Geosystems Mathematics

Series Editors
Willi Freeden
Kaiserslautern, Germany

M. Zuhair Nashed
Orlando, Florida, USA

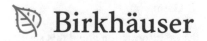

This series provides an ideal frame and forum for the publication of mathematical key technologies and their applications to geo-scientific and geo-related problems. Current understanding of the highly complex system Earth with its interwoven subsystems and interacting physical, chemical, and biological processes is not only driven by scientific interest but also by the growing public concern about the future of our planet, its climate, its environment, and its resources. In this situation mathematics provides concepts, tools, methodology, and structures to characterize, model, and analyze this complexity at various scales. Modern high speed computers are increasingly entering all geo-disciplines. Terrestrial, airborne as well as spaceborne data of higher and higher quality become available. This fact has not only influenced the research in geosciences and geophysics, but also increased relevant mathematical approaches decisively as the quality of available data was improved.

Geosystems Mathematics showcases important contributions and helps to promote the collaboration between mathematics and geo-disciplines. The closely connected series *Lecture Notes in Geosystems Mathemactics and Computing* offers the opportunity to publish small books featuring concise summaries of cutting-edge research, new developments, emerging topics, and practical applications. Also PhD theses may be evaluated, provided that they represent a significant and original scientific advance.

Edited by

- Willi Freeden (University of Kaiserslautern, Germany)
- M. Zuhair Nashed (University of Central Florida, Orlando, USA)

In association with

- Hans-Peter Bunge (Munich University, Germany)
- Roussos G. Dimitrakopoulos (McGill University, Montreal, Canada)
- Yalchin Efendiev (Texas A&M University, College Station, TX, USA)
- Andrew Fowler (University of Limerick, Ireland & University of Oxford, UK)
- Bulent Karasozen (Middle East Technical University, Ankara, Turkey)
- Jürgen Kusche (University of Bonn, Germany)
- Liqiu Meng (Technical University Munich, Germany)
- Volker Michel (University of Siegen, Germany)
- Nils Olsen (Technical University of Denmark, Kongens Lyngby, Denmark)
- Helmut Schaeben (Technical University Bergakademie Freiberg, Germany)
- Otmar Scherzer (University of Vienna, Austria)
- Frederik J. Simons (Princeton University, NJ, USA)
- Thomas Sonar (Technical University of Braunschweig, Germany)
- Peter J.G. Teunissen, Delft University of Technology, The Netherlands and Curtin University of Technology, Perth, Australia)
- Johannes Wicht (Max Planck Institute for Solar System Research, Göttingen, Germany).

More information about this series at http://www.springer.com/series/13389

Nicolae Suciu

Diffusion in Random Fields

Applications to Transport in Groundwater

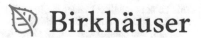 Birkhäuser

Nicolae Suciu
Department of Mathematics
Friedrich-Alexander University
of Erlangen-Nürnberg
Erlangen, Germany

Tiberiu Popoviciu Institute
of Numerical Analysis
Cluj-Napoca Branch
of the Romanian Academy
Cluj-Napoca, Romania

ISSN 2510-1544 ISSN 2510-1552 (electronic)
Geosystems Mathematics
ISBN 978-3-030-15083-9 ISBN 978-3-030-15081-5 (eBook)
https://doi.org/10.1007/978-3-030-15081-5

Mathematics Subject Classification (2010): 60J60, 60G60, 65C10, 65M75, 65C05, 76S05, 86A05

This book is published under the imprint Birkhäuser, www.birkhauser-science.com by the registered company Springer Nature Switzerland AG.
The registered company address is: Gewerbestrasse 11, 6330 Cham, Switzerland

In memory of Călin Vamoş (1955–2017)

Preface

The aim of this book is to provide an accessible and self-consistent theory of diffusion in random velocity fields together with robust numerical simulation approaches. The focus is on transport processes in natural porous media, with applications to contaminant transport in groundwater.

Stochastic modeling has been a leading paradigm in studies of complex systems for several decades. Random media, random environments, or random fields are central topics for thousands of research papers in physics, technology, geophysics, and life sciences. For instance, a search for the topic "random media" (with quotes) in Web of Science (seen online in January 2019) returned 5030 results with 18.64 citations per item in the last two decades, with a strong increasing trend. A similar dynamics (4466 results with 15.54 citations per item) shows for the same period the topic "groundwater contamination," which is one of the investigation directions where the "randomness" paradigm is intensively used.

Mathematical models of transport in random environments (e.g., continuous diffusion processes with random coefficients or random walks with random jump probabilities) are often used for phenomena which are not reproducible experimentally under macroscopically identical conditions or in cases where the incomplete knowledge of the physical parameters precludes deterministic descriptions. To the first class belongs the turbulence, characterized by an intrinsic randomness, which is modeled by random velocity fluctuations. Also in plasma physics the turbulent state of the system of charged particles is described by random electric potentials and magnetic fields. Transport in groundwater belongs to the second class. The way randomness enters modeling in hydrogeology is through stochastic parameterizations of incompletely known hydraulic conductivity fields which induce random Darcy velocity fields.

A common feature of transport processes in random environments is the apparent increase of the diffusion coefficients with the scale of observation. In hydrogeology, the increase from Darcy scale to laboratory and to field scale of the diffusion coefficients inferred from measurements through different approaches (by fitting concentrations with solutions of advection-diffusion equations, by computing spatial moments of tracer concentrations, or by analysis of concentration series recorded at different travel distances from the source) has been called "scale effect." Similar scale dependence characterizes the so-called running diffusion coefficient in plasma physics and the "turbulent diffusivity" in turbulence.

Another characteristic of transport in random media is the presence of various memory effects associated with the departure of the transport process from a genuine Gaussian diffusion. In turbulence and in plasma physics, memory effects manifested by non-Markovian evolution were explicitly associated with the stochastic nature of the environment. In the frame of stochastic subsurface hydrology, the departure from Fickian, linear-time behavior of the second moment of the solute plume can be interpreted as a memory effect. This type of memory effect is usually associated with Markovian diffusion processes and is omnipresent in stochastic models of transport in groundwater. The prototype memory-free process is the Wiener process with independent increments. Therefore, a direct quantification of such memory effects is provided by correlations of increments of the transport process.

The groundwater is contained in aquifer systems consisting of spatially heterogeneous hydrogeological formations. The scarcity of direct measurements of their hydraulic conductivity is compensated by spatial interpolations and correlations. Based on such empirical models, the hydraulic conductivity is further modeled by space random functions. The groundwater flow driven by piezometric pressure gradients is usually modeled by Darcy's law, and the randomness of the hydraulic conductivity induces the randomness of the flow velocity. Contaminant solutes are transported by advection, are diluted by diffusion and hydrodynamic dispersion, and undergo various chemical reactions. Under simplifying assumptions, also supported by experiments, the hydrodynamic dispersion is approximated by a Gaussian diffusion and summing up the molecular diffusion at the pore scale one arrives at a local-scale diffusive model with diffusive flux governed by Fick's law. Hence, the primary mechanism governing the fate of contaminants in groundwater can be described as a diffusion in random velocity fields.

For fixed realizations of the random velocity field, concentrations and transition probability densities of the diffusion process are governed by parabolic partial differential equations local in time and space. However, in case of statistically nonhomogeneous velocity fields, theoretical investigations and numerical simulations show that the evolution of the ensemble average concentration is non-Fickian and has to be described by integro-differential equations nonlocal in both time and space. A model nonlocal in time but local in space of the ensemble average concentration is the "continuous time random walk" process, with uncorrelated polydisperse features consisting of a random walk with waiting times uniformly sampled from a probability distribution. Non-locality also occurs in modeling the local dispersion if the hydraulic parameters of the medium display a fractal structure. Since nonlocal and non-Fickian behavior may arise from either normal or anomalous local-scale diffusion models, it is difficult to extract information on the true nature of the stochastic transport process from experiments. Moreover, if the hydraulic conductivity and the velocity field are characterized by power-law correlations, the model of diffusion in random fields naturally leads to anomalous diffusive behavior of the transport process. Thus, diffusion in random velocity fields remains competitive with respect to other models of non-Fickian behavior.

Unlike models of ensemble-averaged observables of the transport process, the model of diffusion in random fields allows straightforward Monte Carlo (MC) estimates of prediction errors and ergodicity assessments. They can be obtained by comparing results for fixed realizations of the random field, corresponding to the observed transport process, to their ensemble averages. Last but not least, the model of diffusion in random velocity fields is formulated as a Fokker–Planck equation with random coefficients which is appropriate and facilitates the development of methods similar to those used in turbulence studies for the probability density function (PDF) of the random concentration.

The Fokker–Planck structure of the model equations is of great importance in numerical simulations. The solution of the Fokker–Planck equation is essentially the probability density of an Itô diffusion process. This property provides the basis for constructing various "particle methods," such as random walks on lattices or grid-free "particle tracking" (PT) approaches. The latter are actually solutions of Itô stochastic differential equations. Particle densities estimated from ensembles of random walk or PT solutions provide the numerical solution of the Fokker–Planck equation. These approaches are generalized by the "global random walk" (GRW) algorithm. It consists of a superposition of many weak Euler schemes for the Itô equation projected on a regular lattice. The associated system of computational particles evolves globally, by simultaneous jumps of all the particles from a lattice site to neighboring sites according to the random walk rule.

The book consists of seven chapters. Following the introductory Chap. 1, which presents an overview of the problems and the model equations, and Chap. 2, which introduces the basic notions of random variables, random functions, and diffusion processes, Chap. 3 is devoted to numerical simulations of diffusion processes and GRW algorithms. Chapter 4 presents the mathematical model of diffusion in random fields in relation to applications to stochastic modeling in subsurface hydrology. Memory effects and ergodicity issues are investigated by MC simulations in Chap. 5. A mathematical frame of PDF approaches and numerical solutions of PDF evolution equations obtained with the GRW algorithm are presented in Chap. 6. Chapter 7 concludes the book by discussing the relation between model and measurement scales and introduces a new approach which accounts for measurement scale through spatiotemporal averages of GRW solutions. Some technical details are deferred to Appendices A–F.

The moderate level of difficulty of the presentation and the minimum necessary information on stochastic processes provided in the first three chapters make the book accessible to readers with an undergraduate background in engineering or physics. More challenging issues presented in Chaps. 4–7, such as the correlation structure of the process of diffusion in random fields, relation between memory effects and ergodic properties, and derivation and parameterizations of PDF equations, could be of interest for researchers with an advanced mathematical and engineering background and could serve as the basis for further developments in stochastic modeling of groundwater systems.

Nürnberg, Germany   Nicolae Suciu
January 2019

Acknowledgements

Parts of this book present results obtained during several research projects supported by the Deutsche Forschungsgemeinschaft–Germany under Grants SU 415/1-1, and SU 415/1-2, SU 415/2-1, SU 415/4-1. Massive Monte Carlo simulations have been supported by the Jülich Supercomputing Centre–Germany as part of the Project JICG41. Fruitful cooperation with colleagues from Tiberiu Popoviciu Institute of Numerical Analysis of the Romanian Academy, Agrosphere Institute of the Research Centre Jülich, Chair of Applied mathematics 1 of the Friedrich-Alexander University Erlangen-Nürnberg, Helmholtz Centre for Environmental Research—UFZ, Leipzig, and the Department of Mathematics of the University of Bergen is gratefully acknowledged.

Contents

1 Introduction 1
  1.1 Motivation 1
  1.2 Diffusion Equations 4
    1.2.1 Stationary Diffusion and Flow in Porous Media 4
    1.2.2 Advection–Diffusion Processes 4
    1.2.3 Advection–Diffusion–Reaction Processes 5
  1.3 Numerical Methods 6
    1.3.1 Solutions of Deterministic PDEs 6
    1.3.2 Solutions of Stochastic PEDs 7
  References 8

2 Preliminaries 11
  2.1 Random Variables, Processes, and Fields 11
    2.1.1 Random Variables 11
    2.1.2 Random Functions and Stochastic Processes 14
    2.1.3 Densities of Finite Dimensional Probability
        Distributions 20
  2.2 Markov and Diffusion Processes 22
    2.2.1 Markov Processes 22
    2.2.2 Diffusion Processes and Fokker–Planck Equations 25
    2.2.3 Itô Equation and Stochastic-Lagrangian Framework 32
  References 46

3 Stochastic Simulations of Diffusion Processes 47
  3.1 Random Sequences 47
    3.1.1 Programming Tools.. 47
    3.1.2 Convergence 49
  3.2 Numerical Solutions of Itô Equations 57
    3.2.1 Strong and Weak Solutions 57
    3.2.2 Strong and Weak Euler Schemes 58

3.3   Global Random Walk   62
       3.3.1   Weak Approximations by Global Random Walk   62
       3.3.2   Unbiased GRW   67
       3.3.3   Biased GRW   77
       3.3.4   GRW Solutions for Flow and Reactive Transport...........   84
References   88

4   Diffusion in Random Velocity Fields..................................   91
4.1   Classical Stochastic Theories Revisited   91
       4.1.1   Taylor's Theory of Diffusion by Continuous Movements ...   91
       4.1.2   Advection–Dispersion Models...............................   99
       4.1.3   The Eulerian Statistics of the Travel-Time   104
       4.1.4   The Local Dispersivity Tensor   107
       4.1.5   First Order Approximations for Small Variance
                of the Log-Hydraulic Conductivity   108
4.2   Diffusion with Space Variable Drift   112
       4.2.1   Fokker–Planck Equation with Variable Drift................   112
       4.2.2   Dispersion and Memory Terms   114
       4.2.3   Memory Effects and Transition Probabilities   117
4.3   Diffusion in Random Fields Model of Passive Transport   118
       4.3.1   Ensemble-, Effective-, and Center of Mass-Dispersion......   118
       4.3.2   Statistical Homogeneity Properties   120
       4.3.3   Anomalous Diffusion, Ergodicity, and Self-averaging   122
4.4   First Order Approximations...   126
       4.4.1   Lagrangian and Eulerian Representations of Diffusion
                in Random Velocity Fields   126
       4.4.2   Explicit First Order Results for Transport in Aquifers.......   129
       4.4.3   First Order Results for Power-Law Correlated ln K
                Fields...   133
References   135

**5   Monte Carlo GRW Simulations of Passive Transport
     in Groundwater**   139
5.1   Numerical Investigations on Memory Effects and Ergodicity........   139
       5.1.1   Ergodicity of the Center of Mass   140
       5.1.2   Dependence on Initial Conditions   142
       5.1.3   Non-ergodic Effective Dispersion at Finite Times   143
       5.1.4   Loss of Memory and Asymptotic Ergodicity   144
5.2   Numerical Simulations of Transport in Aquifers
       with Evolving Scale Heterogeneity...................................   144
       5.2.1   Quantifying Anomalous Diffusion by Memory Terms   145
       5.2.2   Ensemble and Memory Coefficients.........................   147
       5.2.3   Breakthrough Curves and Cross-Section Concentrations ...   149
       5.2.4   Anomalous Diffusion Behavior..............................   152
References   154

6   Probability and Filtered Density Function Approaches 157
   6.1   PDF/FDF Evolution Equations 157
         6.1.1   Background on PDF/FDF Methods........................... 157
         6.1.2   PDF/FDF Equations for Reactive Transport 160
         6.1.3   The Fokker–Planck Approach 162
   6.2   Spatial Coarse-Graining and FDF Simulations 169
         6.2.1   PDF/FDF Problem for Passive Transport in Aquifers 169
         6.2.2   Mixing Models 172
         6.2.3   Upscaled Velocity Fields and Diffusion Coefficients 177
         6.2.4   Coarse-Grained Simulations of Transport................... 178
   6.3   GRW Solutions for PDF/FDF Equations............................. 182
         6.3.1   Convergence of the Mean Concentration.................... 182
         6.3.2   Convergence of FDF Solutions to PDF Solutions 183
   6.4   Issues and Future Developments of PDF/FDF Approach 185
         6.4.1   Looking for Appropriate Mixing Models 185
         6.4.2   Perspectives of the FDF Approach 188
   References 189

7   Model, Scale, and Measurement 193
   7.1   Sampling Volume and Sampling Time 193
         7.1.1   Sampling Volume Approach 193
         7.1.2   Spatio-Temporal Upscaling 194
   7.2   Coarse-Grained Spatio-Temporal Upscaling......................... 196
         7.2.1   Coarse-Grained Space–Time Averages...................... 196
         7.2.2   Continuous Fields 200
         7.2.3   CGST Average Concentration 201
   References 202

A   Numerical Simulation of Diffusion Processes 205
   A.1   Diffusion Processes Constructed with i.i.d. Variables 205
   A.2   Itô–Euler Schemes.. 206
   A.3   Global Random Walk 209
         A.3.1   One-Dimensional Unbiased GRW Algorithms 209
         A.3.2   Two-Dimensional Unbiased GRW Algorithm
                 for Advection-Diffusion Processes 212
         A.3.3   Biased GRW Algorithms................................... 215

B   GRW Solutions of Fokker–Planck Equations............................. 221
   B.1   GRW Approximations for Continuous Diffusion Processes 221
   B.2   Strict Equivalence Between GRW and the Weak Euler
         Scheme for Constant Velocity 222
   B.3   Biased GRW Approximations for Continuous Diffusion
         Processes.. 222

C Numerical Generation of Random Fields 225
   C.1   Homogeneous Gaussian Random Fields 225
        C.1.1   Randomization Method 225
        C.1.2   Analytic Properties of the Samples 226
   C.2   Filtered Kraichnan Fields 229
   C.3   Kraichnan Field Generators 234
        C.3.1   Hydraulic Conductivity Fields 234
        C.3.2   Kraichnan Approximations of Velocity Fields 239

D Correlation Structure of the Itô Process 245
   D.1   Mean Value and Covariance Components 245
   D.2   Insights from Itô–Taylor Expansions 249
   D.3   Weak Solutions by Successive Approximations 250

E Derivation of PDF Equations by δ-Function Method 253
   E.1   The PDF Equation 253
   E.2   The Fokker–Planck Equation 257

F Upscaled Dispersion Coefficients 261
   F.1   Self-averaging Estimations of Diffusion Coefficients 261
   F.2   Coarse-Grained Dispersion Coefficients 262
   References 263

Index 265

Chapter 1
Introduction

Abstract Stochastic approaches for transport processes in heterogeneous media are motivated by the need to use stochastic parameterizations of the model equations. Essentially, this results in modeling diffusion processes in random fields. For instance, in stochastic subsurface hydrology, random hydraulic conductivity parameters generate random groundwater flow velocity fields and solute transport is modeled by diffusion equations with random drift coefficients. Technically, modeling approaches are based on equivalent Fokker–Planck and Itô representations of the diffusion in random fields, that is, through trajectories of molecules or computational particles and the corresponding continuous fields.

1.1 Motivation

In physics, diffusion denotes the collective behavior of systems of micro-particles which spread from regions with high concentrations to regions with low concentrations and approach to a uniform spatial distribution. The concentration of the particles is governed by a partial differential equation (PDE) called diffusion equation. The relation between the macroscopically observable diffusion process and the irregular, apparently random movement of micro-particles was established in a simple and elegant presentation by Albert Einstein in his 1905 paper on Brownian motion [11]. This seminal paper, where Einstein explains the irregular motion of small particles suspended in a liquid from the point of view of molecular-kinetic theory of heat, contains many of the basic concepts which have been developed since then in the theory of stochastic processes [13], such as Markov property, Chapman–Kolmogorov equation, Fokker–Planck equation, and the principle of constructing stochastic particle trajectories. The latter, which is central to this book, consists of increasing the coordinates of the individual particles by mutually independent increments at consecutive time intervals [11, Sect. 4]. The resulting trajectory is a particular Itô integral, solution of the Itô stochastic differential equation, which will be presented in detail in Chap. 2. On the same principle of summing up independent increments are based the numerical schemes for the integration of the Itô equation presented in Chap. 3. From the frequency law that the independent increments have

© Springer Nature Switzerland AG 2019

N. Suciu, *Diffusion in Random Fields*, Geosystems Mathematics,
https://doi.org/10.1007/978-3-030-15081-5_1

to satisfy, Einstein derived the diffusion equation verified by the concentration of the suspended particles, which is a particular Fokker–Planck equation. This concludes the connection between the microscopic and the macroscopic descriptions of the diffusion process.

The mathematical model of the random trajectory process on which Einstein based his molecular-kinetic explanation for the movement of the suspended particles is since then referred to as Brownian motion, or also as Wiener process (which is precisely a standardized Brownian motion [16]). The discrete time Brownian motion model proposed by Einstein can be regarded as a numerical scheme to solve the Itô equation in continuous space. If one considers independent increments with constant length, the discrete space Brownian motion becomes a random walk on a regular lattice, in case of discrete time variable, or a continuous time random walk governed by a master equation, in case of continuous time variable. Examples of discrete and continuous Brownian motions are discussed in Chap. 2. Brownian motion models have a broad applicability beyond statistical physics and transport phenomena, in modeling stock markets, birth-dead processes, or electronic systems, among other areas [13].

In numerical approaches to diffusion, the term "particle" denotes either an entity modeled by a random trajectory (molecule, asset price on financial markets, prey/predator in birth-death processes, electric pulses, etc.) or the trajectory itself. In the latter case one also uses the term "computational particle." Thus, an ensemble of computational particles (trajectories of the Brownian motion) can be used to make statistical inferences, in particular to obtain numerical approximations of the probability density of the process [16], which is proportional to the density of the system of physical particles and solves the same diffusion equation [11]. If a deterministic drift is superposed on the Brownian motion, one obtains an advection–diffusion process. For physical systems consisting of molecules, considered in this book, the drift coefficients occurring in the trajectory stochastic equation, as well as in the deterministic PDE solved by the concentration of the molecules, are components of a velocity field. If, moreover, the physical system contains several species of interacting molecules, its evolution is described by an advection–diffusion–reaction process. The associated PDEs governing molecular species concentrations contain, in addition to drift and diffusion coefficients, reaction terms and the random trajectory equations are supplemented by terms similar to those of the master equations for birth-dead processes [13].

If the medium where the system of molecules evolves is highly heterogeneous, it is convenient to model the transport process as diffusion in a random environment. The model consists of two components: the environment, specified by a set of random parameters, and the diffusion process defined by these parameters [3, 4, 31]. For a fixed set of parameters given by a realization of the random environment one obtains a deterministic solution of the diffusion problem. The stochastic solution of the diffusive transport problem is further defined by the ensemble of deterministic solutions indexed by the realizations of the random environment. For example, if the random environment is specified by the coefficients of the diffusion PDE (drift and diffusion coefficients, or reaction rates) modeled as random fields (i.e., space random functions) one defines a model of diffusion in random fields [4]. The ensemble of deterministic solutions obtained for an ensemble of realizations of the random field defines in a consistent manner the solution of the stochastic PDE. It can be effectively constructed in two steps by solving the associated equation for the trajectory of the diffusion process. First, an ensemble of trajectories is computed for a fixed realization of the random field to estimate the realization of the concentration field, then the procedure is repeated for an ensemble of realizations of the random field to construct the ensemble of concentration fields. This is the so-called Monte Carlo (MC) solution method, illustrated in Chap. 5. For discrete diffusion processes, the environment can be defined, for instance, by an ensemble of jump probabilities on a lattice. One obtains in this way a model of random walk in random environments [3]. This is also the principle of the numerical method used to compute MC ensembles of solutions presented in Chap. 5, where concentrations estimated from ensembles of random walk trajectories for fixed realizations of the random field are obtained with the global random walk (GRW) algorithm introduced in Chap. 3.

The theory of diffusion in random fields presented in this book is illustrated for solute transport in groundwater. Hydrogeological systems are modeled as heterogeneous porous media. The hydraulic pressure driving flows in saturated porous media is described by an elliptic equation with hydraulic conductivity tensor coefficients usually modeled as realizations of random space functions. Further, Darcy's constitutive law is used to compute the flow velocity by a post-processing procedure or, more accurately, within a mixed finite element method (MFEM) approach [5, 26]. The velocity field enters as a distributed parameter into the numerical finite element method (FEM) solution of the transport equations [26]. Alternatively, flow solutions can be imported into a GRW simulation of the transport process [32]. The uncertainty of the computed concentration fields, induced by that of the hydraulic conductivity or other randomness sources, as, for instance, random sources in the pressure equation [25] or random parameterizations of storage coefficients, in case of transient flows [1], is finally evaluated through the first two statistical moments or, more completely, by the first order probability density function (PDF) [22, 33].

The success of such a stochastic numerical approach to contaminant transport in groundwater is conditioned by the accuracy of the solutions of the parabolic and elliptic equations on which it is based. Also, the usefulness of the approach for real life problems is conditioned by the adequacy of the descriptions for the space–time support scales of the hydrological observations. Spatial scales can be accounted for by estimating PDFs via spatial averaging procedures, leading to parabolic evolution equations in the so- called filtered density function (FDF) approach [33]. The PDF/FDF approaches for transport in groundwater are discussed in Chap. 6. Temporal scales are relevant for the interpretation of the measurements [10] and are explicitly considered in some special models of the measurement, such as in case of quantifying average concentrations from time dependent pumping well data [2]. The possibility to describe space–time averages by evolution equations is shortly discussed in Chap. 7, at the end of the book.

1.2 Diffusion Equations

The equations intensively used in stochastic subsurface hydrology approaches are particular forms of diffusion equations. General advection–diffusion–reaction processes are described by parabolic equations of the form

$$
\frac{\partial c_v}{\partial t} + \underbrace{V_i \frac{\partial c_v}{\partial x_i}}_{advection} = \underbrace{\frac{\partial}{\partial x_i}(D_{ij} \frac{\partial c_v}{\partial x_j})}_{diffusion} + \underbrace{S_v}_{reaction/source} ,
\tag{1.1}
$$

where $i = 1, \ldots, d$, d is the spatial dimension, and the Einstein summation convention is used for repeated indices. The local balance Eq. (1.1) are derived from global (integral) mass balance equations and constitutive relations for the diffusion flux [14, 23]. These equations govern the behavior of a system of reacting chemical species described by concentrations $c_v(\mathbf{x}, t) \in Y_c \subset \mathbb{R}_+^{\mathcal{N}}$, $\mathbf{x} \in Y_x \subset \mathbb{R}^d$, $t \in \mathbb{R}_+$, $v = 1, \ldots, \mathcal{N}$, related by nonlinear reaction terms $S_v(c_1, \ldots, c_{\mathcal{N}})$. V_i and D_{ij} are the components of the velocity vector and diffusion tensor, respectively.

1.2.1 Stationary Diffusion and Flow in Porous Media

Dropping the time derivative and the advection term in (1.1) one obtains an elliptic equation describing a stationary diffusion. The same elliptic equation is verified by the hydraulic head $h(\mathbf{x})$ driving stationary flows in saturated porous media,

$$
-\frac{\partial}{\partial x_i}(K_{ij} \frac{\partial h}{\partial x_j}) = f,
\tag{1.2}
$$

where K_{ij} is the hydraulic conductivity tensor and f is a source term. The flow velocity in porous media is obtained from the solution of Eq. (1.2) and Darcy's law, $V_i = -K_{ij} \frac{\partial h}{\partial x_j}$.

1.2.2 Advection–Diffusion Processes

Dropping the reaction terms in (1.1) one obtains an advection–diffusion equation. Note that the diffusion term in (1.1) is in the so-called Stratonovich form, consistent to Fick's law [13]. With the relations

$$
V_i^* = V_i + \partial D_{ij}/\partial x_j
\tag{1.3}
$$

one defines new velocity components V_i^*, the diffusion term takes the Itô form, and the advection–diffusion equation is equivalent to

$$\frac{\partial c_v}{\partial t} + \frac{\partial}{\partial x_i}(V_i^* c_v) = \frac{\partial^2}{\partial x_i \partial x_i}(D_{ij} c_v). \qquad (1.4)$$

Thus, by using (1.3), the advection–diffusion equation takes the form of the Fokker–Planck equation (1.4) for the probability density of the diffusion process with trajectories governed by the system of Itô equations

$$X_i(t) = X_{0i} + \int_{t_0}^{t} V_i^*(\mathbf{X}(t'))dt' + \int_{t_0} dW_i(\mathbf{X}(t')), \qquad (1.5)$$

where W_i is a non-standard Wiener process with mean $E\{W_i(\mathbf{X}(t'))\} = 0$ and variance $E\{W_i^2(\mathbf{X}(t'))\} = 2\int_{t_0}^{t} D_{ii}(\mathbf{X}(t'))dt'$ [16].

The Fokker–Planck form (1.4) of the advection–diffusion equation is often preferred because its numerical solutions can be obtained within a MC approach from an ensemble of solutions of the associated Itô equation (1.5). The relation between Fokker–Planck equation and Stratonovich PDE is discussed in more detail at the end of Chap. 2

1.2.3   Advection–Diffusion–Reaction Processes

The full equation (1.1) describes the transport of reacting molecular species [17]. The groundwater velocity is a random field determined by the random hydraulic conductivity. The diffusion coefficients and the reaction rates are obtained by appropriate material laws.

In deriving (1.1) Fick's law is assumed to be valid. Since anomalous diffusion in heterogeneous groundwater systems can be modeled by considering long-range correlations of the velocity field [12, 30], keeping the normal behavior of the local diffusion process is not a restrictive assumption. According to Scheidegger's parameterization [28], the diffusion coefficient in (1.1) is the sum of molecular diffusion coefficient and a hydrodynamical dispersion. The latter is a second order tensor expressed through a fourth order dispersivity tensor as a linear combination of velocity components [28]. While this complex structure is relevant for processes at Darcy or laboratory scale [21], only two coefficients, the longitudinal and the transverse dispersivities [24], or even constant anisotropic [29] or isotropic [9] dispersion coefficients are enough to capture important features of the transport process at the field scale.

The system of equations (1.1) has to be completed by algebraic equations to describe equilibrium reactions and by ordinary differential equations if immobile species from the solid matrix of the porous medium react with mobile species transported and dispersed by the fluid flow [18]. If charged particles are also considered, the behavior of electrolyte solutions is modeled by the Darcy–Poisson–Nernst–Planck system of equations [14].

If the coefficients occurring in (1.1), (1.2), and (1.4) are samples of random functions, then one obtains stochastic PDEs of "multiplicative noise" type. Equation (1.5) is a stochastic integral equation. When used to solve the Fokker–Planck equation (1.4) with random coefficients, the solutions of (1.5) are trajectories of diffusion processes in random environments. These processes are mathematically formulated and analyzed in Chap. 4.

1.3   Numerical Methods

For a fixed realization of the hydraulic conductivity, Eqs. (1.1), (1.2), and (1.4) are PDEs with deterministic coefficients and are solved by usual discretization methods such as finite difference/element/volume methods, by spectral methods, as well as by various particle methods. To an ensemble of realizations of the random function modeling the hydraulic conductivity corresponds an ensemble of solutions, which define the solution of the stochastic PDE as a random function.

1.3.1   Solutions of Deterministic PDEs

Finite element methods (FEM), together with their finite volume version, are among the mostly used solution approaches for PDEs [17]. While finite difference and spectral methods assume the existence of the derivatives or the continuity of the solutions, FEM are integral formulations of the problem which require less regularity conditions. Therefore, FEM are appropriate for solving problems for flow and transport in heterogeneous systems, such as natural groundwater formations. FEM solutions are constructed from those of the system of algebraic equations resulted from the finite element discretization of the PDE problem. The latter are usually solved by locally convergent Newton methods or globally convergent linearization schemes [14, 20, 26, 27]. The solution of the flow problem is often obtained with MFEM, the mixed formulation of the FEM, where the gradient of h is considered as a new unknown and the solution for the flow velocity is obtained at the same time with the solution of (1.2). One avoids in this way the numerical errors produced in post-processing procedures for computing the velocity field. The flow velocity is further used as input for the transport problem [5, 26].

Typical conditions for transport of pollutants in groundwater lead to advection-dominated problems. Inaccurate approximations of the advection term which dominates in (1.1) could result in a non-physical, numerical diffusion which affect the solution obtained with FEM or other discretization schemes. An alternative approach based on Einstein's relationship between diffusion and particle trajectories, introduced by Chorin [6], ensures that the "effective diffusion of the scheme equals the nominal diffusion" [7]. The principle of the method can be summarized as follows: a collection of computational particles is used to represent the transported quantity (vorticity in that case); the position of each particle is advanced in time using a splitting scheme which first moves the particles along the streamlines of the advection velocity, then the advected positions are perturbed by displacements given by realizations of a Gaussian random variable; the value of the transported quantity is approximated by the updated distribution of particles.

The first step is a dynamical system associated with the advection equation, i.e., the left-hand side of (1.1) equated to zero, which is a Liouville equation (see Chaps. 2 and 4). The second step solves a diffusion equation without drift for the initial condition given by the solution of the advection step by using Einstein's representation of the diffusion process through stochastic particle trajectories. Together, the two steps correspond to a numerical scheme for the Itô equation (1.5). When these computational steps are taken at successive times to construct the trajectory of each particle, as in Chorin's approach, the method is a PT algorithm [34].

If instead of constructing individual particle trajectories the advection and diffusion steps are applied simultaneously to the whole system of particles one obtains a GRW algorithm. To enable the global GRW procedure, the Gaussian random variable in continuous space (Wiener process) is replaced by a random walk and the advection velocity is discretized on a regular lattice. With these, all particles from a lattice site undergo advective displacements to neighboring sites followed by diffusive jumps performed simultaneously by spreading the particles according to binomial distributions. The GRW algorithm can thus be thought of as a superposition of weak Euler schemes for systems of Itô equations (1.5), governing the movement of an ensemble of computational particles on regular lattices. The solution of the Fokker–Planck equation (1.4) is approximated by particle densities at lattice sites [34]. Alternatively to discretizing the velocity, advective displacements can be accounted for by using biased random walk rules, in biased GRW algorithms which generalize cellular automata for diffusion with sequential spreading of walkers from lattice sites [15]. GRW algorithms can be used to solve parabolic or elliptic equations. Particle simulations for advection–diffusion–reaction problems can also be designed as splitting schemes with advection–diffusion steps performed by GRW algorithms.

1.3.2 Solutions of Stochastic PEDs

Problems for stochastic PDEs are commonly addressed by MC simulations, which consist of solving Eq. (1.1) for large ensembles of realizations Y of the random coefficients. The ensemble of solutions, e.g., concentration vectors $c(x, t, Y)$, is further used to make statistical estimations (expectations, variances, probability distributions). The deterministic PDEs for fixed realizations of the coefficients can be solved either by classical finite difference/element/volume methods or by particle methods [26]. The latter are mainly useful in advection-dominated problems, because they do not produce numerical diffusion.

A particular MC approach associated with difference schemes is the polynomial chaos expansion (PCE). In this approach, the random coefficients as well as the solutions are expressed as

$$u(\mathbf{x}, t,\ Y) = \sum_{i=0}^{N} a_i(\mathbf{x}, t,\ Y)\psi_i(\ Y)$$

where $\psi_i(\ Y)$ are orthogonal polynomial basis functions constructed, for instance, by using Hermite polynomials [19]. The PCE method requires solving the deterministic PDE (1.1) as many times as the number of terms N of the expansion. Then, equating the solutions with their polynomial expansion one obtains a system of algebraic equations for the coefficients a_i. Statistical estimations are further obtained, similarly to classical MC approaches, by evaluating the solution $\mathbf{c}(\mathbf{x}, t,\ Y)$ for an ensemble of realizations Y of the random polynomials ψ_i. The system of algebraic equations has to be solved for all points $(\mathbf{x}, t)$ where one estimates the statistics of the solution. Therefore, the PCE approach is competitive with respect to classical MC mainly when not the whole field of the stochastic solution but only its statistics at a moderately large number of points is required.

MC simulations can be avoided by solving evolution equations for the PDF of the random solution. The one-point one-time PDF $f(c_1, \cdots, c_{\mathcal{N}}; \mathbf{x}, t)$ of a system of reacting species concentrations is proportional to a one-time PDF $p(c_1, \cdots, c_{\mathcal{N}}, \mathbf{x}, t)$ defined on the concentration-position state space $Y_x \times Y_c$ with dimensions $d + \mathcal{N}$, which solves a Fokker–Planck equation similar to (1.1). The coefficients of this equation depend on higher order PDFs. Thus, they are not in closed form and require modeling. Once model coefficients have been determined, PDF solutions f can by efficiently obtained by solving the Fokker–Planck equation for p with GRW schemes in concentration-position spaces [33].

The qualitative behavior of the numerical solution can be investigated through analyses of time series obtained by integrating the PDE solution over all but one spatial coordinates. The approach is mainly relevant in analyzing noisy solutions of diffusion equations with random coefficients [8, 33]. In particular, the analysis of such time series reveals the structure of the unclosed coefficients of the PDF evolution equations. Time series analyses can also be applied to Itô stochastic equations, for instance, to investigate some anomalous diffusion features of the trajectories generated by Eq. (1.5) with random coefficients [35].

References

1. Alzraiee, A.H., Baú, D., Elhaddad, A.: Estimation of heterogeneous aquifer parameters using centralized and decentralized fusion of hydraulic tomography data from multiple pumping tests. Hydrol. Earth Syst. Sci. Discuss. **11**(4), 4163–4208 (2014)

2. Bayer-Raich, M., Jarsjö, J., Liedl, R., Ptak, T., Teutsch, G.: Average contaminant concentration and mass flow in aquifers from time-dependent pumping well data: analytical framework. Water Resour. Res. **40**(8), W08303 (2004)
3. Bogachev, L.V.: Random walks in random environments. In: Françoise, J.-P., Naber, G., Tsou, S.T. (eds.) Encyclopedia of Mathematical Physics, vol. 4, pp. 353–371. Elsevier, Oxford (2006)
4. Bouchaud, J.-P., Georges, A.: Anomalous diffusion in disordered media: statistical mechanisms, models and physical applications. Phys. Rep. **195**, 127–293 (1990)
5. Brunner, F., Radu, F.A., Bause, M., Knabner, P.: Optimal order convergence of a modified BDM1 mixed finite element scheme for reactive transport in porous media. Adv. Water Resour. **35**, 163–171 (2012)
6. Chorin, A.J.: Numerical study of slightly viscous flow. J. Fluid Mech. **57**(4), 785–796 (1973)
7. Chorin, A.J.: Vortex sheet approximation of boundary layers. J. Comput. Phys. **27**(3), 428–442 (1978)
8. Craciun, M., Vamos, C., Suciu, N.: Analysis and generation of groundwater concentration time series. Adv. Water Resour. **111**, 20–30 (2018)
9. de Barros, F.P.J., Fiori, A.: First-order based cumulative distribution function for solute concentration in heterogeneous aquifers: theoretical analysis and implications for human health risk assessment. Water Resour. Res. **50**(5), 4018–4037 (2014)
10. Destouni, G., Graham, W.: The influence of observation method on local concentration statistics in the subsurface. Water Resour. Res. **33**(4), 663–676 (1997)
11. Einstein, A.: On the movement of small particles suspended in stationary liquids required by the molecular kinetic theory of heat. Ann. Phys. **17**, 549–560 (1905)
12. Fiori, A.: On the influence of local dispersion in solute transport through formations with evolving scales of heterogeneity. Water Resour. Res. **37**(2), 235–242 (2001)
13. Gardiner, C.: Stochastic Methods. Springer, Berlin (2009)
14. Herz, M.: Mathematical modeling and analysis of electrolyte solutions. Ph.D. thesis, Erlangen-Nuremberg University (2014). http://www.mso.math.fau.de/fileadmin/am1/projects/PhD_Herz.pdf
15. Karapiperis, T., Blankleider, B.: Cellular automaton model of reaction-transport processes. Physica D **78**, 30–64 (1994)
16. Kloeden, P.E., Platen, E.: Numerical Solutions of Stochastic Differential Equations. Springer, Berlin (1999)
17. Knabner, P., Angermann, L.: Numerical Methods for Elliptic and Parabolic Partial Differential Equations. Springer, New York (2003)
18. Kräutle, S., Knabner, P.: A new numerical reduction scheme for fully coupled multicomponent transport-reaction problems in porous media. Water Resour. Res. **41**(9), W09414 (2005)
19. Li, J., Xiu, D.: A generalized polynomial chaos based ensemble Kalman filter with high accuracy. J. Comput. Phys. **228**, 5454–5469 (2009)
20. List, F., Radu, F.A.: A study on iterative methods for solving Richards' equation. Comput. Geosci. **20**(2), 341–353 (2016)
21. Liu, Y., Kitanidis, P.K.: A mathematical and computational study of the dispersivity tensor in anisotropic porous media. Adv. Water Resour. **62**, 303–316 (2013)
22. Meyer, D.W., Jenny, P., Tchelepi, H.A.: A joint velocity-concentration PDF method for tracer flow in heterogeneous porous media. Water Resour. Res. **46**, W12522 (2010)
23. Müller, I.: Thermodynamics. Pitman, Boston (1985)
24. Naff, R.L., Haley, D.F., Sudicky, E.A.: High-resolution Monte Carlo simulation of flow and conservative transport in heterogeneous porous media 1. Methodology and flow results. Water Resour. Res. **34**(4), 663–677 (1998)
25. Pasetto, D., Guadagnini, A., Putti, M.: POD-based Monte Carlo approach for the solution of regional scale groundwater flow driven by randomly distributed recharge. Adv. Water Resour. **34**(11), 1450–1463 (2011)
26. Radu, F.A., Suciu, N., Hoffmann, J., Vogel, A., Kolditz, O., Park, C.-H., Attinger, S.: Accuracy of numerical simulations of contaminant transport in heterogeneous aquifers: a comparative study. Adv. Water Resour. **34**, 47–61 (2011)

27. Radu, F.A., Nordbotten, J.M., Pop, I.S., Kumar, K.: A robust linearization scheme for finite volume based discretizations for simulation of two-phase flow in porous media. J. Comput. Appl. Math. **289**, 134–141 (2015)
28. Scheidegger, A.E.: General theory of dispersion in porous media. J. Geophys. Res. **66**(10), 3273–3278 (1961)
29. Srzic, V., Cvetkovic, V., Andricevic, R., Gotovac, H.: Impact of aquifer heterogeneity structure and local-scale dispersion on solute concentration uncertainty. Water Resour. Res. **49**(6), 3712–3728 (2013)
30. Suciu, N.: Spatially inhomogeneous transition probabilities as memory effects for diffusion in statistically homogeneous random velocity fields. Phys. Rev. E **81**, 056301 (2010)
31. Suciu, N.: Diffusion in random velocity fields with applications to contaminant transport in groundwater. Adv. Water Resour. **69**, 114–133 (2014)
32. Suciu, N., Radu, F.A., Prechtel, A., Brunner, F., Knabner, P.: A coupled finite element–global random walk approach to advection-dominated transport in porous media with random hydraulic conductivity. J. Comput. Appl. Math. **246**, 27–37 (2013)
33. Suciu, N., Schüler, L., Attinger, S., Knabner, P.: Towards a filtered density function approach for reactive transport in groundwater. Adv. Water Resour. **90**, 83–98 (2016)
34. Vamoş, C., Suciu, N., Vereecken, H.: Generalized random walk algorithm for the numerical modeling of complex diffusion processes. J. Comput. Phys. **186**(2), 527–244 (2003)
35. Vamoş, C., Crăciun, M., Suciu, N.: Automatic algorithm to decompose discrete paths of fractional Brownian motion into self-similar intrinsic components. Eur. Phys. J. B **88**, 250 (2015)

Chapter 2
Preliminaries

Abstract In this chapter, basic concepts used in stochastic modeling of transport processes are revisited. Random fields and stochastic processes will be introduced as particular random functions. The hierarchy of finite dimensional distributions will be introduced and particularized for Markov and diffusion processes. Itô and Fokker–Planck descriptions of the diffusion process will be used to introduce the stochastic-Lagrangian framework.

2.1 Random Variables, Processes, and Fields

2.1.1 Random Variables

Stochastic descriptions of physical systems associate probabilities to states and to transitions between states. Intuitively, the probability is defined as the limit for large number of trials of the relative frequency of occurrence of an event. Further, the probability can be axiomatically defined as a positive and additive function of sets of events, so that the occurrence of any one event from the set of all possible events, i.e., the "sure event," has the probability one. A parameter describing the state of the system with the associated probability is modeled by a random variable. The random variable is the basic mathematical object used to build stochastic descriptions of real processes. Stochastic processes, random fields, and dynamical systems are random variables, defined on particular spaces of events.

The mathematical model of random variables uses σ-algebras of sets and measures defined on σ-algebras. The σ-algebra $\mathscr{A}$ is a set of subsets of a generic set Ω which is closed with respect to intersection and countable reunion operations. The pair $(\Omega, \mathscr{A})$ is a *measurable space;* the set $A \subset \Omega$ is a *measurable set* with respect to $\mathscr{A}$ if $A \in \mathscr{A}$. A *measure* P on a σ-algebra $\mathscr{A}$ is defined as a positive and countably additive function of sets on $\mathscr{A}$, $P : \mathscr{A} \longmapsto \mathbb{R}_+$. The triplet $(\Omega, \mathscr{A}, P)$ defines a *measure space*. When $P(\Omega) = 1$ then P is a normalized measure and $(\Omega, \mathscr{A}, P)$ becomes a *probability space*.

© Springer Nature Switzerland AG 2019

N. Suciu, *Diffusion in Random Fields*, Geosystems Mathematics,
https://doi.org/10.1007/978-3-030-15081-5_2

Let $(\Omega, \mathscr{A}, P)$ be a probability space and $(X, \mathscr{B})$ a measurable space with property $\{x\} \in \mathscr{B}, \forall x \in X$. A *random variable* is an $(\mathscr{A}, \mathscr{B})$-measurable application $\eta : \Omega \longmapsto X$, i.e., endowed with the property

$$\{\eta \in B\} \in \mathscr{A}, \forall B \in \mathscr{B}, \qquad (2.1)$$

where $\{\eta \in B\} = \{\omega \in \Omega \mid \eta(\omega) \in B\}$.

Hence, the counter-images of measurable sets through measurable applications η are measurable sets. Measurability is the analog of continuity of applications on topological spaces. If Ω and X are topological spaces and $\mathscr{A}$ and $\mathscr{B}$ are Borel σ-algebras (generated by open sets) *the continuity of an application from Ω to X implies its $(\mathscr{A}, \mathscr{B})$ measurability*: counter-images of open sets are open sets and belong to the Borel σ-algebra [18, p. 10].

In physics terminology, X is the *phase space*, the sets $B \in \mathscr{B}$ are realizations, and Ω is the *space of elementary events* (see Fig. 2.1).

The sets $A \in \mathscr{A}$ are called *events* and $P(A)$ is the *probability of occurrence of the event A* [8, 27].

The *distribution* of the random variable η is a measure $P_\eta : \mathscr{B} \longmapsto \mathbb{R}_+$, defined by

$$P_\eta(B) = P(\{\eta \in B\}), \ \forall B \in \mathscr{B}. \qquad (2.2)$$

The distribution of the realization B is thus defined as the probability of occurrence of the event $\{\eta \in B\}$. In particular, $P_\eta(X) = P(\{\eta \in X\}) = P(\Omega) = 1$, which shows that the distribution (2.2) organizes the phase space X as a probability space, $(X, \mathscr{B}, P_\eta)$.

The notion of distribution allows the connection of the mathematical model of random variable with experimental data and numerical models. Consider, for example, N measurements of the positions of a particle at different distances along a straight line. The number of occurrences N_{occ} of a measured value inside the interval Δx centered at x yields estimates of the relative frequency N_{occ}/N, which approximates the distribution of a random variable describing the measurement.

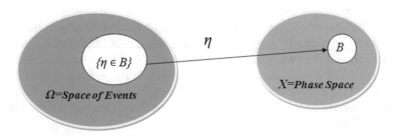

Fig. 2.1 The random variable (2.1) is a measurable application from the space of elementary events Ω to the phase space X

Probabilistic descriptions of practical interest are given by the *stochastic average, or mean* defined as a Lebesgue integral,

$$M(\eta) = \int_\Omega \eta(\omega) P\,(d\omega)\,, \qquad (2.3)$$

the *variance*, $M((\eta - M(\eta))^2)$, *n order moments*, $n > 2$, $M((\eta - M(\eta))^n)$, and *correlation functions* of two variables η and ς, $Corr(\eta, \varsigma) = M(\eta\varsigma) - M(\eta)M(\varsigma)$.

When $B = X$ then $\{\eta \in X\} = \Omega$ and a change of variables theorem shows that the average over the space of elementary events, M_Ω, equals the average over the phase space, M_X,

$$M_\Omega\,(f) = \int_\Omega f(\eta(\omega)) P\,(d\omega) = \int_X f(x) P_\eta(dx) = M_X\,(f)\,, \qquad (2.4)$$

where $x = \eta(\omega)$ [18, p. 180]. Both averages (2.4) are mathematical models of the "average over the statistical ensemble," also called "ensemble mean" or "expectation." Alternative notations for $M(\eta)$, which will also be used in this book, are $E(\eta)$, from "expectation," and angular brackets $\langle\eta\rangle$.

The sufficient and necessary condition of measurability for real functions $\eta :$ $\Omega \to \mathbb{R}$ *is* [13, p. 274],

$$\{\eta < x\} \in \mathscr{A}, \ \forall x \in \mathbb{R}, \ \text{where}\{\eta < x\} = \{\omega \in \Omega \mid \eta\,(\omega) < x\}.$$

For real random variables η one defines the *distribution function* (also known as *repartition function*)

$$F_\eta\,(x) = P(\{\eta < x\}), \ \forall x \in \mathbb{R}.$$

F_η is non-decreasing, left-continuous, measurable, satisfies $F_\eta(-\infty) = 0$, $F_\eta(\infty) = 1$, and generates the Lebesgue–Stieltjes measure

$$P_\eta(\{\eta \in (x_1, x_2)\}) = F_\eta(x_2) - F_\eta(x_1).$$

Since F_η is monotonically non-decreasing it has bounded variation and can be decomposed, up to an additive constant, as sum of a jump function, an absolutely continuous function, and a singular function. Accordingly, the measure P_η is the sum of a discrete measure, an absolutely continuous measure, and a singular measure [13, pp. 342–352]. Illustrative examples of simple distribution functions can be found in [21, Chap. 4].

For real random variables the average can be expressed with the Stieltjes integral with respect to the function F_η and (2.4) takes the particular form

$$M(\eta) = \int_\Omega \eta(\omega) P(d\omega) = \int_{-\infty}^{\infty} x \, dF_\eta(x).$$

2.1.2 Random Functions and Stochastic Processes

Let Λ be a set of parameters, $(\Omega, \mathscr{A}, P)$ a probability space and $(Y, \mathscr{B})$ a measurable space. A *random function* is defined as a random variable on Ω taking values in the "Cartesian product" space Y^Λ,

$$\eta : \Omega \longmapsto Y^\Lambda, \quad \eta(\omega) = y^\omega, \quad y^\omega \in Y^\Lambda, \quad \forall \, \omega \in \Omega. \tag{2.5}$$

If $\mathscr{B}^\Lambda$ is a σ-algebra on Y^Λ, the $(\mathscr{A}, \mathscr{B}^\Lambda)$-measurability of the random variable η is given by the condition $\{\eta \in C\} \in \mathscr{A}$ for every set $C \in \mathscr{B}^\Lambda$. The *distribution of the random function* is defined according to (2.2) by

$$P_\eta(C) = P(\{\eta \in C\}), \quad \forall C \in \mathscr{B}^\Lambda, \tag{2.6}$$

and organizes the function space Y^Λ as a probability space, $(Y^\Lambda, \mathscr{B}^\Lambda, P_\eta)$. For fixed ω, the function $\eta(\omega) = y^\omega$, $y^\omega : \Lambda \longmapsto Y$ is called a *sample* (or *realization*) of the random function. Its values, $y^\omega(\lambda) = y_\lambda$, are points in the space Y, called the *state space* of the random function. For fixed λ and $\omega \in \Omega$, the function $\eta_\lambda : \Omega \longmapsto Y$ $\eta_\lambda(\omega) = y^\omega(\lambda)$ is a random variable with phase space Y.

The *phase space* for the random function η is the space of functions Y^Λ and the realizations are sets of functions y^ω in Y^Λ (see Fig. 2.2). The definition of the random function as a random variable on a space of functions, originally introduced in [6, Chap. 2], is referred to as the definition *in the sense of Doob* [10]. When $\Lambda \subseteq \mathbb{R}$ and λ has the meaning of physical time, the random function is called a *stochastic process*. The samples of a stochastic process are called *trajectories*. When $\Lambda \subset \mathbb{R}^d$, the random function is a *d-dimensional random field* (this is the mathematical object used as model for the velocity field in turbulent diffusion and transport in heterogeneous porous media [16, 19, 25, 27].

The distribution of the random function is completely defined by a *hierarchy of consistent finite dimensional distributions*. The *n-dimensional distributions* are the distributions of the random vectors $\{\eta_{\lambda_1}, \ldots, \eta_{\lambda_n}\}$, obtained from η for fixed values of the parameter λ,

$$P_{\lambda_1 \ldots \lambda_n}(B_1 \times \cdots \times B_n) = P(\{\eta_{\lambda_1} \in B_1, \ldots, \eta_{\lambda_n} \in B_n\}), \quad B_1, \ldots, B_n \in \mathscr{B}.$$
$$\tag{2.7}$$

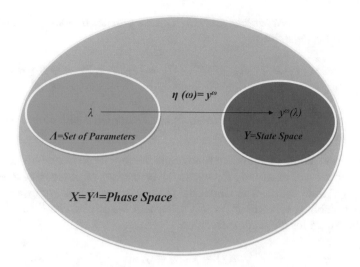

Fig. 2.2 Random function: set of parameters, state space, and phase space

The finite dimensional distributions (2.7) are called *consistent* if they verify the conditions

$$P_{\lambda_{i_1}\ldots\lambda_{i_n}}(B_{i_1} \times \cdots \times B_{i_n}) = P_{\lambda_1\ldots\lambda_n}(B_1 \times \cdots \times B_n), \tag{2.8}$$

for all permutations $\{i_1, \ldots, i_n\}$ of indices $\{1, \ldots, n\}$ and any $B_1, \ldots, B_n \in \mathscr{B}$, and

$$P_{\lambda_1\ldots\lambda_n\lambda_{n+1}}(B_1 \times \cdots \times B_n \times Y) = P_{\lambda_1\ldots\lambda_n}(B_1 \times \cdots \times B_n). \tag{2.9}$$

The distribution $P_{\lambda_1\ldots\lambda_n}(B_1 \times \ldots \times B_n)$ is a measure on the measurable space $(Y^n, \mathscr{B}^n)$, endowed with the σ-algebra $\mathscr{B}^n$ generated by Cartesian products of sets from $\mathscr{B}$, $B = B_1 \times \ldots \times B_n$. The σ-algebra $\mathscr{B}^\Lambda$ on the function space Y^Λ is a system of sets of the form $C_n = \{y_{\lambda_1} \in B_1, \ldots, y_{\lambda_n} \in B_n \mid \lambda_1, \ldots, \lambda_n \in \Lambda, B_1, \ldots, B_n \in \mathscr{B}\}$ [27, p. 80]. The values $P_\eta(C_n)$ of the distribution of the random function η are thus uniquely determined through (2.7) by measures of products of sets from $\mathscr{B}$.

The existence of the process in the sense of Doob (2.5) and the definition of the distribution P_η for the real state space $Y = \mathbb{R}$ with Borel σ-algebra $\mathscr{B}$ are stated in the theorem below [27].

Theorem 2.1 (Kolmogorov Theorem on Finite Dimensional Distributions) *Let $P_{\lambda_1\ldots\lambda_n}$ be probability measures on $(\mathbb{R}^n, \mathscr{B}^n)$, associated with finite sets of distinct parameters $\lambda_1, \ldots, \lambda_n \in \Lambda$. If $P_{\lambda_1\ldots\lambda_n}$ verify the consistency conditions (2.8) and*

(2.9), then, there exists a random function η on the probability space $(\mathbb{R}^{\Lambda}, \mathscr{B}^{\Lambda}, P_{\eta})$, with distribution defined by

$$P_{\eta}(C_n) = P_{\lambda_1 \ldots \lambda_n}(B), \quad \forall C_n \in \mathscr{B}^{\Lambda}, \quad C_n = \{(y_{\lambda_1}, \ldots, y_{\lambda_n}) \in B \mid B \in \mathscr{B}^n\},$$
(2.10)

and reciprocally, finite dimensional distributions associated with a random function by (2.10) are consistent.   □

The simplest example of a random function is the infinite sequence of independent random variables. Let $\Lambda = \mathbb{Z}$, where $\mathbb{Z}$ is the set of integer numbers, and $(Y, \mathscr{B}, P_0)$ a state space. The Cartesian product space $Y^{\mathbb{Z}}$ is endowed with the product probability measure $P = \bigotimes_{-\infty}^{\infty} P_0$ and the triplet $(Y^{\mathbb{Z}}, \mathscr{B}^{\mathbb{Z}}, P)$ is a probability space for the random function

$$\eta : Y^{\mathbb{Z}} \longmapsto Y^{\mathbb{Z}}, \quad \eta(\mathbf{y}) = \mathbf{y}', \quad y_i' = y_{i+1}, \quad -\infty < i < \infty,$$

defining a translation along the components of infinite vectors from $Y^{\mathbb{Z}} = \{(y_n)_{n \in \mathbb{Z}} \mid y_n \in Y\}$. It is straightforward that P verifies (2.10) and the finite dimensional distributions are consistent.

The first random function build by the use of Theorem 2.1 was the *Wiener process* [27, pp. 83–84]. The one-dimensional Wiener process is the random function $W : \Omega \longmapsto \mathbb{R}^{[0,\infty)}$, $w(\omega) = y^{\omega}$, with trajectories on real axis, $y^{\omega} : [0, \infty) \longmapsto \mathbb{R}$, defined by the following three properties:

(w1) $W_0(\omega) = 0$, $\forall \omega \in \Omega$, (the process starts from 0),
(w2) for fixed times, $0 \leq t_0 \leq t_1 \leq \cdots \leq t_n$, the random variables $W_{t_1} - W_{t_0}$, $W_{t_2} - W_{t_1}, \ldots, W_{t_n} - W_{t_{n-1}}$, are independent (independent increments), and
(w3) the distributions of the random variables $W_t - W_s$, $0 \leq s \leq t$, are Gaussian with zero mean and variance $(t - s)$,

$$P_{(W_t - W_s)}(B) = [2\pi (t - s)]^{-1/2} \int_{B \subset \mathbb{R}} \exp\left[-\frac{(y_t - y_s)^2}{2(t - s)}\right] dy_s.$$
(2.11)

The distribution of the sequence of independent increments from (w2) is a product of distributions (2.11). By the means of a linear transformation of the vector of independent increments, one obtains the finite dimensional distribution of the random vector $(W_{t_1}, \ldots, W_{t_n})$ as,

$$P_{W_{t_1} \ldots W_{t_n}}(B) = \prod_{i=1}^{n} [2\pi (t_i - t_{i-1})]^{-1/2} \int_{B \subset \mathbb{R}^n} \exp\left[-\sum_{i=1}^{n} \frac{(y_i - y_{i-1})^2}{2(t_i - t_{i-1})}\right] dy_1 \ldots dy_n,$$
(2.12)

where $t_0 = 0$ and $y_0 = 0$ [27, p. 13].

The Wiener process is a random variable with one-dimensional parameters $t \in [0, \infty]$ defined on the probability space $(\mathscr{C}_0(\mathbb{R}^{[0,\infty)}), \mathscr{B}^{[0,\infty)}, P_{\mathbf{w}})$, where $\mathscr{C}_0(\mathbb{R}^{[0,\infty)})$ is the set of continuous real functions defined on $[0, \infty)$, starting at $t = 0$ from the point 0, and $P_{\mathbf{w}}$ is the distribution completely defined by finite dimensional distributions (2.12) through (2.10) and Theorem 2.1. Averages $M(f)(t)$ over the phase space, as well as higher order moments, are obtained, according to (2.4), by integration with respect to the probability measures (2.12).

2.1.2.1 Equivalent Representations of Random Functions

The value of y^ω for a fixed parameter λ defines a random variable, $\eta_\lambda(\omega) = y^\omega(\lambda)$, $\eta_\lambda : \Omega \longmapsto Y$. The phase space of η_λ coincides with the state space Y. The family of random variables $\{\eta_\lambda \mid \eta_\lambda : \Omega \longmapsto Y\}_{\lambda \in \Lambda}$ defines a random function in the *sense of Wiener*. It is equivalent to a random function in the sense of Doob if the corresponding distributions are projections of P_η,

$$P_{\eta_\lambda}(B) = P(\{\eta_\lambda \in B\}), \ \forall \lambda \in \Lambda.$$

A family of random variables $\{\eta_\lambda \mid \eta_\lambda : \Omega \longmapsto \mathbb{R}\}_{\lambda \in \Lambda}$ can be represented as *function of the parameter* λ with values in the set of random variables defined on Ω with values in Y, denoted by $\Lambda \xrightarrow{\eta_\lambda} L^0((\Omega, \mathscr{A}), (Y, \mathscr{B}))$. This representation is used in the "random analysis," where random functions are studied with methods of functional analysis [18, 27]. In particular, the random function can be seen as *a trajectory in the Hilbert space* $L^2(\Omega, \mathscr{A}, P)$ of square integrable complex functions with scalar product on Ω defined by $(\eta_1, \eta_2) = M(\eta_1 \bar{\eta}_2)$, where the overbar denotes the complex conjugate and M is the stochastic average (2.3). This representation is mainly useful in *correlation theory of second order stationary functions* [10, 28]. The covariance function, for instance, is given by

$$Cov(\lambda_1, \lambda_2) = (\eta_{\lambda_1}, \overline{\eta_{\lambda_2}}) - (\eta_{\lambda_1}, 1)(1, \overline{\eta_{\lambda_2}}).$$

Another useful representation of the random function is a *function of two variables*, $\eta_\lambda(\omega) = \eta(\lambda, \omega)$, $\eta : \Lambda \times \Omega \to Y$. The sufficient condition of measurability is that for every fixed ω the function η be (left or right) continuous as function of λ [27, p. 22]. From (2.3) and Fubini's theorem the stochastic average of the measurable random function η is given by

$$\int_\Lambda \int_\Omega \eta(\lambda, \omega) P_\Lambda(d\lambda) P(d\omega) = \int_\Lambda M[\eta](\lambda) P_\Lambda(d\lambda) = M\left[\int_\Lambda \eta(\lambda, \omega) P_\Lambda(d\lambda)\right],$$

where $P_\Lambda(d\lambda)$ is a measure on Λ. For instance, if one considers the Lebesgue measure on real axis and the uniform probability measure on $[0, T]$ given by $P_T(dt) = dt/T$ (the assumption of time homogeneity from non-relativistic mechanics), one obtains the useful property of permutation between stochastic and time averages

$$\frac{1}{T} \int_0^T M\,[\eta]\,(t)dt = M\left[\frac{1}{T} \int_0^T \eta\,(t,\omega)\,dt\right].$$

General conditions of equivalence of these definitions are presented in [10, p. 164]. While the Wiener and two variables function definitions are useful in specific applications, the Doob definition remains the main tool to build a coherent frame for a general theory of Markov and diffusion processes.

2.1.2.2 Remarks

1. Two random variables defined on the same probability space with values in the same phase space, $\eta : \Omega \to X$ and $\zeta : \Omega \to X$, are called P-a.s. (almost sure) equivalent, $\eta \underset{P\text{-}a.s.}{\sim} \zeta$, if $P(\{\eta \neq \zeta\}) = 0$, where $\{\eta \neq \zeta\} = \{\omega \in \Omega \mid \eta(\omega) \neq \zeta(\omega)\}$. If $\eta \underset{P\text{-}a.s.}{\sim} \zeta$, then the corresponding distributions coincide, $P_\eta \equiv P_\zeta$ [23, p. 11].

   The converse statement is not true. Let $\eta, \zeta : (0, 1) \to \{0, 1\}$, $\eta = 1$ for $\omega \leq 0.5$, $\eta = 0$ for $\omega > 0.5$ and $\zeta = |1 - \eta|$. The basic probability space Ω is now the interval $(0, 1)$, the corresponding σ-algebra $\mathscr{A}$ is the Borel σ-algebra, and the probability measure is the Lebesgue measure $P[(x_1, x_2)] = x_2 - x_1$, $\forall (x_1, x_2) \in \mathscr{A}$. The σ-algebra $\mathscr{B}$ on the phase space $X = \{0, 1\}$ contains the non-empty sets

$$B_1 = \{1\},\ B_2 = \{0\},\ \text{and}\ B_3 = \{0, 1\},$$

and the corresponding counter-images through η and ζ are

$$\{\eta \in B_1\} = (0, 0.5],\ \{\zeta \in B_1\} = (0.5, 1),$$

$$\{\eta \in B_2\} = (0.5, 1),\ \{\zeta \in B_2\} = (0, 0.5],$$

$$\{\eta \in B_3\} = \{\zeta \in B_3\} = (0, 1).$$

Thus, using the definition (2.2) of the distribution, $P_\eta (B) = P(\{\eta \in B\})$, $\forall B \in \mathscr{B}$, we find that the random variables η and ζ have the same distribution. However, because $P(\{\eta \neq \zeta\}) = P[(0, 1)] = 1$, η and ζ are not P-a.s. equivalent.

2. Even though not P-a.s. equivalent, η and ζ from the example above have identical mean and moments, because they have the same distribution. This

shows that random variables with identical distributions belong to a larger equivalence class.

The σ-algebra $\mathscr{A}_\eta$ on Ω generated by sets $\{\eta \in B\}$,

$$\mathscr{A}_\eta = \sigma\{\{\eta \in B\} \mid B \in \mathscr{B}\},$$

is called a *minimal σ-algebra*. Since η is $(\mathscr{A}, \mathscr{B})$ measurable, $\mathscr{A}_\eta \subset \mathscr{A}$. The triplet $(\Omega, \mathscr{A}_\eta, P_{\mathscr{A}_\eta})$, where $P_{\mathscr{A}_\eta}$ is the restriction of the measure P to $\mathscr{A}_\eta$, is called *minimal probability space*. It is shown that *the minimal probability space constructed for a random variable η is isomorphic to the phase space,* $(\Omega, \mathscr{A}_\eta, P_{\mathscr{A}_\eta}) \sim (X, \mathscr{B}, P_\eta)$ [23, p. 15].

Random variables with identical distributions defined on the same measurable space $(X, \mathscr{B})$ have the same phase space $(X, \mathscr{B}, P_\eta)$ and their minimal probability spaces are isomorphic. More generally, the random variables $\eta_1 :$ $\Omega_1 \rightarrow X_1$ and $\eta_2 : \Omega_2 \rightarrow X_2$ are called P_η-*equivalent* if their minimal probability spaces are isomorphic, $(\Omega_1, \mathscr{A}_{\eta_1}, P_{\mathscr{A}_{\eta_1}}) \sim (\Omega_2, \mathscr{A}_{\eta_2}, P_{\mathscr{A}_{\eta_2}})$. The random variables η and ζ from the example given at remark 1 above are P_η-*equivalent* representatives of the equivalence class specified by their phase space.

Given a phase space $(X, \mathscr{B}, P_\eta)$, the simplest representative is the identity function defined on X. Another representative of the same class is a surjective function $\eta : X \longrightarrow X$, which satisfies $P_\eta(B) = P_\eta(\eta^{-1}B)$, $\forall B \in \mathscr{B}$. The latter is the measure-preserving endomorphism on phase spaces studied in ergodic theory [3].

The probability measures constructed according to Kolmogorov's theorem (Theorem 2.1) rather define equivalence classes than representatives. The random functions in the sense of Doob are classes of P_η equivalence.

In applications, P-*a.s.* equivalence characterizes the *strong solutions* of the stochastic differential Itô equation. The *weak solutions*, which only guarantee the correctness of the expected values, are representatives of classes of P_η-equivalent random functions.

3. The conditions of Theorem 2.1 can be too restrictive for some applications. Let us consider an example from [27, p. 19]: a random variable $\tau : \Omega \rightarrow \mathbb{R}_+$ and two random functions $\eta, \zeta : \Omega \rightarrow [0, 1]^{\mathbb{R}_+}$, defined by $\eta_\lambda(\omega) \equiv 0$, $\zeta_\lambda(\omega) = 0$ for $\omega \in \{\tau \neq \lambda\}$, $\zeta_\lambda(\omega) = 1$ for $\omega \in \{\tau = \lambda\}$, $\forall \lambda \in \mathbb{R}_+$.

Since $P\{\eta_\lambda \neq \zeta_\lambda\} = P\{\tau = \lambda\} = 0$ the two random functions are P-*a.s.* equivalent and their distributions coincide (see remark 1). According to Theorem 2.1 η and ζ belong to the same equivalence class constructed with the measurable space $([0, 1]^{\mathbb{R}_+}, \mathscr{B}^{\mathbb{R}_+}_{[0,1]})$.

Essential in this construction was the Borel σ-algebra, on which the distributions of the two variables coincided. If instead one considers a σ-algebra which contains sets of bounded functions, the two random functions are no longer P_η-equivalent. Indeed, in this case $P(\{\sup_{0\le\lambda\le 1} \eta_\lambda \le 1/2\}) = 1$ and $P(\{\sup_{0\le\lambda\le 1} \zeta_\lambda \le 1/2\}) = 0$.

2.1.3   Densities of Finite Dimensional Probability Distributions

Let $(Y, \mathscr{B})$ be a state space, μ a measure on $\mathscr{B}$, and P_η the distribution of the random variable $\eta : \Omega \longmapsto Y$. If the measure P_η is absolutely continuous with respect to μ, i.e., $P_\eta(B) = 0$ for every $B \in \mathscr{B}$, with $\mu(B) = 0$, then the Radon–Nikodym theorem defines the *probability density* p_η of the distribution P_η:

$$\int_B p_\eta(y)\mu(dy) = P_\eta(B). \tag{2.13}$$

By (2.13), one associates to the set function P_η a point function p_η, which is Lebesgue integrable with respect to the measure μ and uniquely defined up to sets of null Lebesgue measure, i.e., it belongs to the set of integrable functions, $p_\eta \in L^1(Y, \mathscr{B}, \mu)$. Also, as a density of a probability measure, p_η has the normalization property $\int_Y p_\eta(y)dy = 1$. In the case of real random variables, i.e., $Y \subseteq \mathbb{R}$, $\mu(dy) = dy$ is the Lebesgue measure on the Borel σ-algebra $\mathscr{B}$ (defined on the set of all open sets from Y) and, from (2.13), one obtains $P_\eta(dy) = p_\eta(y)dy$. The notions introduced here as well as in the next section for the one-dimensional state space Y can be readily extended to higher dimensions by considering $Y \subseteq \mathbb{R}^d$, $d > 1$, and Cartesian products of sets from the Borel σ-algebra $\mathscr{B}$ defined on the d-dimensional state space Y.

From Radon–Nikodym theorem one obtains also the densities of higher order finite dimensional distributions. Let $\eta : \Omega \longmapsto Y^\Lambda$, $\eta(\omega) = y^\omega$, be a random function defined on the probability space $(\Omega, \mathscr{A}, P)$ and $\eta_\lambda : \Omega \longmapsto Y$, $\eta_\lambda(\omega) = y^\omega(\lambda)$, be random variables on the measurable space $(Y, \mathscr{B})$, $Y \subseteq \mathbb{R}$, obtained by fixing the parameter λ. Considering the set $B = B_1 \times \cdots \times B_n$, $B_i \in \mathscr{B}$, and the characteristic function of the set B_i, $1_{B_i}(y)$ (defined by $1_{B_i}(y) = 1$ if $y \in B_i$ and $1_{B_i}(y) = 0$ if $y \notin B_i$), the n-dimensional distribution (2.7) can be written as

$$P_{\lambda_1,\ldots,\lambda_n}(B) = \int_\Omega 1_{B_1}(\eta_{\lambda_1}(\omega)) \ldots 1_{Bn}(\eta_{\lambda_n}(\omega)) P(d\omega). \tag{2.14}$$

The value of the function 1_{B_i} at $y_i = \eta_{\lambda i}(\omega)$ can formally be written using the Dirac function [13, p. 200],

$$1_{Bi}(\eta_{\lambda i}(\omega)) = \int_\mathbb{R} \delta(y_i - \eta_{\lambda_i}(\omega))1_{B_i}(y_i)dy_i = \int_{B_i} \delta(y_i - \eta_{\lambda_i}(\omega))dy_i, \tag{2.15}$$

With (2.15), (2.14) becomes

$$P_{\lambda_1,\ldots,\lambda_n}(B) = \int_{B_1} dy_1 \ldots \int_{Bn} M_\Omega[\delta(y_1 - \eta_{\lambda_1}(\omega)) \ldots \delta(y_n - \eta_{\lambda_n}(\omega))]dy_n, \tag{2.16}$$

where M_Ω is the averaging operator (2.4). If B is a set of null measure in $\mathscr{B}^n$, then at least one of the integrals from (2.16) vanishes, thus $P_{\lambda_1,...,\lambda_n}(B)$ is an absolutely continuous measure with respect to the Lebesgue measure on $\mathscr{B}^n$. With this condition of the Radon–Nikodym theorem fulfilled, (2.16) defines the *n-dimensional density* (i.e., the density of the n-dimensional distribution $P_{\lambda_1,...,\lambda_n}$), as an integrable function from $L^1(Y^n)$, through

$$p(y_1, \lambda_1; \ldots; y_n, \lambda_n) = M_\Omega[\delta(y_1 - \eta_{\lambda_1}(\omega))\ldots\delta(y_n - \eta_{\lambda_n}(\omega))]. \qquad (2.17)$$

The formula (2.17) is used in [26, Chap. 3] to introduce the hierarchy of finite dimensional distributions associated with a stochastic process and is often used in studies on turbulent diffusion [16]. For $n = 1$ and fixed λ, (2.17) gives the one-dimensional density

$$p(y, \lambda) = M_\Omega[\delta(y - \eta_\lambda(\omega))] = \int_\Omega \delta(y - \eta_\lambda(\omega))P(d\omega). \qquad (2.18)$$

The density (2.18) verifies the normalization condition $\int_Y p(y, \lambda)dy = 1$. Indeed, because $\eta_\lambda(\omega) \in Y, \ \forall\, \omega \in \Omega$ we have $1_Y(\eta_\lambda(\omega)) \equiv 1$. Then,

$$\int_Y p(y, \lambda)dy = \int_\Omega P(d\omega) \int_{\mathbb{R}} 1_Y(y)\delta(y - \eta_\lambda(\omega))dy$$

$$= \int_\Omega P(d\omega)1_Y(\eta_\lambda(\omega)) = \int_\Omega P(d\omega) = 1.$$

From (2.17) it follows that $p(y_1, \lambda_1; \ldots; y_n, \lambda_n)$ is an invariant function with respect to permutations of pairs (y_i, λ_i) and it verifies the relation

$$\int_Y p(y_1, \lambda_1; \ldots; y_n, \lambda_n)dy_n = p(y_1, \lambda_1; \ldots; y_{n-1}, \lambda_{n-1}). \qquad (2.19)$$

Thus, for given random functions with real trajectories the finite dimensional densities defined by (2.17) are consistent [26, Chap. 3].

Conversely, given a hierarchy of consistent densities, the distributions

$$P_{\lambda_1,...,\lambda_n}(B) = \int_{B_1} dy_1... \int_{Bn} p(y_1, \lambda_1; \ldots; y_n, \lambda_n)dy_n \qquad (2.20)$$

verify the consistency conditions (2.8) and (2.9), and Theorem 2.1 ensures the existence of the random function η.

2.2 Markov and Diffusion Processes

2.2.1 Markov Processes

Let us consider a stochastic process with real trajectories, $\eta : \Omega \longmapsto Y^T$, $Y \subseteq \mathbb{R}$, $T \subseteq \mathbb{R}_+$, and the sets

$$A_1 = B_1 \times \cdots \times B_r \underbrace{\times Y \cdots \times Y}_{(n-r)\ times}, \qquad A_2 = \underbrace{Y \cdots \times Y}_{r\ times} \times B_{r+1} \times \cdots \times B_n,$$

and their intersection $A_1 \bigcap A_2 = B_1 \times \cdots \times B_n$, where $B_i \in \mathcal{B}$, and $\mathcal{B}$ is the Borel σ-algebra on Y. The distributions $P_{t_1,\dots,t_n}(A_1 \bigcap A_2)$ and $P_{t_1,\dots,t_n}(A_2)$ are related to the conditional probability $P_{t_1,\dots,t_n}(A_1|A_2)$ through the Bayes' theorem,

$$P_{t_1,\dots,t_n}(A_1 \bigcap A_2) = P_{t_1,\dots,t_n}(A_1|A_2) P_{t_1,\dots,t_n}(A_2).$$

Using (2.20) one obtains the corresponding *conditional probability density*

$$p(y_1, t_1; \dots; y_r, t_r \mid y_{r+1}, t_{r+1}; \dots; y_n, t_n) = \frac{p(y_1, t_1; \dots; y_n, t_n)}{p(y_{r+1}, t_{r+1}; \dots; y_n, t_n)}. \quad (2.21)$$

Markov processes are characterized by the dependence of conditional probabilities on only one of the previous states but not on the whole history of the process η. For $t_1 > \cdots > t_n$, the Markov property is expressed by

$$p(y_1, t_1; \dots; y_r, t_r \mid y_{r+1}, t_{r+1}; \dots; y_n, t_n) = p(y_1, t_1; \dots; y_r, t_r \mid y_{r+1}, t_{r+1}). \quad (2.22)$$

From (2.21) and (2.22) one obtains

$$p(y_1, t_1; \dots; y_n, t_n) = p(y_1, t_1 \mid y_2, t_2) \dots p(y_{n-1}, t_{n-1} \mid y_n, t_n) p(y_n, t_n). \quad (2.23)$$

From relation (2.23) and Theorem 2.1 it follows that Markov processes are completely described by the one-dimensional probability and the conditional probability for two states, called *transition probability*. For $t_1 > t_2 > t_3$ and using (2.23), the consistency condition (2.19) written for Markov processes becomes the *Chapman–Kolmogorov equation*,

$$p(y_1, t_1 \mid y_3, t_3) = \int_Y p(y_1, t_1 \mid y_2, t_2) p(y_2, t_2 \mid y_3, t_3) dy_2. \quad (2.24)$$

The property (2.23) of Markov processes establishes a connection between the one-dimensional distributions at successive times. From the definition of finite dimensional densities (2.7) and the consistency property (2.9), it follows that $P_{t_1}(B) = P(\eta_{t_1} \in B) = P(\eta_{t_1} \in B; \eta_{t_2} \in Y)$, for every $B \in \mathcal{B}$. With formula (2.20) and relation (2.23), written for two successive times, we obtain

$$P_{t_1}(B) = \int_B p(y_1, t_1) dy_1 = \int_B dy_1 \int_Y p(y_1, t_1; y_2, t_2) dy_2$$
$$= \int_B dy_1 \int_Y p(y_1, t_1 \mid y_2, t_2) p(y_2, t_2) dy_2.$$

Thus, the one-dimensional density at t_1 can be calculated from the corresponding density at t_2 and the transition probability with the relation

$$p(y_1, t_1) = \int_Y p(y_1, t_1 \mid y_2, t_2) p(y_2, t_2) dy_2 = (K_{t_1,t_2} p)(y_1, t_1). \qquad (2.25)$$

Equation (2.25) introduces the *Markov operator*, $K_{t_1,t_2} : L^1(Y) \longmapsto L^1(Y)$, where L^1 is the space of Lebesgue integrable functions such that $\int_Y |f(y)| dy < \infty$. Since the density of the transition probability $p(y_1, t_1 \mid y_2, t_2)$ fulfills the positivity and normalization conditions,

$$(k1) \qquad p(y_1, t_1 \mid y_2, t_2) \geq 0,$$

$$(k2) \int_Y p(y_1, t_1 \mid y_2, t_2) dy_1 = 1,$$

it is a *stochastic kernel*. The *kernel-type Markov operator* K_{t_1,t_2} is a linear operator on $L^1(Y, \mathcal{B}, \mu)$ and, as follows from (k1-k2), has the properties [14, p. 101],

$$(M1) \ K_{t_1,t_2} f \geq 0,$$
$$(M2) \ \| K_{t_1,t_2} f \|_{L^1} = \|f\|_{L^1}, \ \forall f > 0, \ f \in L^1(Y).$$

Using Markov operators, the Chapman–Kolmogorov equation (2.24) takes the form

$$K_{t_1,t_3} f = K_{t_1,t_2} K_{t_2,t_3} f, \ \forall f \in L^1(Y), \ t_1 > t_2 > t_3.$$

Stationary Markov processes are defined by

$$p(y_1, t_1 \mid y_2, t_2) = p_s(y_1, \tau \mid y_2) \text{ and } p(y, \tau) = p_s(y), \text{ where } \tau = t_1 - t_2.$$

Stationary Markov operators form a semigroup of operators, $\{K_\tau\}_{\tau \geq 0}$,

$$K_\tau f(y) = \int_Y p(y, \tau \mid y_0) f(y_0) dy_0,$$

$$K_{\tau_1 + \tau_2} f = K_{\tau_1} K_{\tau_2} f, \ \forall f \in L^1(Y), \ \forall \tau_1, \tau_2 \geq 0,$$

$$K_0 f(y) = \int_Y \delta(y - y_0) f(y_0) dy_0 = f(y), \ \forall f \in L^1(Y).$$

The *strong ergodicity* property of stationary Markov processes

$$p_s(y_1, t_1 - t_2 \mid y_2) \longrightarrow p_s(y_1), \ \text{for } (t_1 - t_2) \longrightarrow \infty$$

ensures the equality of time and stochastic averages (it is a sufficient condition for vanishing variance of mean and correlation function estimated through time averages [8, Sect. 3.7.3]).

Nonstationary processes obtained by a "preparation of the initial state" using strong ergodic Markov processes,

$$p(y, t) = p_s(y, t - t_0 \mid y_0) \text{ and } p(y, t \mid y', t') = p_s(y, t - t' \mid y'), \ \forall t > t' > t_0,$$

possess the following convergence property:

$$\|K_t p - p_s\|_{L^1} \to 0, \ \text{for } t \to \infty,$$

which is a form of *exactness property* for semigroups of operators used to model thermodynamic irreversibility in ergodic theory [22–24]. Exactness implies the *mixing* of K_t and, as a consequence, the decay of correlations

$$Cor[f(y(t_1))f(y(t_2))] = \{M_Y[f(y(t_1))f(y(t_2))] - M_Y[f(y(t_1))]M_Y[f(y(t_2))]\} \to 0.$$

Markov operators provide a unitary frame for both Markov processes and dynamical systems, viewed as degenerated Markov processes.

Dynamical systems $\{S_t \mid S_t : Y \longmapsto Y, t \in \mathbb{R}\}$ can be viewed as families of $(\mathscr{B}, \mathscr{B})$ measurable random variables defined on the measure space $(Y, \mathscr{B}, P)$, where the measure P has a density p with respect to the Lebesgue measure [22, 23]. For f and g integrable functions, the Frobenius–Perron operator U_t is defined as the adjoint of the Koopman operator $U_t^* g(y) = g(S_t(y))$ induced by the dynamical system,

$$\int_Y f(y) U_t^* g(y) dy = \int_Y g(y) U_t f(y) dy.$$

This relation implies, with the action of the Koopman operator expressed as $U_t^* g(y) = g(S_t(y)) = \int_Y g(y')\delta(y' - S_t(y))dy'$, that the Frobenius–Perron operator is explicitly defined by $U_t f(y') = \int_Y \delta(y' - S_t(y)) f(y)dy$.

The Frobenius–Perron U_t is stationary and verifies the properties M1 and M2 of the Markov operator. It acts on one-dimensional densities like a stationary Markov operator,

$$U_t p(y) = \int_Y p(y, t \mid y_0) p(y_0) dy_0,$$

with singular density of the transition probability localized on the trajectory of the dynamical system

$$p(y, t \mid y_0) = \delta(y - S_t(y_0)).$$

Note that for non-degenerated Markov processes, the transition probability constructed by (2.21) and (2.17) has a support of non-zero Lebesgue measure. The reason is that, while a dynamical system is a function which maps the point $y \in Y$ into $S_t(y) \in Y$ for every fixed t, a stochastic process consists of an infinity of trajectories $\eta_t(\omega)$, indexed by $\omega \in \Omega$, passing through every point $y \in Y$.

2.2.2   Diffusion Processes and Fokker–Planck Equations

2.2.2.1   Definitions, Properties, Examples

Let $\eta : \Omega \longmapsto Y^T$, $Y \subseteq \mathbb{R}^d$, $T \subseteq \mathbb{R}$, be a Markov process defined on the probability space $(\Omega, \mathscr{A}, P)$ valued into the trajectories space $\{y^\omega \in Y^T \mid y^\omega : T \longmapsto Y, y^\omega = \eta(\omega), \omega \in \Omega\}$, and $(Y, \mathscr{B})$ its measurable state space endowed with the Borel σ-algebra $\mathscr{B}$. For this d-dimensional process, the densities of the finite dimensional distributions absolutely continuous with respect to Lebesgue measure and the corresponding transition densities are defined similarly to one-dimensional processes by using (2.17) and (2.21).

The Markov process defined above is a *diffusion process* in the sense of Kolmogorov [8, 11, 27] if for any $\varepsilon > 0$ the transition probabilities satisfy, uniformly in x and t as $\Delta t \longrightarrow 0$, the conditions

(i) $\lim\limits_{\Delta t \to 0} \frac{1}{\Delta t} \int\limits_{|y-x| \geq \varepsilon} p(y, t + \Delta t \mid x, t)dy = 0$,

(ii) $\lim\limits_{\Delta t \to 0} \frac{1}{\Delta t} \int\limits_{|y-x| < \varepsilon} (y_i - x_i)p(y, t + \Delta t \mid x, t)dy = A_i(x, t) + \mathscr{O}(\varepsilon)$,

(iii) $\lim\limits_{\Delta t \to 0} \frac{1}{\Delta t} \int\limits_{|y-x| < \varepsilon} (y_i - x_i)(y_j - x_j)p(y, t + \Delta t \mid x, t)dy = 2B_{ij}(x, t) + \mathscr{O}(\varepsilon)$.

The condition (i) ensures *the continuity with probability* 1 for the trajectories of the Markov process [8, p. 46], [27, p. 167]. This property means that for almost all values of ω (excepting sets of null measure, $A \in \mathscr{A}$, $P(A) = 0$) the trajectories y^ω are continuous time functions. Continuous Markov processes are sometimes called "diffusion processes in a large sense."

If there exist partial derivatives of A_i, $B_{i,j}$ and of the transition probabilities p, then, from (i)–(iii) in the limit $\varepsilon \longrightarrow 0$ one derives the *Fokker–Planck* equation

$$\partial_t p(x,t \mid x_0, t_0) = -\nabla[Ap(x,t \mid x_0, t_0)] + \nabla^2[\tilde{B}(x,t) \, p(x,t \mid x_0, t_0)], \quad (2.26)$$

where A is the *drift vector* and $\tilde{B}$, the *diffusion tensor*. If $\tilde{B}$ is positively defined the Eq. (2.26) can be solved if the initial condition $p(x,t \mid x_0, t) = \delta(x - x_0)$ and suitable boundary conditions are assumed [8, Sect. 3.5.2].

The integrals in (ii)–(iii) have the meaning of local averages, $M_{|y-x|<\varepsilon}[f(y) \mid x, t](t + \Delta t; x, t)$, of some functions $f(y)$ defined on Y. These averages are calculated inside a sphere of radius ε, at the time moment $t + \Delta t$ and conditioned by the value x taken by the trajectories of the process at the time t. Using the mean value theorem and the condition (i), one estimates the average over the state space Y at $t + \Delta t$ as follows:

$$M[f(y) \mid x, t](t + \Delta t; x, t) = \int_{|y-x|<\varepsilon} f(y)p(y, t + \Delta t \mid x, t)dy$$

$$+ M_{|y-x|\geq\varepsilon}[f(y) \mid x, t](t + \Delta t; x, t) \int_{|y-x|\geq\varepsilon} p(y, t + \Delta t \mid x, t)dy \quad (2.27)$$

$$= M_{|y-x|<\varepsilon}[f(y) \mid x, t](t + \Delta t; x, t) + o\,(\Delta t).$$

Thus, for continuous processes the local average is of the same order of magnitude for $\Delta t \to 0$ as the average over the entire state space (here and above, o and $\mathcal{O}$ stand for the Euler's relations "of smaller order" and "of the same order," see, e.g., [9, Sect. 1.2]). In (ii)–(iii) there is no relation between ε and Δt. However, since (i)–(iii) hold for any $\varepsilon > 0$, the order relation $\varepsilon = o(\Delta t)$ can be assumed [27, Chap. 11]. With this assumption, from (2.27), (ii), and (iii), one obtains

$$A_i(x, t) = \frac{d}{ds} M[(y_i - x_i) \mid x, t](t + s; x, t) \mid_{s=0}, \quad (2.28)$$

respectively,

$$B_{ij}(x, t) = \frac{1}{2} \frac{d}{ds} M[(y_i - x_i)(y_j - x_j) \mid x, t](t + s; x, t) \mid_{s=0}. \quad (2.29)$$

This proves that for continuous processes (i.e., verifying the condition (i)) the existence of the coefficients (2.28) and (2.29) under the order of magnitude hypothesis $\varepsilon = o(\Delta t)$ is equivalent with the diffusion conditions (ii) and (iii). The coefficients (2.28) and (2.29) are "forward derivatives" with respect to time of the mean value and variance of the displacement $(y_i - x_i)$, conditioned by the initial state (x, t) (the variance is obtained when the product $M[(y_i - x_i) \mid x, t]M[(y_j - x_j) \mid x, t]$, which is of the order $o(s)$ for $s \to 0$, is subtracted from the mean in (2.29)). It is only under these conditions that diffusion coefficients can be defined as time derivatives of displacement variances, as, for instance, in [5, 25].

The typical diffusion process is the one-dimensional *Gaussian process* specified by the average x_0, the variance $\sigma^2 = 2D(t-t_0)$, where $D > 0$ is a constant diffusion coefficient, and the Gaussian probability density

$$p(x, t \mid x_0, t_0) = [4\pi D(t - t_0)]^{-\frac{1}{2}} \exp[-(x - x_0)^2/4D(t - t_0)]. \tag{2.30}$$

Since the variance is a linear time function, the Gaussian process with constant coefficients has a remarkable reproducibility property: processes starting at any initial time have the same variance after the same time interval. It was shown that their images in the state space are fractal sets [14, 15]. An important consequence of this behavior is that over infinitesimal time intervals Δt the increments of the process are of order $\sqrt{\Delta t}$ and so the quotient defining the time derivative of the process is of order $1/\sqrt{\Delta t}$ and its limit as $\Delta t \longrightarrow 0$ does not exist. This heuristic argument is completed by a theorem proved by Paley, Wiener, and Zygmund which states that the sample paths of the Gaussian diffusion process with constant coefficients are nowhere differentiable with probability one [20].

Using (2.28) and (2.29), one obtains the coefficients for the Fokker–Planck equation which corresponds to the process with transition density (2.30). The result is the *diffusion equation*

$$\partial_t p = D\partial_x^2 p. \tag{2.31}$$

More generally, for a non-vanishing constant drift coefficient u one obtains the *advection–diffusion equation*

$$\partial_t p + u\partial_x p = D\partial_x^2 p \tag{2.32}$$

and the transition probability is given by (2.30) with x_0 replaced by $x_0 + u(t - t_0)$.

The solutions of the parabolic equations like (2.31)–(2.32) are uniquely determined by initial and boundary conditions [2, 4, 8, 15]. For instance, the transition probability (2.30) is a solution of Eq. (2.31) with the initial condition $p(y, t_0 \mid x, t_0) = \delta(y - x)$. The one-dimensional probability density

$$p(x, t) = \int_{\mathbb{R}} p(x, t \mid x_0, t_0)p(x_0, t_0)dx_0$$

also verifies Eq. (2.31), with the initial condition $p(x, t_0) = p(x_0, t_0)$. This proves that the transition probability density $p(x, t \mid x_0, t_0)$ is the fundamental solution of Eq. (2.31).

The continuity condition (i) can be generalized as

$$(\mathrm{i}')\ \lim_{\Delta t \to 0} \frac{1}{\Delta t} \int_{|y-x| \geq \varepsilon} p(y, t + \Delta t \mid x, t) dy = \Psi(y \mid x, t),$$

where Ψ is the *jump probability* from state x into state y at time t [8, 26]. Instead of (2.26) one obtains, under the same conditions as for the Fokker–Planck equation, an *integro-differential form of the Chapman–Kolmogorov equation*,

$$\partial_t p(x, t \mid x_0, t_0) =$$

$$- \nabla[\mathbf{A} p(x, t \mid x_0, t_0)] \hspace{5cm} a)\ \text{Liouville}$$

$$+ \nabla^2[\tilde{\mathbf{B}}(x, t)\, p(x, t \mid x_0, t_0)] \hspace{4cm} b)\ \text{diffusion}$$

$$+ \int_Y dy[\Psi(x \mid y, t) p(y, t \mid x_0, t_0) - \Psi(y \mid x, t) p(x, t \mid x_0, t_0)]. \hspace{0.5cm} c)\ \text{master}$$

$$(2.33)$$

If $\Psi \geq 0$, almost everywhere (a.e.), then the integro-differential equation (2.33) has solutions for the same initial and boundary conditions as for the Fokker–Planck equation [8, Sect. 3.7]. Equation (2.33) shows that generally the trajectory of the Markov process can be represented as sum of three functions: a continuous and derivable function, the determinist component, with a probability density function solution of the Liouville equation (2.33a), when $\tilde{B} \equiv \tilde{0}$ and $\Psi \equiv 0$, a continuous and nowhere derivable function, with a probability density function solution of the diffusion equation (2.33b), when $A \equiv 0$ and $\Psi \equiv 0$, and a jump function (discontinuous at all times), with a probability density function solution of the *master equation* (2.33c), when $A \equiv 0$ and $\tilde{B} \equiv \tilde{0}$. Under some conditions, the master equation approximates the diffusion equation [8, Chap. 7] and, reciprocally, there are situations when the master equation is approximated by a diffusion equation [26, Chap. 10].

The asymptotic behavior of the solutions is governed by the following theorem [8, Sect. 3.7.3].

Theorem 2.2 *If the one-dimensional stationary density $p_s(x) \neq 0$ a.e., the state space Y is simply connected, and the diffusion tensor $\tilde{B}$ and jump probability Ψ are a.e. positively defined, then the solutions of the integro-differential Chapman–Kolmogorov equation (2.33) tend to the stationary solution p_s,*

$$\|p(x, t) - p_s(x)\|_{L^1(Y)} \longrightarrow 0, \quad \text{for } t \to \infty. \hspace{2cm} \square$$

2.2.2.2 Examples

1. Random Walk Intuitively, the one-dimensional random walk is an undecided movement with left-right steps of equal length taken with equal probabilities. Generally, the transition and the one-dimensional probabilities of the random walk process are solutions of the *master equation* (2.33c) particularized to a discrete state space $Y \subseteq \mathbb{Z}$,

$$\partial_t P(n, t) = d[P(n - 1, t) + P(n + 1, t) - 2P(n, t)].$$

The master equation describes the balance of transitions from a state n to first neighbor states, with jump probabilities $\Psi(n|n - 1, t) = \Psi(n - 1|n, t) = \Psi(n|n + 1, t) = \Psi(n + 1|n, t) = d$, and $\Psi(n|m, t) = 0$ for $n = m$ or $|n - m| > 1$.

After discretization of the time derivative,

$$\partial_t P(n, t) = [P(n, (k + 1)\delta t) - P(n, k\delta t)]/\delta t,$$

the master equation takes the form of a finite difference (FD) scheme for the heat equation in continuous space, $\partial_t P = D\partial_x^2 P$, with x represented as $x = n\delta x$ and diffusion coefficient defined by $D = d\delta x^2$.

It is also noticeable that, using the discretization of the time derivative and defining $d = 1/(2\delta t)$, the master equation takes the form of the traditional intuitive description of the random walk in *discrete time* [8, Sect. 3.8.2],

$$P(n, (k + 1)\delta t) = \frac{1}{2}[P(n - 1, k\delta t) + P(n + 1, k\delta t)].$$

Both equations can be solved by using the *characteristic function*, defined as Fourier transform of the probability distribution,

$$G(s, t) = M(e^{ins}) = \sum_{n \in Y} P(n, t)e^{ins}$$

Since $\sum_{n \in Y} P(n, t)e^{ins} = \sum_{n \in Y} P(n \pm 1, t)e^{i(n \pm 1)s} = G(s, t)$, one obtains for master and discrete time equations, respectively,

$$\partial_t G_{mast}(s, t) = d[e^{is} + e^{-is} - 2]G_{mast}(s, t), \text{ and}$$

$$G_{discr}(s, k\delta t) = d[e^{is} + e^{-is}]G_{discr}(s, k\delta t).$$

For the initial condition $G(s, 0) = 1$, corresponding to $P(0, 0) = 1$ $P(n \neq 0, 0) = 0$, the solutions are, respectively,

$$G_{mast}(s, t) = \exp[(e^{is} + e^{-is} - 2)td], \text{ and } G_{discr}(s, k\delta t) = \left[\frac{1}{2}(e^{is} + e^{-is})\right]^k.$$

Considering $d\delta t = 1/2$ and $\delta t = t/k$, one obtains

$$\lim_{k\to\infty} G_{discr}(s, k\delta t) = \lim_{k\to\infty}\left[1 + \frac{1}{k}td(e^{is} + e^{-is} - 2)\right]^k$$

$$= \exp[(e^{is} + e^{-is} - 2)td] = G_{mast}(s, t),$$

and the same for the corresponding probability distributions, $\lim_{k\to\infty} P_{discr}$ $(n, k\delta t) = P_{mast}(n, t)$.

The probability distributions can be obtained from characteristic functions by discrete Fourier inversion. For instance, in case of discrete time one obtains the Bernoulli distribution which gives the probability of n heads in tossing an unbiased coin k times [8, p. 70],

$$P_{discr}(n, k\delta t) = \left(\frac{1}{2}\right)^k k! \left[\left(\frac{k-n}{2}\right)! \left(\frac{k+n}{2}\right)!\right]^{-1}.$$

As discrete state-space stochastic process, random walks provide natural discretization schemes for continuous diffusion processes. The relation with the diffusion process and the convergence properties of the random walk are at the hearth of the particle methods for transport equations.

2. Wiener Process The one-dimensional distribution of the Wiener process can be retrieved as limit of continuous space of the random walk distribution $P_{mast}(n, t)$.

For $x = n\delta x$,

$$\Phi_{mast}(s, t) = M(e^{isn\delta x}) = G_{mast}(s\delta x, t) = \exp[(e^{is\delta x} + e^{-is\delta x} - 2)td].$$

Expanding $e^{is\delta x}$ and $e^{-is\delta x}$ up to second order in δx and defining $D = \lim_{\delta x\to 0} \delta x^2 d$ yields

$$\lim_{\delta x\to 0} \Phi_{mast}(s, t) = \exp(-s^2 tD) = \Phi_W(s, t).$$

Φ_W solves $\partial_t \Phi_W = -Ds^2 \Phi_W$ for $\Phi_W(s, 0) = 1$, i.e., it is the characteristic function of the Wiener process with density

$$p(x, t) = \frac{1}{\sqrt{4\pi Dt}} \exp\left(-\frac{x^2}{4Dt}\right),$$

solution of Fokker–Planck equation $\partial_t p = D\partial_x^2 p$ with $p(x, 0) = \delta(x)$. The limit of the discrete time random walk, $\lim_{\delta x\to 0, \delta t\to 0} \Phi_{discr}(s, t)$, gives the same result if $D = \lim_{\delta x\to 0, \delta t\to 0} \delta x^2/(2\delta t)$.

3. Poisson Process The one-sided random walk with jump probabilities $\Psi(n + 1|n, t) = d$ and $\Psi(n|m, t) = 0$ for any other m and n is governed by the master equation

$$\partial_t P(n, t) = d[P(n - 1, t) - P(n, t)].$$

The characteristic function $G(s, t) = \exp[(e^{is} - 1)td]$ solves

$$\partial_t G(s, t) = d[e^{is} - 1]G(s, t), \quad G(s, 0) = 1.$$

Expanding in powers of e^{is} one obtains the characteristic function

$$G(s, t) = e^{(-td)}[1 + (td)e^{is} + \frac{1}{2!}(td)^2(e^{is})^2 + \cdots],$$

and identifying with the definition $G(s, t) = \sum_{n \in Y} P(n, t)e^{ins}$ one obtains the Poisson probability distribution

$$P(n, t) = e^{(-td)}(td)^n/n!$$

For $x = n\delta x$, and taking the limit of continuous space $\delta x \to 0$, with $\delta x d \equiv v$ held fixed, one obtains

$$\lim_{\delta x \to 0} G(s, t) = \lim_{\delta x \to 0} \exp[(e^{is\delta x} - 1)td] = \lim_{\delta x \to 0} \exp[(1 + is\delta x + \cdots - 1)td] = \exp(isvt).$$

By inverse Fourier transform one obtains the probability density

$$p(x, t) = \frac{1}{2\pi} \int_{-\infty}^{\infty} \exp(isvt)\exp(-isx)ds = \frac{1}{2\pi} \int_{-\infty}^{\infty} \exp[-is(x - vt)]ds = \delta(x - vt).$$

In the same limit, from the master equation $\partial_t P(n\delta x, t) = d[P((n - 1)\delta x, t) - P(n\delta x, t)]$ one obtains the Liouville equation $\partial_t p(x, t) = -v\partial_x p(x, t)$.

If one expands the characteristic function up to second order in δx,

$$G(s, t) = \exp[(e^{is\delta x} - 1)td] \simeq \exp[(1 + is\delta x + \frac{1}{2}s^2\delta x^2 + \cdots - 1)td] \simeq \exp[(isv - s^2 D)t],$$

where $D = \delta x^2 d/2$. This is the characteristic function of the Gaussian process of mean vt, variance $2Dt$, and probability density

$$p(x, t) = (4\pi Dt)^{1/2} \exp[-(x - vt)^2/(4Dt)],$$

solution of the Fokker–Planck equation

$$\partial_t p(x, t) = -v\partial_x p(x, t) + D\partial_x^2 p(x, t).$$

The latter can also be obtained by expanding the master equation of the Poisson process up to order δx^2 [8, p. 72]. Note that this is an approximation not a limit: with well-defined $v = \lim_{\delta x \to 0} \delta x d$, the "diffusion coefficient" $D = \delta x^2 d/2 = v\delta x \to 0$ as $\delta x \to 0$ and one retrieves the Liouville equation.

While Examples 1 and 2 show how the master equation approximates the diffusion equation, Example 3 illustrates a situation where the master equation can be approximated by a diffusion equation. This is a very simple case of van Kampen's system size expansion [26, Chap. 7] and is also the principle of "upscaling" advection (Liouville) equations to diffusion equations in stochastic hydrology [5, 22].

2.2.3 Itô Equation and Stochastic-Lagrangian Framework

2.2.3.1 Gaussian White Noise and Wiener Process

The solution of the advection–diffusion equation (2.32) is of the form (2.30) where the mean value x_0 is replaced by $x_0 + u(t - t_0)$ (see also Example 3 above). If the derivatives of the coefficients of the Fokker–Planck equation (2.26) are negligible compared to those of the probability density p (e.g., evolutions over small time intervals close to a sharp initial condition [8, Sect. 3.5.2]), the diffusion process behaves locally as a Gaussian process with constant coefficients. Let us consider the one-dimensional process described by the advection–diffusion equation (2.32). For infinitesimal time intervals dt, the transition dx between the states of this process can be interpreted using the definitions (2.28) and (2.29) of the coefficients u and D as a superposition between a translation udt and a fluctuation with mean amplitude $\sqrt{\sigma^2} = \sqrt{2Ddt}$, i.e.,

$$dx(t) = udt + \gamma\sqrt{2D}\sqrt{dt}, \tag{2.34}$$

where γ is a realization of a random variable of mean zero and variance one. Assuming that Eq. (2.34) is integrable, the trajectory $x(t)$ is continuous and nowhere differentiable (because of the $\sqrt{dt}$ term which makes dx/dt infinite for $dt \to 0$).

The second term in the right-hand side of (2.34) is significant as compared with the first term only when $\sqrt{dt} = \mathcal{O}(dt)$. This order of magnitude relation comes in conflict with the usual rules of differential calculus. That is why new rules, called "Itô calculus," are introduced. Heuristically, one looks for a stochastic process ζ, $\zeta : \Omega \longmapsto Z^I$, $Z \subseteq \mathbb{R}$, $I \subseteq \mathbb{R}$, so that, at fixed time moments t, $\zeta(t, \omega) = \zeta_t(\omega)$ is a random variable describing the "fluctuation velocity" in (2.34), with realizations completely determined by those of the random variable γ,

$$\zeta_t dt = \gamma\sqrt{dt}. \tag{2.35}$$

Using (2.35), Eq. (2.34) can be rewritten in the integral form as

$$X(t, \omega) = X_0 + \int_{t_0}^{t} u \, dt' + \sqrt{2D} \int_{t_0}^{t} \zeta(t', \omega) \, dt'. \tag{2.36}$$

Hereafter, we shall use capital letters for processes and random variables and lowercase letters for their realizations. Hence, we distinguish the stochastic process $X(t, \omega)$, represented as function of two variables (see Sect. 2.1.2.1), from its realization $x(t)$ described by Eq. (2.34).

The solution of (2.36) corresponds to the process with transition probabilities verifying the Fokker–Planck equation (2.32) only if it has the same mean value and variance, i.e.,

$$M[X(t, \omega)] = x_0 + u(t - t_0),$$

$$M[X(t, \omega)^2] - M[X(t, \omega)]^2 = 2D \int_{t_0}^{t} dt' \int_{t_0}^{t} dt'' M[\zeta(t', \omega)\zeta(t'', \omega)]$$

$$= 2D(t - t_0). \tag{2.37}$$

The first relation (2.37) is a consequence of (2.35). The last relation holds only if the correlation function of the random variables $\zeta(t', \omega)$ and $\zeta(t'', \omega)$ has the singular form of a Dirac function,

$$M[\zeta(t', \omega)\zeta(t'', \omega)] = \delta\left(t' - t''\right). \tag{2.38}$$

The process ζ with zero mean and singular Dirac correlation is called *Gaussian white noise* [26, Chap. 8].

Since the variance given by the second relation (2.37) is linear in time, the process X has uncorrelated increments [30]. Moreover, there exists a wide sense Gaussian process version of X with independent increments [6] (a more detailed discussion on this topic is given in Chap. 4, Sect. 4.2.3). In particular, for $x_0 = 0, t_0 = 0, u = 0$, $D = 1/2$, and making the change of variables

$$dW(t, \omega) = \zeta(t, \omega) dt, \tag{2.39}$$

one obtains the process

$$X(t, \omega) = \int_{0}^{t} dW(t', \omega) = W(t, \omega).$$

The process W is a wide version of a Gaussian process with independent increments which starts from origin, thus it possesses the properties (w1)–(w3) of the Wiener process defined in Sect. 2.1.2.

Sometimes, the relation (2.39) is used to define the white noise as "derivative" of the Wiener process [14, p. 293], but this has only an intuitive meaning because, as we have seen in Sect. 2.2.2.1 the Wiener process, as well as all Gaussian processes, is not differentiable. The consistent mathematical interpretation of the white noise is obtained via (2.39): displacements produced by a white noise velocity are increments of a Wiener process. With this interpretation of the second integral in (2.36), the trajectories of the diffusion process are constructed in the sense of Einstein (see beginning of Chap. 1) by summing up independent increments of the Wiener process.

2.2.3.2 Itô Stochastic Differential Equation

The representation of the Wiener process as a function of two variables (Sect. 2.1.2.1), $W(t, \omega)$, $W : [0, \infty] \times \Omega \longmapsto Y$, $Y \subset \mathbb{R}^d$, and (2.39) allow the generalization of (2.34) to the *Itô stochastic differential equation*. The d-dimensional Itô equation with variable coefficients is usually written componentwise as

$$dX_i(t) = a_i(t, \mathbf{X}(t))dt + b_{ij}(t, \mathbf{X}(t))dW_j(t), \qquad (2.40)$$

where $i, j = 1 \cdots d$, the components of the Wiener process are mutually independent and, for the sake of simplicity, we drop the dependence on ω. If for every fixed ω there exists a unique solution of (2.40), $x_i(t) = X_i(t, \omega, \mathbf{x}_0, t_0)$, which satisfies the initial condition $\mathbf{x}(t_0) = \mathbf{x}_0$, $\mathbf{x}_0 \in Y$, then the time functions $\mathbf{x}(t)$ are trajectories of a stochastic process defined in the Cartesian product space $\Omega \times Y$, $\chi : \Omega \times Y \longmapsto Y^{[0,\infty]}$. For $a_i = 0$ and $b_{ij} = 1/2\delta_{ij}$, the Itô equation (2.40) describes a d-dimensional Wiener process, also known as continuous time Brownian motion.

To complete this picture of the diffusion process, we have to show that the probability densities of the process described by the Itô equation with variable coefficients verify a Fokker–Planck equation. To proceed, we need a relation between the stochastic average with respect to the density of the transition probability from the definition (i)–(iii) of the diffusion process and the average over the ensemble of trajectories generated by Eq. (2.40).

Considering the density of initial states given by $p(x_0, t_0)$ and using the definition (2.18) we obtain the one-dimensional probability density of the process χ at the moment t,

$$p(x, t) = M_{\Omega \times Y}[\delta(x - X(t, \omega, x_0, t_0))]$$

$$= \int_Y M_\Omega[\delta(x - X(t, \omega, x_0, t_0))]p(x_0, t_0)dx_0,$$

and comparing with the Markov operator (2.25), we find that the density of the transition probability of the process χ is given by

$$p(x, t \mid x_0, t_0) = M_\Omega[\delta(x - X(t, \omega, x_0, t_0))]. \tag{2.41}$$

The average of a function $f(x)$, conditioned by the initial state (x_0, t_0), is an integral weighted by the density of the transition probability (2.41),

$$M[f(x) \mid x_0, t_0](t; x_0, t_0) = \int_Y f(x) p(x, t \mid x_0, t_0) dx$$

$$= \int_\Omega P(d\omega) \int_Y f(x)\delta(x - X(t, \omega, x_0, t_0)) dx$$

$$= M_\Omega[f(X(t, \omega; x_0, t_0))]. \tag{2.42}$$

As shown by the last equality in (2.42), the conditional average can thus be computed as average over the ensemble of trajectories starting from (x_0, t_0).

Further, let us consider the explicit finite difference approximation of the Itô equation (2.40),

$$\Delta X_i(t + s, \omega; \mathbf{x}, t) = X_i(t + s, \omega) - x_i = a_i(t, \mathbf{x})s + b_{ik}(t, \mathbf{x})\Delta W_k(t), \tag{2.43}$$

where $\Delta W_j(t) = [W_j(t + s, \omega) - W_j(t, \omega)]$ and $\mathbf{x} = \mathbf{X}(t)$ is a non-random position at the fixed time moment t. Applying the averaging operator (2.42) and using the properties of the Wiener process with mutually independent increments, one obtains

$$M_\Omega[\Delta X_i] = a_i s,$$

and

$$M_\Omega[\Delta X_i \Delta X_j] = b_{ik} b_{jl} M_\Omega[\Delta W_k \Delta W_l] = b_{ik} b_{jl} \delta_{kl} s = b_{ik} b_{jk} s.$$

With these, one finds that the coefficients (2.28) and (2.29) of the Fokker–Planck equation can be related to the coefficients of the Itô equation (2.40) by

$$A_i(\mathbf{x}, t) = a_i(\mathbf{x}, t), \quad \text{and} \quad B_{ij}(\mathbf{x}, t) = \tfrac{1}{2}(\tilde{b}\tilde{b}^{\mathrm{T}})_{ij}(\mathbf{x}, t). \tag{2.44}$$

Thus, if the solutions of the Itô equation are computed in a way consistent with the explicit discretization (2.43), χ is a diffusion process, also called *Itô process*. The probability density of χ verifies the Fokker–Planck equation (2.26) with coefficients given by (2.44).

The description realized by the Fokker–Planck equation and probability densities as space-time functions is the so-called *Eulerian statistics*, while the complementary description which uses trajectories of the diffusion process governed by the Itô equation and averages over realizations $\omega \in \Omega$ as in (2.42) is called *Lagrangian statistics* [1].

Since the derivative of the term $M_\Omega[\Delta X_i] M_\Omega[\Delta X_j] = \mathscr{O}(s^2)$ vanishes for $s \to 0$, the diffusion coefficients (2.29) can be recast using the Lagrangian statistics as

$$B_{ij}(x, t) = \frac{1}{2} \frac{d}{ds} \tilde{\sigma}_{ij}^2(t + s; x, t)\,|_{s=0},\tag{2.45}$$

where $\tilde{\sigma}_{ij}^2(t + s; \mathbf{x}, t)$ are the coefficients of the variance tensor computed at $t + s$, over the trajectories starting from $(\mathbf{x}, t)$ (conditional averages (2.42)) [23].

Itô Integral

The differential Itô equation (2.40) is interpreted as a stochastic integral equation. Let us consider the one-dimensional Itô equation

$$dX_t = a(t, X_t)dt + b(t, X_t)dW_t$$

and its equivalent integral form

$$X_t = X_0 + \int_0^t a(t', X_{t'})dt' + \int_0^t b(t', X_{t'})dW_{t'},\tag{2.46}$$

where X_t is a convenient notation for $X(t)$ in one-dimensional case and the dependence on ω is not written explicitly.

The first integral in (2.46) is the usual Riemann integral. The second integral is a stochastic integral with respect to the differential of the Wiener process,

$$I_t(b) = \int_0^t b(t', X_{t'})dW_{t'},$$

formally written as a Stieltjes integral. However, the integral I_t cannot be interpreted as a Riemann–Stieltjes integral because the sample paths of the Wiener process are not of bounded variation [12]. Itô's definition of the stochastic integral is essentially based on the explicit finite difference (2.43) used above to compute the coefficients of the Fokker–Planck equation.

To fix the ideas, let us consider a partition $0 = t_1 < t_2 < \cdots < t_{n+1} = t$ and the explicit finite difference

$$X_{t_{i+1}} = X_{t_i} + a(t_i, X_{t_i})(t_{i+1} - t_i) + b(t_i, X_{t_i})(W_{t_{i+1}} - W_{t_i}).\tag{2.47}$$

Unlike in (2.43), where one starts with a non-random position, now X_{t_i} is random and depends recursively on the successive increments of the Wiener process up to $(W_{t_i} - W_{t_{i-1}})$. But, since the increments of the Wiener process over nonoverlapping time intervals are statistically independent, X_{t_i} and the functions a and b are independent of $(W_{t_{i+1}} - W_{t_i})$. In general, a function $f(t) = f(t, X_t)$ which depends on the solution X_t of the Itô equation and is independent of $(W_s - W_t)$ for all $t < s$ is called *nonanticipating function* [8, 12]. Assuming that f is continuous for all $\omega \in \Omega$, mean-square integrable, and that it can be represented as a limit of step functions, the partial sums

$$S(f)(\omega) = \sum_{i=1}^{n} f_{t_i}(\omega)[W_{t_{i+1}}(\omega) - W_{t_i}(\omega)], \qquad (2.48)$$

converge in mean square as $n \to \infty$ to the Itô integral $I_t(f)(\omega)$ [12].

The nonanticipativeness of the integrand in (2.48) is assured by evaluating it at the left of discretization intervals, in accordance with (2.47). For any square integrable function f, the independence of the increments dW_t of the Wiener process leads to the cancelation of the expectation of the Itô integral,

$$E[I_t(f)] = E\left[\int_0^t f(t', X_{t'})dW_{t'}\right] = 0, \qquad (2.49)$$

and because the variance of the increments is related to the time increment by $E(dW_t^2) = dt$, one obtains [8, 12]

$$E\left[(I_t(f))^2\right] = E = \int_0^t f^2(t', X_{t'})dt'. \qquad (2.50)$$

Vector valued Itô integrals are defined similarly to the one-dimensional case by interpreting the vector Itô equation as a stochastic integral and assuming that the increments of the components of the Wiener process are pairwise independent,

$$E[(W_{i,s} - W_{i,t})(W_{j,s} - W_{j,t})] = (t - s)\delta_{ij}.$$

Itô Formula

The Taylor expansion up to the second order of a function of $f(t, x)$ evaluated on the trajectories (2.46) of the Itô process, $f(t, x) = f(t, X_t)$, yields

$$df(t, X_t) = \frac{\partial f(t, X_t)}{\partial t}dt + \frac{\partial f(t, X_t)}{\partial x}dX_t + \frac{1}{2}\frac{d^2 f(t, X_t)}{\partial x^2}dX_t^2$$

$$= \frac{\partial f(t, X_t)}{\partial t}dt + \frac{\partial f(t, X_t)}{\partial x}[a(t, X_t)dt + b(t, X_t)dW_t]$$

$$+ \frac{1}{2}\frac{d^2 f(t, X_t)}{\partial x^2}\left[b^2(t, X_t)dW_t^2 + 2a(t, X_t)b(t, X_t)dtdW_t + a^2(t, X_t)dt^2\right].$$

By formally considering $dW_t^2 = dt$ (which is only accurate in the mean, $E(dW_t^2) = dt$), collecting terms of first order in dt and dW and neglecting the terms $\mathcal{O}(dt^{3/2})$ and $\mathcal{O}(dt^2)$, one obtains the stochastic chain rule of differentiation known as the *Itô formula* (see, e.g., [12, Section 3.3]),

$$df(t, X_t) = \frac{\partial f(t, X_t)}{\partial t} dt + \left[a(t, X_t) \frac{\partial f(t, X_t)}{\partial x} + \frac{1}{2} b^2(t, X_t) \frac{\partial^2 f(t, X_t)}{\partial x^2} \right] dt$$

$$+ b(t, X_t) \frac{\partial f(t, X_t)}{\partial x} dW_t. \tag{2.51}$$

The Itô formula is often used in its integral form,

$$f(t, X_t) = f(0, X_0) + \int_0^t \frac{\partial f(t', X_{t'})}{\partial t'} dt'$$

$$+ \int_0^t \left[a(t', X_{t'}) \frac{\partial f(t', X_{t'})}{\partial x} + \frac{1}{2} b^2(t', X_{t'}) \frac{\partial^2 f(t', X_{t'})}{\partial x^2} \right] dt'$$

$$+ \int_0^t b(t', X_{t'}) \frac{\partial f(t', X_{t'})}{\partial x} dW_{t'}, \tag{2.52}$$

and followed by a stochastic average. For instance, considering $f(x) = X_t$ in (2.52), by applying the expectation operator E and using the properties (2.49) and (2.50) of the Itô integral one obtains the first moment, or the mean, of the Itô process,

$$m_1(t) = E(X_t) = X_0 + \int_0^t E\left(a(t', X_{t'}) \right) dt', \tag{2.53}$$

and for $f(x) = X_t^2$ one obtains the second moment

$$m_2(t) = E(X_t^2) = X_0^2 + 2 \int_0^t E\left(a(t', X_{t'}) X_{t'} \right) dt' + \int_0^t b^2(t', X_{t'}) dt'. \tag{2.54}$$

In general, the moments can be explicitly computed by (2.53) and (2.54) only if the solutions of the Itô equation are known. However, this is not necessary in the special situation of linear equation with constant coefficients, $dX_t = c_1 X_t dt + c_2 X_t dW_t$. In this case (2.53) and (2.54) become

$$m_1(t) = X_0 + c_1 \int_0^t m_1(t') dt' \quad \text{and} \quad m_2(t) = X_0^2 + (2c_1 + c_2^2) \int_0^t m_2(t') dt',$$

or, in differential form,

$$\frac{dm_1(t)}{dt} = c_1 m(t), \text{ and } \frac{dm_2(t)}{dt} = (2c_1 + c_2^2) m_2(t). \tag{2.55}$$

Equations (2.55) uniquely determine the first two moments of the Itô processes governed by linear equations, like those considered in Sect. 3.2.2, for deterministic initial conditions $m_1(0) = X_0$ and $m_2(0) = X_0^2$.

Itô-Taylor expansions are obtained by iterating the Itô formula [12, Chap. 5]. Using the operators

$$L_1(t, X_t) = \frac{\partial}{\partial t} + a(t, X_t)\frac{\partial}{\partial x} + \frac{1}{2}b(t, X_t)\frac{\partial^2}{\partial x^2} \quad \text{and} \quad L_2(t, X_t) = b(t, X_t)\frac{\partial}{\partial x}$$

after the first iteration of the Itô formula (2.52) one obtains

$$f(X_t) = f(X_0) + \int_0^t L_1 f(X_r)dr + \int_0^t L_2 f(X_r)dW_r$$

$$= f(X_0) + \int_0^t \left[L_1 f(X_0) + \int_0^r L_1 L_1 f(X_s)ds + \int_0^r L_2 L_1 f(X_s)dW_s \right]dr$$

$$+ \int_0^t \left[L_2 f(X_0) + \int_0^r L_1 L_2 f(X_s)ds + \int_0^r L_2 L_2 f(X_s)dW_s \right]dW_r$$

$$= f(X_0) + L_1 f(X_0) \int_0^t dr + L_2 f(X_0) \int_0^t dW_r + \mathcal{R}_1.$$

By further expanding the remainder $\mathcal{R}_1$,

$$\mathcal{R}_1 = L_1 L_1 f(X_0) \int_0^t dr \int_0^r ds + L_2 L_1 f(X_0) \int_0^t dr \int_0^r dW_s$$

$$+ L_1 L_2 f(X_0) \int_0^t dW_r \int_0^r ds + L_2 L_2 f(X_0) \int_0^t dW_r \int_0^r dW_s + \mathcal{R}_2,$$

one obtains the Itô-Taylor expansion of the second order,

$$\begin{aligned}
f(X_t) = &\ f(X_0) \\
&+ L_1 f(X_0) I_{(0),t} + L_2 f(X_0) I_{(1),t} \\
&+ L_1 L_1 f(X_0) I_{(0,0),t} + L_2 L_1 f(X_0) I_{(1,0),t} + L_1 L_2 f(X_0) I_{(0,1),t} \\
&+ L_2 L_2 f(X_0) I_{(1,1),t} + \mathcal{R}_2,
\end{aligned} \tag{2.56}$$

where $I_{\alpha,t}$ denotes multiple Itô integrals of multi-index $\alpha = (\alpha_1, \alpha_2, \cdots, \alpha_i, \cdots)$, where α_i takes on either the value 0, which corresponds to a Riemann integral $\int_0^t ds$, or the value 1, for an Itô integral $\int_0^t dW_s$ [11]. The order of the expansion is that of the multiplicity of the Itô integrals, i.e., that of the dimension of the multi-index α. Thus, a sufficiently smooth function of the Itô process can be expanded as a linear combination of multiple Itô integrals $I_{\alpha,t}$, with constant integrands, and a remainder consisting of a finite sum of multiple Itô integrals with non-constant integrands.

If $f(X_t) = X_t$, one obtains truncated Itô-Taylor expansions of the Itô process which, for integrable remainders and more or less strong additional assumptions, converge to X_t in the mean-square sense or with probability one [12, Propositions 5.9.1. and 5.9.2].

Itô-Taylor expansions are a basic tool in designing and analyzing the numerical schemes for Itô stochastic differential equations. For instance, the first order in (2.56) yields the strong Euler scheme of order $\beta = 0.5$ [12, Theorem 10.2.2],

$$X_t = X_0 + a(t, X_0)t + b(t, X_0)W_t.$$

The term $L_2 L_2 f(X_0) I_{(1,1),t}$ from the second order expansion (2.56) becomes

$$b(t, X_0) \frac{\partial b(t, X_0)}{\partial x} \int_0^t dW_r \int_0^r dW_s.$$

To evaluate the double integral, one considers the Itô process $X_t = W_t = \int_0^t dW_r$ and one applies the Itô formula (2.52) to the function $f(X_t) = X_t^2$ to obtain $W_t^2 = \int_0^t dr + 2 \int_0^t X_r dW_r = \int_0^t dr + 2 \int_0^t dW_r \int_0^r dW_s$, from where $\int_0^t dW_r \int_0^r dW_s = \frac{1}{2}(W_t^2 - t)$. By adding the term $L_2 L_2 f(X_0) I_{(1,1),t}$ to the Euler scheme one obtains the well-known *Milstein scheme*

$$X_t = X_0 + a(t, X_0)t + b(t, X_0)W_t + \frac{1}{2}b(t, X_0) \frac{\partial b(t, X_0)}{\partial x}(W_t^2 - t),$$

which has an increased order of strong convergence of $\beta = 1$ [12, Theorem 10.6.3].

Connection with the Fokker–Planck Equation

Let $f(x)$ be an arbitrary function with compact support in $\mathbb{R}$ and continuous first and second order derivatives which is evaluated on trajectory of the Itô process X_t. The evolution of its expectation under the action of X_t is described, according to Itô formula (2.51), by the equation

$$\frac{d}{dt} E[f(X_t)] = E\left[a(X_t) \frac{\partial f(X_t)}{\partial x} + \frac{1}{2}b^2(X_t) \frac{\partial^2 f(X_t)}{\partial x^2} \right].$$

Further, computing the expectations as averages with respect to the one-dimensional probability density $p(x, t)$,

$$\frac{d}{dt} \int_{\mathbb{R}} f(x)p(x, t)dx = \int_{\mathbb{R}} f(x) \frac{\partial p(x, t)}{\partial t} dx$$

$$= \int_{\mathbb{R}} \left[a(x) \frac{\partial f(x)}{\partial x} + \frac{1}{2}b^2(x) \frac{\partial^2 f(x)}{\partial x^2} \right] p(x, t)dx,$$

integrating by parts and discarding the terms $f(x)p(x, t)$ evaluated for $x \to \infty$, one obtains

$$\int_{\mathbb{R}} f(x) \left\{ \frac{\partial p(x, t)}{\partial t} + \frac{\partial [a(x)p(x, t)]}{\partial x} - \frac{1}{2} \frac{\partial^2 [b^2(x)p(x, t)]}{\partial x^2} \right\} dx = 0.$$

Hence, since the $f(x)$ is arbitrary, the last relation is equivalent to the Fokker–Planck equation verified by $p(x, t)$,

$$\frac{\partial p(x, t)}{\partial t} + \frac{\partial [a(x)p(x, t)]}{\partial x} = \frac{1}{2} \frac{\partial^2 [b^2(x)p(x, t)]}{\partial x^2}. \tag{2.57}$$

As seen in Sect. 2.2.2.1, the transition probability of the Itô process verifies the same Fokker–Planck equation. The drift and diffusion coefficients of the Fokker–Planck equation (2.57) are particular cases for scalar Itô processes of the coefficients (2.44). This derivation is however somewhat redundant because it has been already shown that Itô processes are diffusion processes which verify a Fokker–Planck equation with coefficients given by (2.44).

2.2.3.3 Stratonovich Stochastic Differential Equation

If instead of explicit discretization (2.47) one uses an implicit midpoint approximation of the drift and diffusion coefficients,

$$X_{t_{i+1}} = X_{t_i} + a\left(t_i, \tfrac{1}{2}(X_{t_{i+1}} + X_{t_i})\right)(t_{i+1} - t_i) + b\left(t_i, \tfrac{1}{2}(X_{t_{i+1}} + X_{t_i})\right)(W_{t_{i+1}} - W_{t_i}). \tag{2.58}$$

one obtains the Stratonovich interpretation of the stochastic differential equation (2.40). With the implicit discretization (2.58), a and b are no longer nonanticipating functions, the Taylor expansion of a function $f(t, x)$ evaluated on trajectory X_t truncated at the first order no longer contains a term involving the second derivative of b, which would be of the order dW_t^2, and instead of Itô formula (2.51) one obtains

$$df(t, X_t) = \frac{\partial f(t, X_t)}{\partial t} dt + \left[a(t, X_t) \frac{\partial f(t, X_t)}{\partial x} + +b(t, X_t) \frac{\partial f(t, X_t)}{\partial x} dW_t \right] dt. \tag{2.59}$$

The differentiation rule (2.59) is that of ordinary calculus. The corresponding Stratonovich equation is readily obtained by comparing (2.51) and (2.59),

$$dX_t = \alpha(t, X_t)dt + \beta(t, X_t)dW_t, \tag{2.60}$$

where $\alpha(t, X_t) = a(t, X_t) - \frac{1}{2}\frac{\partial b(x)}{\partial x}$ and $\beta(t, X_t) = b(t, X_t)$. Using these relations between coefficients, the Fokker–Planck equation (2.57) can be expressed in terms of coefficients α and β of the Stratonovich equation (2.60) as

$$\frac{\partial p(x, t)}{\partial t} + \frac{\partial[\alpha(x) p(x, t)]}{\partial x} = \frac{1}{2}\frac{\partial}{\partial x}\left[\beta^2(x)\frac{\partial p(x, t)}{\partial x}\right]. \tag{2.61}$$

The PDE (2.61) associated with the Stratonovich equation (2.60) is an advection–diffusion equation consistent with Fick's law, which states that the diffusion flux is proportional to the gradient of a macroscopic field, e.g., $-\beta^2(x)\frac{\partial p(x,t)}{\partial x}$.

If the diffusion coefficient is a space function, the relevant transport equation is the Stratonovich PDE, where the α and β^2 represent the velocity and the diffusion coefficients (see Eq. (1.1)). But from a computational perspective, the Itô formulation of the diffusion process based on the explicit finite difference discretization (2.47) could be more appropriate. As we shall see in Chap. 3, the relation (2.47) can be used to construct weak solutions of the Itô equation which, rather than approximating trajectories of the diffusion process, approximate the solution of the Fokker–Planck equation by computing densities of random walkers on regular lattices with a GRW algorithm. Therefore, a practical approach is to obtain the solution of the Stratonovich PDE by solving the equivalent Fokker–Planck equation with the same diffusion coefficient and modified drift coefficient $a(t, X_t) = \alpha(t, X_t) + \frac{1}{2}\frac{\partial b(x)}{\partial x}$.

2.2.3.4 Langevin Equation

A particular Itô equation is the *Langevin equation*,

$$dX_t = -kX_t dt + \sqrt{2D}dW_t, \tag{2.62}$$

where $x \in \mathbb{R}$, $k > 0$ and $D > 0$. The solutions of (2.62) are trajectories of the Ornstein–Uhlenbeck process [8, Sect. 4.4.4]. Solving (2.62) with the initial condition $X_{t_0} = x_0$ and using (2.55) one obtains the mean value

$$M_\Omega[X_t] = x_0 \exp(-k(t - t_0)), \tag{2.63}$$

and the variance

$$\sigma^2(X_t) = \frac{D}{k}[1 - \exp(-2k(t - t_0))]. \tag{2.64}$$

According to (2.44) the coefficients of the corresponding Fokker–Planck equation are $A = -kx$ and $B = D$. The equivalence between Langevin and Fokker–Planck descriptions of the Ornstein–Uhlenbeck process has been rigorously proven (see, e.g., [26, p. 241]). Unlike the Gaussian process with variance $\sigma^2 = 2D(t - t_0)$

linearly increasing in time, the Ornstein–Uhlenbeck process with mean (2.63) and variance (2.64) has a stationary state with zero mean and constant variance D/k. This process fulfills the conditions from Theorem 2.2 ($D > 0$), and the probability density of the process (2.62) tends to the stationary state for $(t - t_0) \longrightarrow \infty$.

2.2.3.5 Asymptotic Diffusive Behavior

Diffusion processes with constant coefficients described by Eqs. (2.31), (2.32), and the Ornstein–Uhlenbeck process (2.62) have the remarkable *self-averaging property* [1],

$$\lim_{t \to \infty} \frac{\sigma^2(t)}{t} = 2D^*, \tag{2.65}$$

where D^* is an *effective diffusion coefficient*.

The self-averaging property is common for diffusion processes. For Gaussian diffusion processes described by Eqs. (2.31) and (2.32) the effective coefficient (2.65) coincides with the diffusion coefficient D. For the Ornstein–Uhlenbeck process with variance given by (2.64), the limit (2.65) is $D^* = 0$, which shows that the diffusion ceases when the asymptotic stationary state is reached.

The definition (2.65) can be used to check the existence of *asymptotic diffusive behavior* for more complex processes described by stochastic differential equations even when the correspondence with a Fokker–Planck equation cannot be established and the properties (i)–(iii) defining diffusion processes do not hold.

2.2.3.6 Green–Kubo Formula

The following example is a process with nontrivial asymptotic diffusive behavior, used as model for equilibrium thermodynamics, entropy law, and Gibbs relation for the "canonical ensemble" [17, 22, 23].

Consider a physical system consisting of a single particle, of mass m and moving under the action of a Newtonian force perturbed by a white noise ζ_t,

$$dX_t = V_t dt, \quad dV_t = -\gamma V_t dt + \sqrt{2D} dW_t, \tag{2.66}$$

where X_t is the position and V_t the velocity of the particle, $(X_t, V_t) \in \mathbb{R}^2$, $\gamma > 0$, $D > 0$, and the white noise perturbation is interpreted, according to (2.39), as an increment of the Wiener process, $\zeta_t dt = dW_t$. Note that in this example the physical dimension of D is L^2/T^3, where L stands for length and T for time units, i.e., D is a diffusion coefficient for the velocity process. The second equation (2.66) is the Langevin equation and its solutions are trajectories of an Ornstein–Uhlenbeck process describing the particle velocity. According to Theorem 2.2, the density of the transition probability tends to a Gaussian stationary density,

$$p(v, t - t_0 \mid v_0) \xrightarrow[t-t_0 \to \infty]{} p_s(v) = \left[\frac{2\pi D}{\gamma}\right]^{-\frac{1}{2}} \exp\left[-\frac{v^2}{\frac{2D}{\gamma}}\right]. \qquad (2.67)$$

For each velocity realization, the first equation (2.66) provides a trajectory of the particle, $X_t = \int_{t_0}^t V_{t'} dt'$. The displacement variance is obtained by averaging over the velocity statistics,

$$\sigma^2(X_t) = \int_{t_0}^t \int_{t_0}^t M[(V_{t'} - M(V_{t'}))(V_{t''} - M(V_{t''}))] dt' dt''. \qquad (2.68)$$

The limit (2.65) can be expressed, by using the L'Hopital rule from differential calculus, as the limit of the derivative of the variance σ^2,

$$D^* = \lim_{t \to \infty} \frac{\sigma^2(t)}{2t} = \lim_{t \to \infty} \frac{1}{2} \frac{d}{dt} \sigma^2(t). \qquad (2.69)$$

From (2.68) and (2.69) on obtains

$$D^* = \lim_{t \to \infty} \int_{t_0}^t M[(V_t - M(V_t))(V_{t'} - M(V_{t'}))] dt', \qquad (2.70)$$

and using the Gaussian distribution specified by (2.63) and (2.64) one finds the value $D^* = D/\gamma^2$ of the effective diffusion coefficient (the physical dimensions of D^* are now L^2/T) [23]. Thus, the process describing the position of a particle moving with a velocity modeled as an Ornstein–Uhlenbeck process has the self-averaging property and behaves asymptotically diffusive.

The stationary density from (2.67) describing the equilibrium state of the velocity process has the mean value equal to zero, $[M(V_t)]^e = 0$, and the variance $[M(V_t^2)]^e = \int_{\mathbb{R}} v^2 p_s(v) dv = D/\gamma$. From (2.63), the mean value at t'' of the velocity process staring at t' from v' is $\int_{\mathbb{R}} v'' p(v'', t'' - t' \mid v') dv'' = v' \exp[-\gamma(t'' - t')]$. The velocity correlation in the equilibrium state can be computed as

$$[M(V_{t'} V_{t''})]^e = \int_{\mathbb{R}} \int_{\mathbb{R}} v' v'' p(v'', t'' - t' \mid v') p_s(v') dv' dv''$$

$$= \int_{\mathbb{R}} v' p_s(v') dv' \int_{\mathbb{R}} v'' p(v'', t'' - t' \mid v') dv''$$

$$= \exp[-\gamma(t'' - t')] \int_{\mathbb{R}} (v')^2 p_s(v') dv'.$$

Thus, the equilibrium correlation is given by

$$[M(V_{t'}V_{t''})]^e = \frac{D}{\gamma} \exp[-\gamma(t'' - t')]. \tag{2.71}$$

Inserting the equilibrium velocity correlation (2.71) into (2.70) one obtains the following expression for the effective diffusion coefficient:

$$D^* = \lim_{t \to \infty} \int_{t_0}^{t} [M(V_{t'}V_t)]^e dt' = \frac{D}{\gamma^2}. \tag{2.72}$$

The relation (2.72) is a *Green–Kubo formula* [7, Chap. 4] for transport coefficients.

The difference with respect to the Ornstein–Uhlenbeck position process, described by the Langevin equation (2.62), is that the process described by (2.66), where the velocity is an Ornstein–Uhlenbeck process, is nonstationary and behaves asymptotically diffusive, with an effective diffusion coefficient given by (2.72). This example shows that a *necessary condition* for diffusive behavior is the convergence of the integral (2.70) of the velocity correlation. The exponential decay of correlation (2.71) is a *sufficient condition* for the existence of the finite diffusion coefficient D^*.

The velocity correlation can be characterized by the *correlation time* τ_c,

$$\tau_c = \lim_{t \to \infty} \frac{1}{[M(V_t^2)]^e} \int_{t_0}^{t} [M(V_{t'}V_t)]^e dt'. \tag{2.73}$$

A finite correlation time is a necessary condition for the asymptotic diffusive behavior. For the process described by (2.66) with velocity correlation given by (2.71), the correlation time is $\tau_c = 1/\gamma$ and from (2.71) one obtains the relation $D^* = [M(V_t^2)]^e \tau_c$. Thus, the effective coefficient is completely determined by the velocity variance and the correlation time.

The model (2.66) is a particular case of diffusion in random velocity fields. The diffusive behavior of the position process X_t is induced by that of the velocity process V_t. The latter is a Lagrangian velocity, uniquely determined by the initial condition and the Langevin equation, uniform in space. The more complex situation of a spatially variable random field will be studied in Chap. 4. In that case, the Lagrangian velocity will be defined as Eulerian velocity viewed from the trajectory of the diffusion process and effective diffusion coefficients will be computed, similarly to the Green–Kubo formula, by integrating Lagrangian velocity correlations.

References

1. Avellaneda, M., Torquato, S., Kim, I.C.: Diffusion and geometric effects in passive advection by random arrays of vortices. Phys. Fluids A **3**(8), 1880–1891 (1991)
2. Carslaw, H.S., Jaeger, J.C.: Conduction of Heat in Solids. Oxford University Press, Oxford (1959)
3. Cornfeld, I., Fomin, S., Sinai, Ya.G.: Ergodic Theory. Springer, New York (1982)
4. Crank, J.: The Mathematics of Diffusion. Oxford University Press, Oxford (1975)
5. Dagan, G.: Flow and Transport in Porous Formation. Springer, Berlin (1989)
6. Doob, J.L.: Stochastic Processes. Wiley, New York (1990)
7. Evans, D.J., Morriss G.P.: Statistical Mechanics of Nonequilibrium Liquids. Academic, London (1990)
8. Gardiner, C.W.: Stochastic Methods. Springer, Berlin (2009)
9. Georgescu, A.: Asymptotic Treatment of Differential Equations. Applied Mathematics and Mathematical Computation, vol. 9. Chapman & Hall, London (1995)
10. Iosifescu, M., Tăutu, P.: Stochastic Processes and Applications in Biology and Medicine. Editura Academiei Bucureşti and Springer, Bucharest (1973)
11. Kloeden, P.E., Platen, E.: Relations between multiple Ito and Stratonovich integrals. Stoch. Anal. Appl. **9**(3), 86–96 (1991)
12. Kloeden, P.E., Platen, E.: Numerical Solutions of Stochastic Differential Equations. Springer, Berlin (1999)
13. Kolmogorov, A., Fomine, S.: Élémentes de la théorie des fonctions et de l'analyse fonctionelle. Mir, Moscou (1975)
14. Lasota, A., Mackey, M.C.: Probabilistic Properties of Deterministic Systems. Springer, New York (1985)
15. Lasota, A., Mackey, M.C.: Chaos, Fractals, and Noise, Stochastic Aspects of Dynamics. Springer, New York (1994)
16. Lundgren, T.S., Pointin, Y.B.: Turbulent self-diffusion. Phys. Fluids **19**(3), 355–358 (1976)
17. Mackey, M.C.: The dynamic origin of increasing entropy. Rev. Mod. Phys. **61**(4), 981–1016 (1989)
18. Malliavin, P.: Integration and Probability. Springer, New York (1995)
19. Monin, A.S., Yaglom, A.M.: Statistical Fluid Mechanics, Volume II: Mechanics of Turbulence. MIT Press, Cambridge (1975)
20. Paley, R.E.A.C., Wiener, N., Zygmund, A.: Notes on random functions. Math. Z. **37**(1), 647–668 (1933)
21. Papoulis, A., Pillai, S.U.: Probability, Random Variables and Stochastic Processes. McGraw-Hill, Singapore (2002)
22. Suciu, N.: Mathematical background for diffusion processes in natural porous media. Forschungszentrum Jülich/ICG-4. Internal Report No. 501800 (2000)
23. Suciu, N.: On the connection between the microscopic and macroscopic modeling of the thermodynamic processes (in Romanian). Ed. Univ. Piteşti (2001)
24. Suciu, N., Georgescu, A.: On the Misra-Prigogine-Courbage theory of irreversibility. Mathematica **44**(67), 215–231 (2002)
25. Taylor, G.I.: Diffusion by continuous movements. Proc. Lond. Math. Soc. **2**(20), 196–212 (1921)
26. van Kampen, N.G.: Stochastic Processes in Physics and Chemistry. North-Holland, Amsterdam (1981)
27. Wentzell, A.D.: A Course in the Theory of Stochastic Processes. McGraw-Hill, New York (1981)
28. Yaglom, A.M.: Correlation Theory of Stationary and Related Random Functions, Volume I: Basic Results. Springer, New York (1987)

Chapter 3
Stochastic Simulations of Diffusion Processes

Abstract Random sequences, their numerical simulation, and convergence properties are shortly introduced as basic tools in solving stochastic differential equations. Strong and weak solutions of Itô equations will be defined and illustrated for the Euler scheme. The global random walk (GRW) algorithm will be introduced as a superposition of arbitrarily large numbers of Euler schemes on regular lattices. Unbiased and biased GRW algorithms will be described and their relation to Fokker–Planck and Itô equations will be discussed.

3.1 Random Sequences

3.1.1 Programming Tools

Pseudo-random number generators are deterministic algorithms to generate sequences of random variables. Generally they have the recursive form

$$X_{n+1} = aX_n + b \pmod{c},$$

with a and c positive integers and b nonnegative integer. For an initial value (called *seed*) X_0 by iterations one obtains a sequence of integer numbers between 0 and $c - 1$ given by the remainders of $aX_n + b$ divided by c. If the modulus c is as large as possible and a is taken prime to c, the numbers

$$U_n = X_n/c$$

are almost uniformly distributed in the unit interval $[0, 1]$ [14].

The sequence of uniformly distributed random numbers $U(0, 1)$ can be used to generate pseudo-random numbers with given distributions. For instance, in Fortran, C, C^{++} programming one uses "ran2()" generator of $U(0, 1)$, and "gasdev()" generator of normally distributed Gaussian random numbers of mean zero and variance one, $N(0, 1)$, from Numerical Recipes. MATLAB provides a wide class

© Springer Nature Switzerland AG 2019
N. Suciu, *Diffusion in Random Fields*, Geosystems Mathematics,
https://doi.org/10.1007/978-3-030-15081-5_3

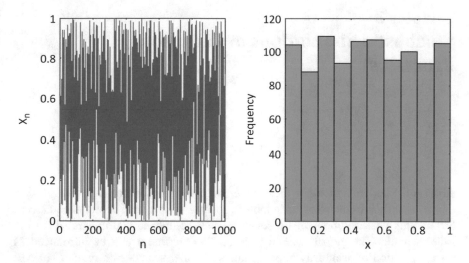

Fig. 3.1 Sample (left) and histogram (right) of a sequence $X = \{X_n\}$, $n = 1 \ldots N$, of $N = 10^3$ of uniformly distributed random numbers $U(0, 1)$ generated by the Matlab function "rand(N,1)," with sample mean $mean(X) = 0.5052$ and sample variance $var(X) = 0.0854$

of random number generators for the supported distributions, as well as specific built-in functions contained in the Statistical Toolbox.

For example, Fig. 3.1 illustrates the use of "rand" function to generate sequences of $U(0, 1)$ numbers and Fig. 3.2 illustrates the use of "randn" function for sequences of $N(0, 1)$ numbers. The histograms from Figs. 3.1 and 3.2 can be transformed into empirical probability densities functions $p(x)$ by dividing the number of occurrences N_{occ} of random numbers in bins by the length N of the sequence and by the length Δx of the bin, $p(x) = N_{occ}/(N\Delta x)$, as shown in Fig. 3.3 for a sequence of $N(0, 1)$ numbers.

Useful in applications are also the *auto-regressive of order 1* (AR(1)) sequences of random variables,

$$X_n = \phi X_{n-1} + Z_n, \tag{3.1}$$

where Z_n is a discrete Gaussian process modeled by $N(0, 1)$ random numbers and ϕ is a real constant $|\phi| < 1$. Figures 3.4 and 3.5 show AR(1) sequences, with the corresponding histograms. One sees that as the AR(1) parameter ϕ approaches to 1, the sequence develops a noisy trend and the probability distribution becomes asymmetric with a heavy tail. This behavior corresponds to the fractal behavior features exhibited by AR1 processes for large ϕ [32].

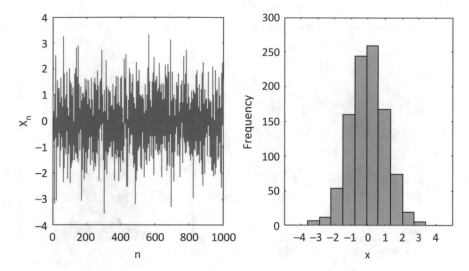

Fig. 3.2 Sample (left) and histogram (right) of a sequence $X = \{X_n\}$, $n = 1 \ldots N$, of $N = 10^3$ of normally distributes random numbers $N(0, 1)$ generated by the Matlab function "randn(N,1)," with sample mean $mean(X) = -0.0368$ and sample variance $var(X) = 1.0801$

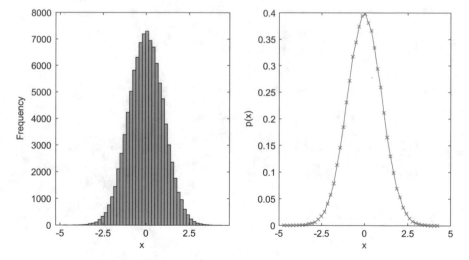

Fig. 3.3 Histogram of a sequence $X = \{X_n\}$, $n = 1 \ldots N$, of $N = 10^5$ normally distributed random numbers $N(0, 1)$, with sample mean $mean(X) = 0.0023$ and sample variance $var(X) = 1.0103$ (left) and the empirical $p(x)$ estimated by using 100 bins of equal length (right)

3.1.2 Convergence

In analyzing infinite sequences of random variables, $X_1, X_2, \cdots, X_n \cdots$, one investigates their asymptotic behavior, i.e., the existence of a random variable X

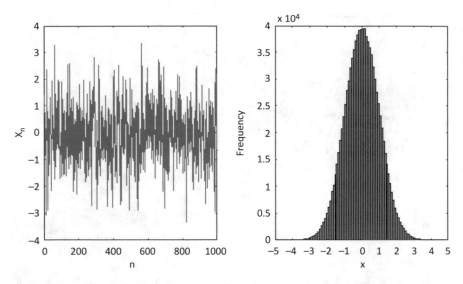

Fig. 3.4 First 10^3 values of an AR(1) sequence $X = \{X_n\}$, $n = 1 \ldots N$, $N = 10^6$, with $\phi = 0.5$ (left) and the corresponding histogram (right)

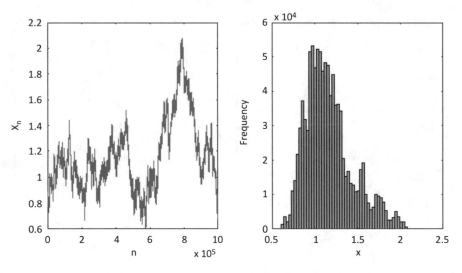

Fig. 3.5 AR(1) sequence $X = \{X_n\}$, $n = 1 \ldots N$, $N = 10^6$, with $\phi = 0.999999$ (left) and the corresponding histogram (right)

which is the limit of the sequence in some sense. Assuming that all random variables are defined on the same probability space, one defines convergence in strong and weak sense [10, 14].

Strong convergence criteria are formulated for the strict convergence of the sequence to a limit random variable with the aid of probability measures or of the expectation of the mean-square distance.

I. *Convergence with probability one* (w.p.1), or *almost sure convergence*:

$$P\left(\left\{\omega \in \Omega : \lim_{n \to \infty} |X_n(\omega) - X(\omega)| = 0\right\}\right) = 1.$$

II. *Mean-square convergence* holds if $E\left(X_n^2\right) < \infty$, for $n = 1, 2, \cdots$, $E\left(X^2\right) < \infty$, and

$$\lim_{n \to \infty} E\left(|X_n - X|^2\right) = 0.$$

III. *Convergence in probability (stochastic convergence)*:

$$\lim_{n \to \infty} P\left(\{\omega \in \Omega : |X_n(\omega) - X(\omega)| \geq \epsilon\}\right) = 0 \text{ for all } \epsilon > 0.$$

Weaker convergence properties are formulated in terms of probability distributions or expectations. The following two criteria for weak convergence use repartition functions $F(x)$ and integration with respect to Stieltjes measures $dF(x)$ (see Sect. 2.1.1).

IV. *Convergence in distribution (convergence in law)*:

$$\lim_{n \to \infty} F_{X_n}(x) = F_X(x) \text{ at all continuity points of } F_X.$$

V. *Weak convergence (convergence in expectation)*:

$$\lim_{n \to \infty} \int_{-\infty}^{\infty} f(x)dF_{X_n(x)} = \int_{-\infty}^{\infty} f(x)dF_X(x)$$

for all test functions with bounded support $f : \mathbb{R} \to \mathbb{R}$.

Convergence w.p.1 (I) and mean-square convergence (II) imply convergence in probability (III). The converse is not true and (I) is not equivalent with (II). Convergence in probability (III) implies convergence in distribution (IV), which in turn implies weak convergence (V).

3.1.2.1   Law of Large Numbers

Let $X_1, X_2, \cdots, X_n$ be a sequence of *independent identically distributed* (i.i.d.) random variables with mean value $\mu = E(X_n)$.

Table 3.1 LLN estimations (3.2) of the mean μ for different sequences of i.i.d. random variables of length $n = 10^6$

	$N(0, 1)$: randn	$U(01)$: rand	RW: $X = 1$ if rand ≤ 0.5 else $X = -1$	Poisson: poissrnd(μ)
μ	0	0.5	0	2
S_n/n	7.9388×10^{-4}	0.4999	3.1000×10^{-4}	1.9975

The *law of large numbers* (LLN) states that the average of the sequence converges,

$$\frac{1}{n} S_n = \frac{1}{n}(X_1 + X_2 + \cdots + X_n) \longrightarrow \mu \text{ as } n \longrightarrow \infty, \tag{3.2}$$

in a stronger (I, II) or weaker sense (III), depending on the version of the law.

Table 3.1 illustrates the LLN convergence for normally and uniformly distributed random variables, as well as for random walk (RW) and Poisson random variables (see Examples 1 and 3 in Sect. 2.2.2.2) generated with MATLAB functions.

This property of i.i.d. random variables is also a form of ergodicity. In particular, for $N(0, 1)$ i.i.d. (3.2) expresses the ergodicity of the discrete white noise. LLN also provides a theoretical justification for the intuitive idea of defining probabilities as limits of relative frequency obtained from repeated experiments [8, Chap. 2].

3.1.2.2 Central Limit Theorem

A stronger result, providing information about the limit distribution, is the *central limit theorem* (CLT): the standardized sum S_n of i.i.d. random variables with mean μ and variance σ^2 from LLN (3.2),

$$Z_n = \frac{(S_n - n\mu)}{\sigma\sqrt{n}}, \tag{3.3}$$

converges in distribution (IV) to an $N(0, 1)$ Gaussian variable Z of mean $E(Z) = 0$ and unit variance $\sigma_Z^2 = 1$.

The convergence of normalized sums of i.i.d. to $N(0, 1)$ Gaussian variables is illustrated in Figs. 3.6, 3.7, 3.8, and 3.9. One can see that normalized sums of quite different i.i.d. variables, as RW (Fig. 3.6) and Poisson variables (Fig. 3.8), can be used to approximate normally distributed variables (Figs. 3.7 and 3.9).

CLTs can also hold for weaker conditions than the i.i.d. assumption. A CTL theorem was, for instance, invoked to argue the Gaussianity of the transport process in aquifers with low heterogeneity [5]. A proof of CLT for a similar problem formulated as diffusion in random environments can be found in [15].

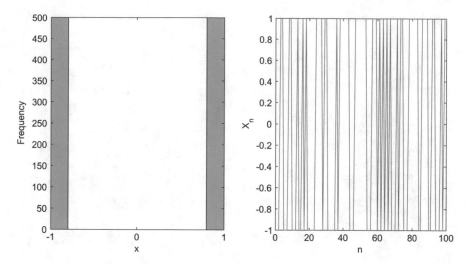

Fig. 3.6 Histogram of a RW sequence $X = \{X_n\}$, $n = 1 \ldots N$, $N = 10^3$ (left) and the first 100 values X_n (right)

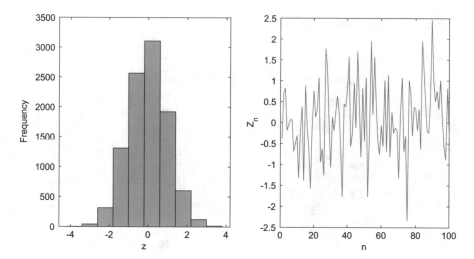

Fig. 3.7 Histogram of 10^4 standardized sums (3.3) of RW random variables from Fig. 3.6 (left) and 100 values Z_n (right)

3.1.2.3   Diffusion

The sum (3.2) of n i.i.d. random variables I_n can be written as a recurrence relation

$$X_n = X_{n-1} + I_n, \quad X_0 = 0. \tag{3.4}$$

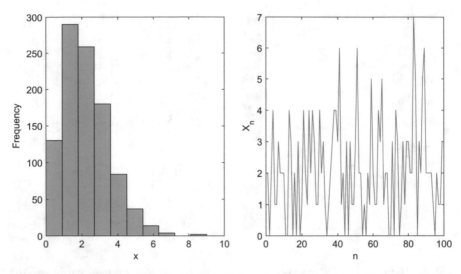

Fig. 3.8 Histogram of a sequence of Poisson i.i.d. random variables, $X = \{X_n\}$, $n = 1 \ldots N$, $N = 10^3$ (left) and the first 100 values X_n (right)

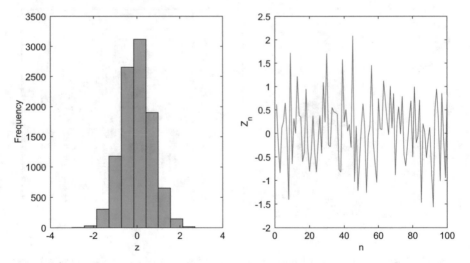

Fig. 3.9 Histogram of 10^4 standardized sums (3.3) of Poisson random variables from Fig. 3.8 (left) and 100 values Z_n (right)

Given the mean value μ and the variance σ^2 of the i.i.d. variable I_n, the mean and the variance of the sum X_n are readily obtained, by virtue of independence, as $\mu_{X_n} = n\mu$ and $\sigma^2_{X_n} = n\sigma^2$. Since the variance is linear as function of the discrete time n, the process is diffusive with diffusion coefficient $D = \sigma^2_{X_n}/(2n) = \sigma^2/2$.

According to (3.3) and CLT, if n is sufficiently large, then the sum X_n is given by

$$X_n = n\mu + \sigma\sqrt{n}Z_n, \tag{3.5}$$

where Z_n approximates a Gaussian variable $N(0, 1)$. From (3.5) one obtains the same mean, $\mu_{X_n} = n\mu + E(Z_n) = n\mu$, variance, $\sigma_{X_n}^2 = E[(X_n - n\mu)^2] = n\sigma^2 E(Z_n^2) = n\sigma^2$, and diffusion coefficient $D = \sigma^2/2$.

Relation (3.5) can be recast as

$$X_{n+1} = X_n + \mu + \sigma(\sqrt{n+1}Z_{n+1} - \sqrt{n}Z_n)$$

and since, according to CLT (3.3), for large n the random variables Z_n are approximately normally distributed with mean zero and verify the relation $\sum_{i=1}^{n} Z_i = \sqrt{n}Z_n$, it takes the equivalent iterative from

$$X_{n+1} = X_n + \mu + \sigma Z_{n+1}, \quad X_0 = 0. \tag{3.6}$$

Note that CLT also implies $E[(\sum_{i=1}^{n} Z_i)^2] = n$, which is a discrete form of the continuous time relations used in Itô calculus (see Eq. (2.50)). The iteration from (3.6) is also similar to relation (3.1) used to define AR(1) sequences, $X_n = \phi X_{n-1} + Z_n$. The latter can be reformulated as follows:

$$X_{n+1} = X_n - (1 - \phi)X_n + Z_{n+1}.$$

Since $|\phi| < 1$ implies $(1 - \phi) > 0$, the AR(1) sequence is a discrete version of the Ornstein–Uhlenbeck process (2.62). For $\phi = 0$ the iteration in the above relation produces a discrete white noise, $X_{n+1} = Z_{n+1}$, and for $\phi = 1$ it yields a discrete time Brownian motion (or a discretization of the Wiener process), $X_{n+1} = X_n + Z_{n+1}$.

A continuous time diffusion process with the same parameters μ and D is a solution of the Itô equation

$$dX_n = \mu dt + \sigma dW(t). \tag{3.7}$$

The Euler approximation of Eq. (3.7) is the discrete stochastic process satisfying the iterative scheme [14, Chap. 9]

$$Y_{n+1} = Y_n + \mu \Delta_n + \sigma \Delta W_n, \tag{3.8}$$

where $Y_n = Y(t_n)$, $\Delta_n = t_{n+1} - t_n$, and $\Delta W_n = W_{n+1} - W_n$ is an $N(0, \Delta_n)$ Gaussian random variable with mean $E(\Delta W_n) = 0$ and variance $E[(\Delta W_n)^2] = \Delta_n$.

We can thus see that (3.6) is the Euler approximation (3.8) of the Itô equation (3.7), for equidistant time steps $\Delta = 1$. While (3.7) is expected to converge

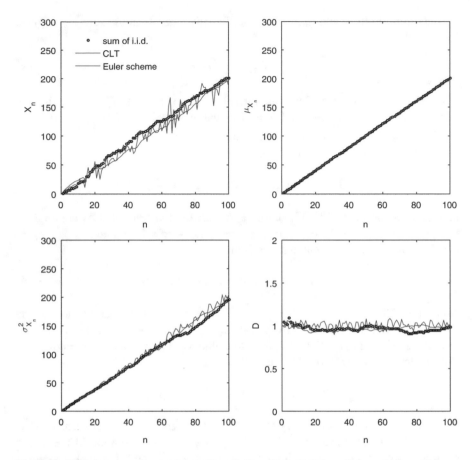

Fig. 3.10 Diffusion processes constructed by relations (3.4)–(3.6) from Poisson i.i.d. variables

to the exact solution of (3.7) for $X_0 = 0$, the processes (3.4), (3.5), and (3.5) can be regarded as weak solutions of the Itô equation.

Figure 3.10 illustrates the equivalence in a weak sense (i.e., the same mean, variance, and diffusion coefficient) of the discrete diffusion processes X_n generated by (3.4)–(3.6) for Poisson i.i.d. variables of mean and variance $\mu = \sigma^2 = 2$, with diffusion coefficient $D = \sigma^2_{X_n}/(2n) = \sigma^2/2 = 1$. The weak equivalence for other i.i.d. random variables can be shown by using the Matlab script given in Appendix A.1.

The derivation from CLT (3.5), passing through (3.6), of the Euler scheme (3.8) for Itô equations with constant coefficients (3.7) highlights the discrete nature of the physical Brownian motion, which essentially is a sum of i.i.d. random variables. If the limit of (3.8) exists as $\Delta_n \to 0$ (i.e., the Euler scheme converges), the probability density of the continuous Itô process (3.7) obeys a Fokker–Planck equation (2.26). This is one of the rare situations where to a discrete description of the physical system one associates a description through continuous fields (normalized concentration of microscopic entities) governed by a closed-form equation. The issue of macroscopic modeling of discrete physical systems will be investigated in more detail in Chap. 7.

3.2 Numerical Solutions of Itô Equations

3.2.1 Strong and Weak Solutions

Let us consider the general one-dimensional Itô equation

$$dX_t = a(t, X_t)dt + b(t, X_t)dW_t. \tag{3.9}$$

The existence of *strong solutions* of the Itô equation (3.9) for deterministic initial conditions is ensured by the assumptions [14, Chap. 4]:

A1. (Measurability): a and b are jointly measurable in $(t, x) \in [0, T] \times \mathbb{R}$
A2. (Lipschitz condition): for all $t \in [0, T]$ and $x, y \in \mathbb{R}$ there exists a constant $M > 0$ such that

$$|a(t, x) - a(t, y)| \leq M |x - y| \text{ and } |b(t, x) - b(t, y)| \leq M |x - y|$$

A3. (Linear growth bound): for all $t \in [0, T]$ and $x, y \in \mathbb{R}$ there exists a constant $M > 0$ such that

$$\left| a(t, x)^2 \right| \leq M^2 \left| 1 + x^2 \right| \text{ and } \left| b(t, x)^2 \right| \leq M^2 \left| 1 + x^2 \right|.$$

The Lipschitz condition ensures both the existence and unicity. For illustration, let us consider the simpler deterministic equation (3.9) with $b = 0$. If the solution $x(t) - x_0 + \int_0^t a(x(t')dt'$ is constructed by successive Picard approximations $x^{(n+1)}(t) = x_0 + \int_0^t a(x^{(n)}(t')dt'$, the existence and uniqueness follow from A2, which yields the relation

$$\left| x^{(n+1)}(t) - x^{(n)}(t) \right| \leq M \int_0^t \left| x^{(n)}(t') - x^{(n-1)}(t') \right|.$$

The linear growth bound condition A3 guarantees that the solution does not explode to infinity at finite times. This happens, for instance, in deterministic case of (3.9) with $a(x) = x^2$ and $b = 0$, when the solution is $x(t) = x_0/(1 - x_0 t))$. The solution blows up at $t = 1/x_0$ because a violates A3.

A proof of the existence of pathwise unique solution, in the sense of almost sure convergence (I., Sect. 3.1.2), is given by Kloeden and Platen [14, Theorem 4.5.3]. If one assumes, in addition to A1–A3, that a and b are continuous, the strong solution is a diffusion process, with coefficients $A = a$ and $B = b^2$ verifying the conditions (i)–(iii) (Sect. 2.2.2.1) [14, Theorem 4.6.1].

Conversely, given a diffusion process with probability densities obeying the Fokker–Planck equation (2.26) with coefficients A and B, the Itô process solution of (3.9) with coefficients $a = A$ and $b = B^{1/2}$, under the assumptions A1–A3 and for any Wiener process W_t, is equivalent to the given diffusion process in distribution, i.e., it has the same probability law (see type IV convergence, Sect. 3.1.2). Thus, the existence of diffusion processes as *weak solutions* of Itô equation only require to specify the coefficients A and B. For the sample path (strong) equivalence of the two processes the Wiener process in (3.9) is no longer arbitrary and has to be constructed, under supplementary conditions of boundedness of a, of b, and of its inverse b^{-1}, as well as of the first order partial derivatives of b [14, Theorem 4.7.1].

Theorems 4.6.1 and 4.7.1 proved in [14] provide a rigorous frame for the equivalence of Fokker–Planck and Itô description of the diffusion processes. The proof of this equivalence was given for the first time, with a more detailed explanation of its meaning, in [6, Chapter VI, Section 3].

3.2.2 Strong and Weak Euler Schemes

In analogy with deterministic equations, the simplest approximation of the solution is given by the Euler scheme. For given equidistant time discretization $0 < \Delta < \cdots < k\Delta \cdots < K\Delta = T$, the Euler approximation for the solution of the Itô equation (3.9) is a discrete time process satisfying the iterative scheme

$$Y_{k+1} = Y_k + a_k \Delta + b_k \delta W_k, \tag{3.10}$$

where $Y_k = Y_{k\Delta}$, $a_k = a(k\Delta, Y_k)$, $b_k = a(k\Delta, Y_k)$, $\delta W_k = W_{k+1} - W_k$ is the increment of the Wiener process, and the deterministic initial condition is the same as for (3.9), $Y_0 = X_0$. According to (w1)–(w3) (Sect. 2.1.2), the increments δW_k of the Wiener process with mean $E(\delta W_k) = 0$ and variance $E(\delta W_k^2) = \Delta$ are Gaussian variables $N(0, \Delta)$ defined in the interval $k\Delta < t < (k + 1)\Delta$. They can be generated by one of the random number generators mentioned in Sect. 3.1.1, for instance, (Δ*gasdev(), Δ*randn, or random(normal,0,Δ)). In physics and engineering applications Euler schemes are often referred to as the PT methods (e.g., [31]).

An approximation Y_t with the step size Δ (or with the maximum step size Δ, for non-equidistant discretization) *converges strongly* to X_t with order $\beta > 0$ at given time t if there exists a constant $C > 0$ such that

$$\lim_{\Delta \longrightarrow 0} E\left(|X_t - Y_t|\right) \leq C\Delta^{\beta}. \tag{3.11}$$

In many applications one needs to approximate only moments $E(X_t^m)$, $m = 1, 2, \cdots$, or functionals $g(X_t)$, where $g : \mathbb{R} \longrightarrow \mathbb{R}$ is some function with compact support. Then, one defines the *weak convergence* with order $\beta > 0$,

$$\lim_{\Delta \longrightarrow 0} |E\left(g(X_t)\right) - E\left(g(Y_t)\right)| \leq C\Delta^\beta. \tag{3.12}$$

The convergence is intimately related to the *consistency* of the numerical scheme with the differential equation. Consistency requires that the increment of the approximation converges to that of the Itô process. It is shown that under suitable strong or weak consistency conditions consistency implies the strong or the weak convergence [14, Theorems 9.6.2 and 9.7.4]. The stability requirement means that errors will remain bounded with respect to the error of the initial condition. *Stochastic numerical stability* is formulated in the sense of stochastic convergence (III, Sect. 3.1.2). The Euler scheme is stable under the sufficient conditions for existence of unique solutions A1–A3.

Under A1–A3 and some supplementary boundedness conditions for coefficients and initial conditions, the Euler scheme converges strongly with order $\beta = 0.5$ and, if a and b are four times continuously differentiable, it converges weakly with order $\beta = 1$ [14, Theorems 10.2.2 and 14.5.1]. Weak convergence with $\beta < 1$ still holds if the coefficients are Hölder continuous. Then, if all the moments of X_t exist and the coefficients are twice continuously differentiable the convergence order will be $\beta = 1$ [14, Theorem 14.1.5.]. The results mentioned above hold generally for systems of Itô equations with dimension greater than one.

To attain the strong pathwise convergence, the Euler scheme (3.10) has to use the same Wiener process as in Itô equation (3.9). For weak convergence, when only the probability distribution is approximated, the increments of the Wiener process can be replaced by random variables ξ with similar moments. The conditions to be fulfilled by moments are rather weak and they follow from Taylor-like expansions of the Itô process and convergence theorems for weak schemes [14, Sections 5.12, 14.1, 14.2]. For the weak Euler scheme of order $\beta = 1$ the first three moments of ξ have to satisfy, for some constant M, the condition

$$|E(\xi)| + \left|E(\xi^3)\right| + \left|E(\xi^2) - \Delta\right| \leq M\Delta^2.$$

Generally one looks for easily generated noise increments which can speed up the computations. For instance, the RW random variable (see Table 3.1, Figs. 3.6 and 3.7) can be used to construct the two-state random variable

$$\xi : \Omega \longrightarrow \{-\sqrt{\Delta}, +\sqrt{\Delta}\}, \quad P(\xi = \pm\sqrt{\Delta}) = \frac{1}{2} \tag{3.13}$$

which satisfies the above condition and has the same mean and variance as the increments of the Wiener process: $E(\xi) = 0$, $E(\xi^2) = \Delta$. Surrogate increments r with amplitude $s = \sqrt{\Delta}$ of the Wiener process constructed from $U(0, 1)$ random variables

```
if (rand<=0.5)
    r=-s;
else
    r=s;
end
```

can be used to obtain weak solutions of Itô equations with the Matlab script given in Appendix A.2.

The Matlab script from Appendix A.2 can be used to compare trajectories, diffusion coefficients, variances, and expectations obtained by strong and weak Euler schemes which solve the linear Itô equation with constant coefficients

$$dX_t = aX_t dt + bX_t dW_t \tag{3.14}$$

in the time interval $[0, 1]$, for the initial condition $X(0) = 1$. Equation (3.14) is explicitly solvable, with the analytical solution [14]

$$X(t) = X(0)exp[(a - \frac{1}{2}b^2)t + bW(t)],$$

which allows direct assessments for the accuracy of the numerical scheme. Figure 3.11 shows results for the particular case $a = -1.0$ and $b = 1.0$, which is just the Ornstein–Uhlenbeck process (2.62). Also implemented are solutions of the linear equation (3.14) with $a = 1.5$ and $b = 1.0$, Itô exponential equation $dX = 0.5Xdt + XdW$, and Itô drift-free equation $dX = XdW$.

The strong Euler scheme yields pathwise approximations of the trajectory of the Ornstein–Uhlenbeck process (illustrated in Fig 3.11 for a time step $\Delta = 2^{-8}$). This is not the case for the trajectory computed with the weak scheme, which does not approach the analytical solution of Eq. (3.14) (see top-left panel in Fig. 3.11). Instead, both strong and weak schemes provide good approximations of the mean, variance, and diffusion coefficient.

The convergence is investigated by successively halving the time step from $\Delta = 2^{-4}$ to $\Delta = 2^{-9}$. To improve the convergence, the error estimates from (3.11) and (3.12) were averaged with respect to t over the unit time interval $[0, 1]$. The precision of the pointwise approximation given by the strong Euler scheme is quantified by the time average ϵ_X of the error estimate from the strong convergence criterion (3.11). The weak convergence is investigated by error estimates ϵ_μ, ϵ_{σ^2},

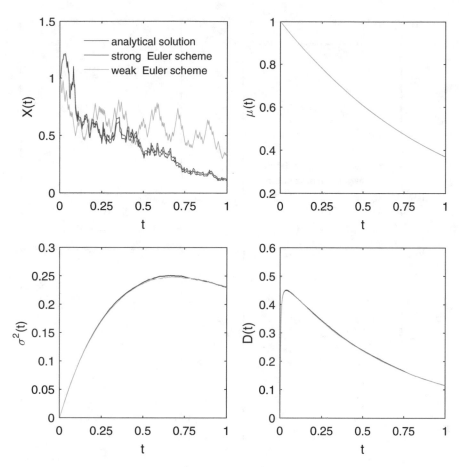

Fig. 3.11 Comparison of trajectories, means, variances, and diffusion coefficients computed by weak and strong Euler schemes for the Itô equation governing the Ornstein–Uhlenbeck process

and ϵ_D, which correspond to functionals $g(X)$ in (3.12) defining the mean $\mu(t) = E[X(t)]$, the variance $\sigma^2(t) = E[(X(t) - \mu(t))^2]$, and an "effective" diffusion coefficient $D(t) = \sigma^2(t)/(2t)$. The results for the Ornstein–Uhlenbeck process presented in Fig. 3.12 indicate that the strong scheme converges pathwise with the strong order $\beta = 0.5$ (top-left panel in Fig. 3.12) and with the weak order $\beta = 1$ (estimates for mean, variance, and diffusion coefficient). The weak scheme not only does not converge strongly but also converges hardly in the weak sense (the results presented in Figs. 3.11 and 3.12 were obtained by averaging over $N = 10^6$ trajectories).

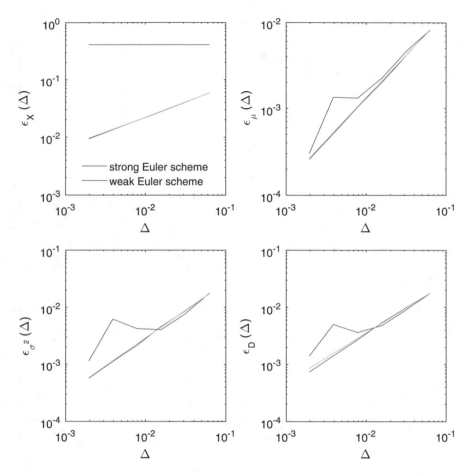

Fig. 3.12 Strong and weak convergence of Euler schemes for the Ornstein–Uhlenbeck process. Green straight lines represent the $\epsilon_X \sim \Delta^{0.5}$ behavior of the pointwise convergence (top-left) and the $\epsilon \sim \Delta$ weak convergence, obtained by fitting with a power law the results obtained by the strong Euler scheme

3.3   Global Random Walk

3.3.1   Weak Approximations by Global Random Walk

We will see in the following that, as far as one approximates probability distributions and their moments, the trajectories of the weak Euler scheme based on two-state random variables (3.13) are in fact not necessary.

The probability distribution of the surrogate random increments of the Wiener process (3.13) is the limit for large numbers of trials N of the relative frequency n/N of occurrence of n heads or tails of an unbiased coin (see Sect. 2.2.2.2, Example

1. Random walk). This is the same as the probability that a random walker takes
unbiased left/right jumps,

$$P(\leftarrow) = P(\rightarrow) = \lim_{N \to \infty} \frac{n^{\leftarrow}}{N} = \lim_{N \to \infty} \frac{n^{\rightarrow}}{N} = \frac{1}{2}. \qquad (3.15)$$

The evaluation of the mean $E(X_t)$ in the numerical implementation of the weak
Euler scheme for Itô processes given in Appendix A.2 consists of an arithmetic
average over N trajectories at a given time. The weak Euler scheme (3.10) for $a_k =
0$ and $b_k = 1$ (Wiener process) constructed with the two-state random variable
(3.13) is a random walk on a regular lattice (see Fig. 3.13). In this case, the mean is
evaluated by $E(X_t) = \sum_{i=1}^{L} i \delta x \frac{n_i}{N}$, where L is the length of the one-dimensional
lattice, δx is the lattice constant, and n_i is the number of random walk trajectories
crossing at the lattice site i (see Fig. 3.14). The latter is related to the number of
unbiased jumps from (3.15) through $n_i = n_{i-1}^{\rightarrow} + n_{i+1}^{\leftarrow}$, where $n_{i-1}^{\rightarrow}$ and $n_{i+1}^{\leftarrow}$ are

Fig. 3.13 Trajectory of the
weak Euler scheme for the
Wiener process

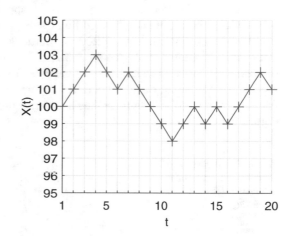

Fig. 3.14 Superposition of
$N = 300$ trajectories of the
weak Euler scheme

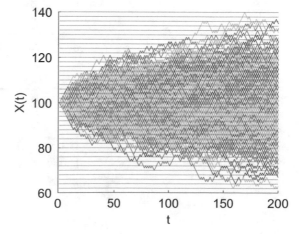

the number of walkers jumping to the site i from the left and from the right first-neighbor sites, respectively. Generally, the m-th order moment of X_t is estimated by

$$E(X_t^m) = \sum_{i=2}^{L-1} (i\delta x)^m \left(\frac{n_{i-1}^{\rightarrow}}{N} + \frac{n_{i+1}^{\leftarrow}}{N} \right). \tag{3.16}$$

For large N, the Bernoulli distributions of $n^{\rightarrow}$ and $n^{\leftarrow}$ can be approximated as follows. If the number n_i of walkers at the grid site i is even, then $n_i^{\rightarrow} = n_i^{\leftarrow} = n/2$. If n_i is odd, then one walker is allocated to either $n_i^{\rightarrow}$ or $n_i^{\leftarrow}$ with probability $1/2$. Since the walkers are globally moved between lattice sites the procedure is called "global random walk" (GRW) algorithm [31].

Figure 3.15 illustrates the evolution of the numbers n_i of random walkers in a GRW simulation of the Wiener process starting with N walkers at the middle of the lattice. One remarks that even and odd modes alternate: even lattice sites are occupied at even time steps and odd sites are occupied at odd time steps. Figure 3.16 shows the final distribution of walkers which approaches a Gaussian shape. This is, precisely, the histogram obtained by counting the number of trajectories from Fig. 3.14 at the final time $t = 200$. In particular, by choosing $N = 1$ in the GRW code presented in Appendix A.3.1.1, the time dependence of the position of the random walker is a trajectory of the weak Euler scheme (e.g., Fig. 3.13) and, repeating the simulations, one obtains the superposition of weak solutions shown in Fig 3.14. The GRW algorithm is thus a superposition of weak Euler schemes on a regular lattice.

Fig. 3.15 Distribution of $N = 300$ random walkers at successive steps in the GRW simulation

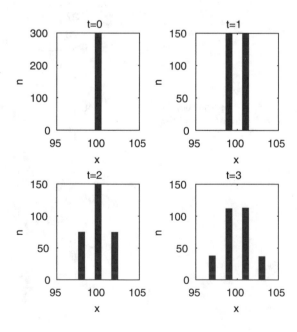

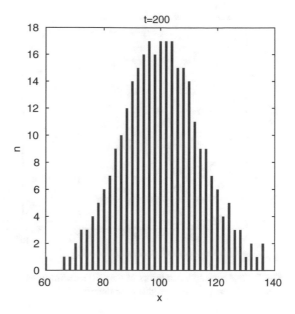

Fig. 3.16 Distribution of $N = 300$ random walkers at the final simulation time

Unlike the Euler scheme, the GRW algorithm is a discrete time–space stochastic scheme. As shown in Example 2 (Wiener process) of Sect. 2.2.2.2, for a given diffusion process, the space and time steps are related to the diffusion coefficient by the relation

$$D = \frac{\delta x^2}{2\delta t}. \tag{3.17}$$

Since the numerical scheme is constrained by (3.17), the GRW algorithm is not affected by numerical diffusion. The GRW scheme is also stable because the number of random walkers N is conserved.

It is also possible to simplify the GRW algorithm by completely removing the randomness from the scheme. This is done by setting $n_{i+1}^{\rightarrow}$ and $n_{i-i}^{\leftarrow}$ to the exact value of $n/2$. In this case N has no longer the meaning of a number of random walkers and can be taken as an arbitrary positive real number, usually equal to 1. This deterministic scheme is equivalent to the FD scheme for the heat equation (see also Sect. 2.2.2.2, Example 2. Wiener process) and converges as δx^2 for $\delta x \longrightarrow 0$. The convergence of the stochastic GRW simulation reaches the same order of convergence if the number of random walkers N is large enough to smooth the random fluctuations of n_i [31].

The convergence with respect to N is illustrated in Fig. 3.17 for a diffusion process with constant coefficient $D = 1$ by results obtained from GRW simulations with $\delta x = 1$, $\delta t = 0.5$, and increasing N. The estimated diffusion coefficient $D_{grw}(t) = [E(X_t^2) - E(X_t)^2]/(2t)$, computed according to (3.16), is compared with the nominal diffusion coefficient D through the absolute error $\left| D_{grw} - D \right|$

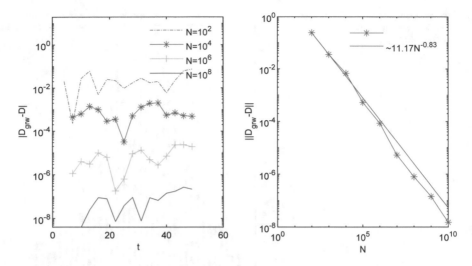

Fig. 3.17 Absolute value errors of the GRW estimations of the diffusion coefficient (left) and the global convergence norm (right) for increasing N

and the global convergence norm $\left\| D_{grw} - D \right\|$ defined by

$$\left\| D_{grw} - D \right\|^2 = \sum_{k=1}^{T/\delta t} \left[D_{grw}(k\delta t) - D \right]^2 .$$

Note that the GRW scheme described above is insensitive to the increase of the number of random walkers N. Assuming that all L grid points contain random walkers at all computation time steps, one needs LT calls of the random number generator $U(0, 1)$ for the entire simulation. Hence, the total computation time is of the order of that for the simulation of a single trajectory of the Itô process by the weak Euler scheme. We have seen that the GRW procedure is equivalent to a superposition of weak Euler schemes on a regular lattice. Hence, the GRW algorithm can be thought as a simultaneous PT procedure for arbitrary large numbers of particles, which performs a global redistribution of the computational particles at lattice sites. In the previous section we have seen that extremely large number of trajectories are necessary to achieve the convergence of the weak Euler scheme. Since the costs of a simulation for N trajectories with the Euler scheme is of the order of NT, the speed-up achieved by GRW is L/N, which can be considerable for finite L and very large N. For instance, the convergence investigation with GRW presented in Fig. 3.17 lasted about 2 s but a similar one with the Euler scheme, shown in Fig. 3.12, needed about 10 min on the same computer, although the number of trajectories, $N = 10^6$, was four orders of magnitude smaller than the number of walkers used in the GRW simulations.

3.3.2 Unbiased GRW

3.3.2.1 One-Dimensional Algorithm

The simple unbiased GRW algorithm introduced in the previous section as a weak approximation of the Wiener process is a particular case of a general algorithm which provides weak solutions of the one-dimensional Itô equation with non-vanishing drift V and diffusion coefficient D,

$$dX_t = V(t, X_t)dt + (2D(t, X_t))^{1/2}dW_t. \tag{3.18}$$

The moments of the process X_t can be computed similarly to (3.16) and the number of particles at lattice sites is proportional to the probability density p which solves the Fokker–Planck equation

$$\partial_t p + \partial_x (Vp) = \partial_x^2 (Dp). \tag{3.19}$$

In this complete one-dimensional GRW algorithm for advection–diffusion processes, the number of computational particles n at lattice sites i and successive time steps k and $k + 1$ is given by the relations

$$n(j, k) = \delta n(j + v_j, j, k) + \delta n(j + v_j - d, j, k) + \delta n(j + v_j + d, j, k), \tag{3.20}$$

$$n(i, k + 1) = \delta n(i, i, k) + \sum_{j \neq i} \delta n(i, j, k), \tag{3.21}$$

where $v_j = [V_j \delta t / \delta x]$ are discrete displacements due to advection by the local velocity field, computed as the integer part $[\cdot]$ of the non-dimensional velocity, δt and δx are the time and the space steps, $j + v_j$ are new positions after advective displacements, and d is an integer describing the amplitude of the diffusive jumps $d\delta x$. The number of particles undergoing diffusion jumps, $\delta n(j + v_j \pm d, j, k)$, and the number of particles waiting at $j + v_j$ over the k time step, $\delta n(j + v_j, j, k)$, are binomial random variables. The space and time steps, δx and δt, are related to the diffusion coefficient D through

$$D = r \frac{(d\delta x)^2}{2\delta t}, \tag{3.22}$$

where r is a rational number, $0 \leq r \leq 1$.

The relation (3.22) is the Kolmogorov's definition of the diffusion coefficient (iii, Sect. 2.2.2.1) projected on the lattice, where the parameter r plays the role of the transition probability. Indeed, according to (3.20), the trajectory of each particle is governed by

$$\hat{X}_{k+1} = \hat{X}_k + v\delta x + \xi, \tag{3.23}$$

where $\hat{X}_k = j\delta x$, $v = v_j$, and the discrete process ξ is an *unbiased random walk* with amplitude $|\xi| = d\delta x$ and transition probabilities

$$P\{\xi = \pm\sqrt{2D\delta t}\} = \frac{r}{2}, \ P\{\xi = 0\} = 1 - r. \tag{3.24}$$

In Appendix B.1 it is proved that the GRW algorithm ((3.20)–(3.22)) yields consistent approximations for continuous diffusion processes with finite first and second moments at finite times. As shown by (3.22) and (B.2), the algorithm is *free of numerical diffusion* by construction. The main source of errors is the truncation of the advective displacement from the last term in (B.1). A priori error estimates are not available for the GRW algorithm. However, a posteriori error estimates obtained by comparisons with a biased-GRW (see Sect. 3.3.3) indicate the decrease of the truncation errors with refining the lattice [25].

The resolution of the velocity field in unbiased GRW simulations is controlled by a parameter U which represents the mean velocity and an integer parameter s such that [24]

$$s = U\delta t/\delta x. \tag{3.25}$$

An empirical recipe to reduce the truncation errors in case of variable velocity is to choose a value of the mean Courant number $U\delta t/d\delta x = s/d$ smaller than one [30]. The parameter s introduces a new constraint in choosing the time and space steps related by relation (3.22). For instance, given U and s one chooses the space step δx and the time step $\delta t = s\delta x/U$ such that $2D\delta t/(d\delta x)^2 = r \leq 1$.

Figures 3.18 and 3.19 show histograms of $N = 10^{24}$ particles, at the beginning of the simulation and at the final time, obtained with the Matlab code for unbiased GRW algorithm for advection–diffusion processes presented in Appendix A.3.1.2. The simulation was carried out for constant coefficients of Eq. (3.19), $V_0 = 1$, $D_0 = 0.25$, and chosen parameters $d = 1$, $s = 1$, $\delta x = 1$. The time step $\delta t = 0.5$ resulted from (3.25) and relation (3.22) yield the sub-unit parameter $r = 0.5$. Repeating the simulations for $N = 300$, $V_0 = 0$, $D_0 = 1$, $d = 1$, $s = 0$, $\delta x = 1$, $r = 1$, and $\delta t = 0.5$, obtained now from (3.22) alone, one retrieves the results from Figs. 3.15 and 3.16. Comparing the results, one remarks the translation of the distributions of particles with the constant advection velocity. One also remarks that the parameter $r < 1$ mixes the even and the odd modes which appeared for $r = 1$, resulting in a smoother approximation of the Gaussian distribution.

Fig. 3.18 Distribution of $N = 10^{24}$ random walkers at successive steps in the GRW simulation

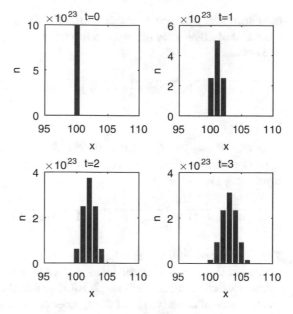

Fig. 3.19 Distribution of $N = 10^{24}$ random walkers at the final simulation time

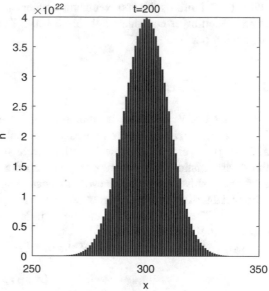

As shown in B.2, for a constant drift V the GRW algorithm is strictly equivalent to a superposition of weak Euler schemes, with convergence order $\mathscr{O}(\delta t)$, which approximates the one-time probability density of the Itô process by the particle density at lattice sites.

The mean of the binomial random variables $\delta n(j + v_j \pm d, j, k)$ with parameters $n(j, k)$ and $r/2$ (see (3.24)), i.e., the mean number of unbiased right/left jumps from the lattice site j at time t, equals $\frac{1}{2}rn(j,k)$ [18, p. 156]. Taking the mean over an ensemble of GRW runs (denoted in the following by an overline) and using (3.20) one obtains

$$\overline{\delta n(j+v_j\pm d,j,k)} = \frac{1}{2}r\,\overline{n(j,k)}, \qquad \overline{\delta n(j,j+v_j,k)} = (1-r)\,\overline{n(j,k)}.$$

$$(3.26)$$

In case of constant D and $V = 0$, according to (3.20) and (3.21), the evolution of the mean number of particles is described by an explicit FD scheme for the diffusion equation $\partial_t c = D\partial_x^2 c$,

$$\overline{n(i,k+1)} = \frac{r}{2}\overline{n(i+d,k)} + (1-r)\overline{n(i,k)} + \frac{r}{2}\overline{n(i-d,k)}. \qquad (3.27)$$

The continuous solution can be approximated by $c(x_i, t_k) = \overline{n(i,k)}/\delta x$ [31], somewhat similarly the approximations by sums of Dirac measures used in weak formulations for particle methods [9, 16]. The initial value problem is well-posed (as consequence of conservation of the number of particles). Since the scheme (3.27) is stable ($r \le 1$ fulfills the von Neumann's criterion), it is also convergent, according to Lax–Richtmyer equivalence theorem [20]. The convergence order is $\mathcal{O}(\delta t)$ in time and $\mathcal{O}(\delta x^2)$ in space.

3.3.2.2 Implementation

The exact GRW algorithm is implemented by extracting the random variables $\delta n(j+v_j\pm d, j, k)$ from the cumulative binomial distribution function (see [31, pp. 532–533]). Several other implementations were also proposed in [31], for instance, the deterministic GRW, where one gives up the particle indivisibility and n are arbitrary positive real numbers evolving according to (3.20), approximations of the binomial distributions by *erf*-functions for large n, or the reduced fluctuations GRW algorithm. The latter proved its efficiency in large-scale simulations of transport in groundwater [27, 28].

In the reduced fluctuations GRW, the number of left jumps is given by

$$\delta n(j+v_j-d, j, k) = \begin{cases} n/2 & \text{if } n \text{ is even} \\ [n/2]+\theta & \text{if } n \text{ is odd}, \end{cases} \qquad (3.28)$$

where $n = n(j,k) - \delta n(j+v_j, j, k)$, $[n/2]$ is the integer part of $n/2$, and θ is a random variable taking the values 0 and 1 with probability $1/2$. The number of right jumps is given by the difference $n - \delta n(j+v_j - d, j, k)$.

In practice, (3.28) is implemented by summing up reminders of division by 2 and multiplication by r of $n(j,k)$ and by assigning a particle to the lattice site where the sum of reminders reaches the unity (see implementation in Appendix A.3.1.2). In this way, one avoids the need to use random number generators [30, p. 3].

3.3.2.3 Convergence Properties

The GRW solution to the initial value problem for a Gaussian diffusion is illustrated in Figs. 3.18 and 3.19. Comparing Fig. 3.19 with Fig. 3.16 one can see that by increasing the number of particles the unaveraged GRW solution approaches the analytical solution described by the Gaussian probability density (2.30). It was found that the GRW algorithm is *self-averaging*, in the sense that if the total number of particles N is large enough, no ensemble averaging over GRW runs is necessary to obtain smooth solutions [31, Fig. 5]. Thus, the GRW solution converges as $\mathcal{O}(\delta x^2) + \mathcal{O}(1/\sqrt{N})$. In case of reduced fluctuations GRW the decay with N of the error norm is a bit faster than for the exact algorithm (Fig. 3.20). The number of particles required for self-averaging increases with the simulation time and the dimension of the spatial domain. For instance, in case of large-scale simulations in groundwater it was found to be $N \sim 10^{10}$ [24].

Compared with sequential PT procedures, usually consisting of ensembles of strong solutions of the Itô equation, GRW has the advantage of providing smooth solutions by using huge numbers of particles, at low computational costs. This is shown in Fig. 3.21 by a comparison of central processing unit (CPU) time used to solve the same problem. While for PT the CPU time increases linearly with N and requires increasing numbers of processors (up to 256 for $N = 10^9$ on a Cray T3E parallel computer), the computing time increases significantly only for more than $N > 10^8$ in case of exact GRW algorithm (GRW0) and is practically constant in case of reduced fluctuations GRW [31].

Fig. 3.20 Convergence with the number of particles of the exact GRW algorithm (GRW0) and of the reduced fluctuations algorithm (GRW), for a one-dimensional Gaussian diffusion problem

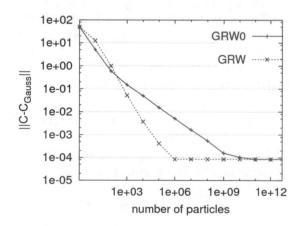

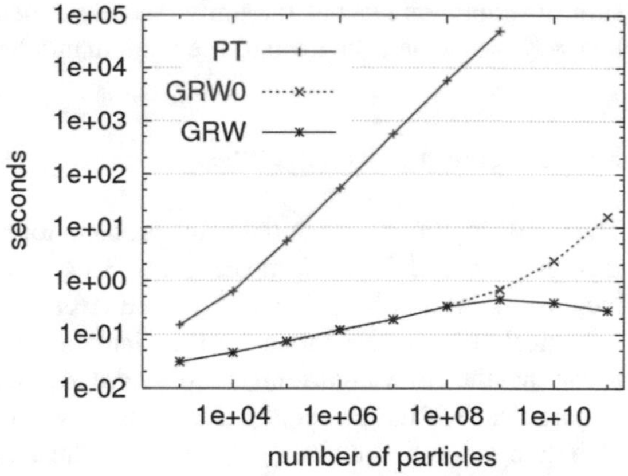

Fig. 3.21 Comparison of
CPU times for simulations
carried out with GRW,
GRW0, and PT of a
three-dimensional Gaussian
diffusion problem over ten
time steps

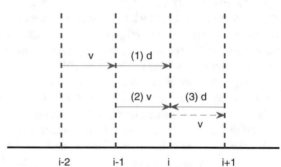

Fig. 3.22 Derivation of FD
scheme (3.29) from GRW
relations (3.21), (3.20), and
(3.26)

3.3.2.4   GRW and Finite Difference Schemes

Giving up particle indivisibility, $n(i, k)$ becomes a real number, relations (3.28) are
verified exactly, without averaging, and one obtains a deterministic FD scheme. For
illustration purposes, let us consider the simple case $v = d = 1$. Then, summing
up the contributions (3.21) to $n(i, k + 1)$ coming from diffusive jumps (d) and
displacements due to velocity field (v), according to (3.20) and (3.26), respectively,
(1)d, (2)v, and (3)d shown in Fig. 3.22, one obtains the explicit FD scheme

$$n(i, k + 1) = (1 - r)n(i - 1, k) + \frac{r}{2}[n(i - 2), k + n(i, k)]. \tag{3.29}$$

The FD scheme (3.29) is equivalent to a splitting FD scheme with the following
advection (a) and diffusion (d) steps:

$$n^a(i, k + \tfrac{1}{2}) = n(i - 1, k), \tag{3.30}$$

$$n^d(i, k + 1) = \frac{r}{2}n^a(i - 1, k + \tfrac{1}{2}) + (1 - r)n^a(i, k + \tfrac{1}{2})$$

$$+ \frac{r}{2}n^a(i + 1, k + \tfrac{1}{2}). \tag{3.31}$$

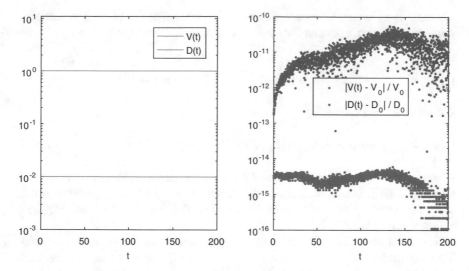

Fig. 3.23 Numerical estimations of velocity and diffusion coefficient computed from the first two spatial moments of $n(i, k)$ approximated by the GRW-FD scheme (left) and the corresponding relative errors (right)

The advection step (3.30) is the discretized version of the exact solution of the advection equation, $p(x, t + \delta t) = p(x - V_0 \delta t, t)$. The diffusion scheme (3.31) is the stable explicit centered FD scheme (3.27) for the diffusion equation.

Figure 3.23 shows numerical estimations and relative errors of the mean velocity $V = E(X_t)/t$ and diffusion coefficient $D = [E(X_t^2) - E(X_t)^2]/(2t)$ defined with the first two moments of the process (3.18). The expectations were computed according to (3.16) from the numerical solution $n(i, k)$ normalized to unity obtained with the GRW-FD scheme (3.29). The coefficients of Eq. (3.19) are chosen as constants, $V_0 = 1$ and $D_0 = 0.01$ (spatial and temporal dimensions are given in meters and days, respectively), which are representative for a typical numerical setup for transport in groundwater [19, 22]. The initial condition is given by the analytical solution of Eq. (3.19), i.e., the normalized Gaussian function

$$p(x, t) = (4\pi D_0 t)^{-1/2} \exp\left[-(x - x_0 - V_0 t)/(4 D_0 t)\right], \tag{3.32}$$

evaluated at $t = 1$ for $x_0 = 1$. The numerical solution is computed with the Matlab code given in Appendix A.3.1.3 for $\delta x = \delta t = 0.1$, which corresponds to a Courant number $Cr = V_0 \delta t / \delta x = v = 1$. The large Peclet number $Pe = V_0 \delta x / D = 10$ indicates advection-dominated transport.

The GRW advection-scheme (3.30) is a particular case of

$$n^a(i, k + \tfrac{1}{2}) = n(i - Cr, k), \tag{3.33}$$

where Cr is an integer Courant number equal to the magnitude of the dimensionless velocity v. One remarks that for $Cr = 1$ this advection-scheme coincides with the "backward" scheme

$$n(i, k + \tfrac{1}{2}) = n(i, k) - Cr(n(i, k) - n(i - 1, k)),$$

and with the "box" scheme

$$n(i, k + \tfrac{1}{2}) = n(i - 1, k) + [(1 - Cr)/(1 + Cr)](n(i, k) - n(i - 1, k + 1)).$$

While the box-scheme is unconditionally stable, the backward scheme is stable only for $Cr \leq 1$ [20]. Since the dimensionless velocity has to be an integer greater than unity, the GRW splitting-scheme (3.30)–(3.31) is defined only for integer Courant numbers, $Cr \geq 1$.

The three advection-schemes can be compared with the aid of the Matlab code from Appendix A.3.1.3 by solving the advection-dominated transport problem described above. One finds that for $Cr < 1$ the backward scheme produces numerical diffusion indicated by the flattening of the numerical solution. The solutions obtained with the box scheme for $Cr \neq 1$ also distort the shape of the analytical solution (3.32), being affected by oscillations which grow with the departure from $Cr = 1$ and are mainly significant for $Cr > 1$. The solutions obtained with the GRW advection-scheme (3.33) closely follow the shape of the theoretical solution for all integer values $Cr \geq 1$. The diffusion coefficient is estimated with relative errors smaller than 10^{-10} and the constant advection velocity is practically estimated with the machine precision (see Fig. 3.23). Since the three schemes coincide for $Cr = 1$, it follows that the numerical diffusion can be avoided by using the GRW discretization (3.33) of the advection step in FD splitting schemes.

3.3.2.5   Two-Dimensional GRW Algorithm

Two- and three-dimensional GRW algorithms are designed by repeating the one-dimensional procedure for each spatial direction, in case of constant diffusion coefficients (Fig. 3.24), or by using independent random walks, in case of variable diffusion coefficients (Fig. 3.25). A two-dimensional GRW for advection–diffusion processes, in case of diagonal diffusion tensor with variable components, is constructed with space–time variable r_x and r_y, $r_x + r_y \leq 1$. For given d_x, d_y, δx and δy, the time step is chosen to satisfy

$$\delta t \leq \left(\frac{2D_x^{\max}}{(d_x \delta x)^2} + \frac{2D_y^{\max}}{(d_y \delta y)^2} \right)^{-1},$$

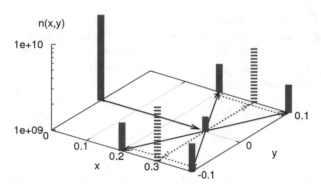

Fig. 3.24 Two-dimensional GRW for constant diffusion coefficient, built as a superposition of two one-dimensional GRW procedures

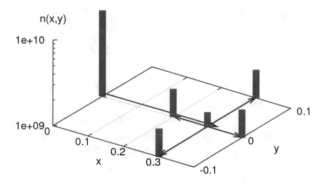

Fig. 3.25 Two-dimensional GRW for variable diffusion coefficients, based on two independent random walks on x- and y-directions (right)

where $D_x^{\max} = \max\{D_x(x, y, t)\}$ and $D_y^{\max} = \max\{D_y(x, y, t)\}$. This implementation yields accurate solutions even if the diffusion coefficients are highly variable and random (for instance, in simulations of diffusion in human skin, modeled as a three-layer two-dimensional model with Gaussian distributed diffusion coefficients [29]).

The two-dimensional GRW illustrated in Figs. 3.24 and 3.25 is defined by the relations

$$n(i, j, k) = \delta n(i + v_i, j + v_j \mid i, j, k) \tag{3.34}$$
$$+ \delta n(i + v_i + d_i, j + v_j \mid i, j, k) + \delta n(i + v_i - d_i, j + v_j \mid i, j, k)$$
$$+ \delta n(i + v_i, j + v_j + d_j \mid i, j, k) + \delta n(i + v_i, j + v_j - d_j \mid i, j, k),$$
$$n(l, m, k + 1) = \delta n(l, m, k) + \sum_{i \neq l, j \neq m} \delta n(l, m \mid i, j, k), \tag{3.35}$$

where $n(i, j, k)$ is the number of particles at the site $(x, y) = (i\delta x, j\delta y)$ at the time $t = k\delta t$ and the δn are binomial random variables describing the spread of $n(i, j, k)$. To the drift and diffusion coefficients of the transport problem, $V_x(i\delta x, j\delta y, k\delta t)$, $D_x(i\delta x, j\delta y, k\delta t)$, $V_y(i\delta x, j\delta y, k\delta t)$, and $D_y(i\delta x, j\delta y, k\delta t)$, one associates dimensionless parameters

$$v_i = V_x \frac{\delta t}{\delta x}, \quad v_j = V_y \frac{\delta t}{\delta y}, \quad r_i = D_x \frac{2\delta t}{(d_i \delta x)^2}, \quad r_j = D_y \frac{2\delta t}{(d_j \delta y)^2}, \tag{3.36}$$

where d_i and d_j are integers describing the amplitude of the diffusive jumps.

The average over GRW runs of the terms from (3.34) are related by

$$\overline{\delta n(i + v_i, j + v_j \mid i, j, k)} = (1 - r_i - r_j)\, \overline{n(i, j, k)},$$

$$\overline{\delta n(i + v_i \pm d_i, j + v_j \mid i, j, k)} = \frac{r_i}{2}\, \overline{n(i, j, k)},$$

$$\overline{\delta n(i + v_i, j + v_j \pm d_j \mid i, j, k)} = \frac{r_j}{2}\, \overline{n(i, j, k)}. \tag{3.37}$$

The distribution of $N = 10^{24}$ particles in a two-dimensional GRW simulation for $V_x = V_y = 0$ and constant diffusion coefficients $D_x = D_y = 0.01$ is presented in Figs. 3.26 and 3.27. Figures 3.28 and 3.29 show results of a two-dimensional GRW simulation, performed with the Matlab code given in Appendix A.3.2, of an advection–diffusion process with the same constant diffusion coefficient and drift coefficients consisting of a realization of a random velocity field computed with the Matlab function given in Appendix C.3.2.2.

Fig. 3.26 Distribution of $N = 10^{24}$ particles at successive steps in a two-dimensional GRW simulation

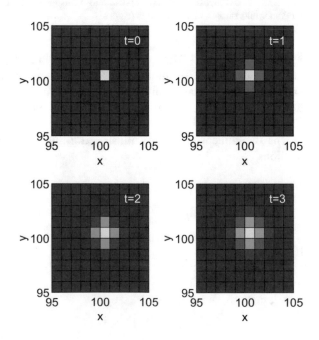

Fig. 3.27 Distribution of
$N = 10^{24}$ particles at $t = 100$
and $t = 200$ in a
two-dimensional GRW
simulation

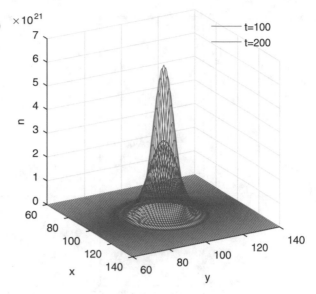

Fig. 3.28 Distribution of
$N = 10^{24}$ particles at
successive steps in a
two-dimensional GRW
simulation

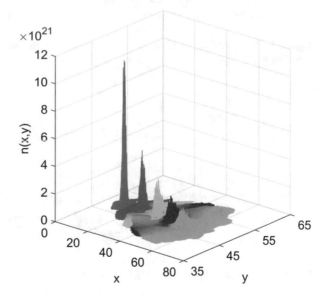

3.3.3   Biased GRW

If the velocity and the diffusion coefficients vary in space, overshooting errors may
occur when the particles jump over more than one lattice site (see Fig. 3.30). This is
mainly the case of diffusion in space-variable velocity fields, when velocity values at
sites lying between the initial and final position of the group of particles during the

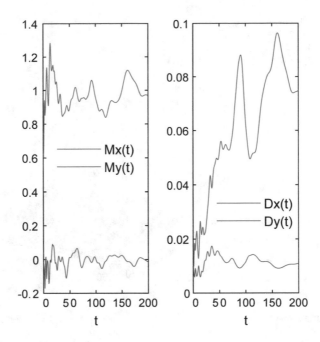

Fig. 3.29 Velocity of the center of mass components and effective diffusion coefficients corresponding to Fig 3.28

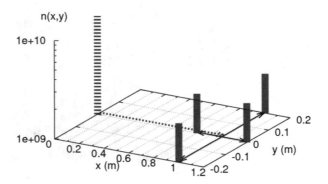

Fig. 3.30 The state of the unbiased GRW at $t = \delta t = 0.5$

advection step may have sharp variations. Overshooting can be avoided if advection is simulated by a bias in the random walk probability and only jumps to the nearest sites are allowed, as shown in Fig. 3.31. This results in a biased global random walk (BGRW) algorithm [25]. Since BGRW moves all the particles lying at a lattice site in a single numerical procedure, N can be as large as necessary to ensure the self-averaging, which is the main difference with respect to the biased-random walks on lattices which move particles sequentially [9, 13]).

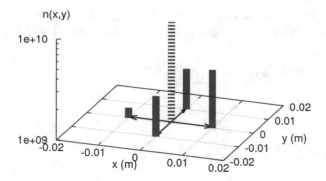

Fig. 3.31 BGRW state at $t = \delta t = 0.0025$, for the same problem as in Fig. 3.30

The two-dimensional BGRW is defined by the relation

$$
\begin{aligned}
n(i, j, k) = {} & \delta n(i, j \mid i, j, k) \\
& + \delta n(i + 1, j \mid i, j, k) + \delta n(i - 1, j \mid i, j, k) \\
& + \delta n(i, j + 1 \mid i, j, k) + \delta n(i, j - 1 \mid i, j, k),
\end{aligned} \tag{3.38}
$$

where $n(i, j, k)$ is the number of particles at the site $(x, y) = (i\delta x, j\delta y)$ at the time $t = k\delta t$ and the δn are binomial random variables describing the number of particles waiting at the initial lattice site or jumping to the first-neighbor sites. To the drift (velocity) and diffusion coefficients of the transport problem, $V_x(x, y, t)$, $V_y(x, y, t)$, $D_x(x, y, t)$ and $D_y(x, y, t)$, one associates the dimensionless parameters similar to those of the biased GRW described in Sect. 3.3.2.5,

$$
v_i = V_x \frac{\delta t}{\delta x}, \quad v_y = V_j \frac{\delta t}{\delta y}, \quad r_i = D_x \frac{2\delta t}{\delta x^2}, \quad r_j = D_y \frac{2\delta t}{\delta y^2}. \tag{3.39}
$$

But instead of (3.37), the average over BGRW runs of the terms in (3.38) are now related by

$$
\overline{\delta n(i, j \mid i, j, k)} = (1 - r_i - r_j)\, \overline{n(i, j, k)},
$$

$$
\overline{\delta n(i \pm 1, j \mid i, j, k)} = \frac{1}{2}(r_i \pm v_i)\overline{n(i, j, k)},
$$

$$
\overline{\delta n(i, j \pm 1 \mid i, j, k)} = \frac{1}{2}(r_j \pm v_j)\overline{n(i, j, k)}. \tag{3.40}
$$

The reduced fluctuations BGRW is implemented similarly to (3.28). As shown in Appendix B.3, the BGRW algorithm fulfills the requirements for diffusion processes with finite first two moments at finite times and, unlike the unbiased GRW, it is free of round-off errors in the representation of the drift coefficients.

Defining the particle density $\rho(x, y, t) = \bar{n}(i, j, k)$, summing up the contributions coming from the first neighbors to a lattice site, and using (3.38)–(3.40) one obtains

$$\frac{\rho(x, y, t + \delta t) - \rho(x, y, t)}{\delta t} +$$

$$\frac{(V_x\rho)(x + \delta x, y, t) - (V_x\rho)(x - \delta x, y, t)}{2\delta x} +$$

$$\frac{(V_y\rho)(x, y + \delta y, t) - (V_y\rho)(x, y - \delta y, t)}{2\delta y} =$$

$$\frac{(D_x\rho)(x + \delta x, y, t) - 2(D_x\rho)(x, y, t) + (D_x\rho)(x - \delta x, y, t)}{\delta x^2} +$$

$$\frac{(D_y\rho)(x, y + \delta y, t) - 2(D_y\rho)(x, y, t) + (D_y\rho)(x, y - \delta y, t)}{\delta y^2}. \tag{3.41}$$

The relation (3.41) is the forward-time centered-space FD scheme for the Fokker–Plank equation

$$\frac{\partial \rho}{\partial t} + \frac{\partial}{\partial x}(V_x\rho) + \frac{\partial}{\partial y}(V_y\rho) = \frac{\partial^2}{\partial x^2}(D_x\rho) + \frac{\partial^2}{\partial y^2}(D_y\rho). \tag{3.42}$$

As follows from (3.40), the BGRW algorithm is subject to the following restrictions:

$$r_x + r_y \leq 1, \quad |v_x| \leq r_x, \quad |v_y| \leq r_y. \tag{3.43}$$

By the last two inequalities in (3.42), the Courant numbers $|V_x|\delta t/\delta x$ and $|V_y|\delta t/\delta y$ are smaller than one, which ensures that the BGRW algorithm is free of overshooting errors. If, in addition, one imposes the conditions $r_x \leq 0.5$ and $r_y \leq 0.5$, the von Neumann's criterion for stability is also satisfied. Thus, the convergence of the scheme (3.41) is implied by the Lax–Richtmyer equivalence theorem [20].

Under the conditions stated above, the numerical solutions of the BGRW algorithm ((3.38)–(3.40)) converge with the order $\mathcal{O}(\delta x^2)$ to the solutions of the Fokker–Planck equation (2.26) for initial value problems.

As shown by (3.41), the BGRW algorithm is equivalent to a FD scheme even if the velocity field is a space–time function, unlike in case of unbiased GRW, for which the equivalence holds only for constant velocity. Instead, since advection is accounted for by biased jump probabilities, BGRW is no longer equivalent to an Euler scheme for the Itô equation.

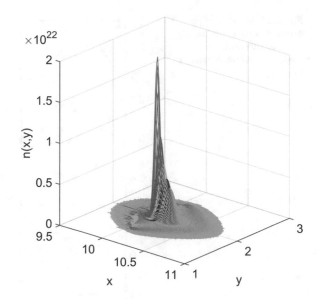

Fig. 3.32 Distribution of $N = 10^{24}$ particles at successive steps in a two-dimensional BGRW simulation

The advection–diffusion equation which corresponds to Fick's law,

$$\frac{\partial \rho}{\partial t} + \frac{\partial}{\partial x}(V_x^* \rho) + \frac{\partial}{\partial y}(V_y^* \rho) = \frac{\partial}{\partial x}\left(D_x \frac{\partial \rho}{\partial x}\right) + \frac{\partial}{\partial y}\left(D_y \frac{\partial \rho}{\partial y}\right), \tag{3.44}$$

is equivalent to the Fokker–Planck equation (2.26) if the drift coefficients are defined by $V_x = V_x^* + \partial D_x/\partial x$ and $V_y = V_y^* + \partial D_y/\partial y$ (see Eq. (1.3)).

The BGRW algorithm is highly accurate but more expensive than the unbiased GRW, because of the restriction of sub-unit Courant numbers in (3.43). Therefore, BGRW is mainly used to validate the faster but less accurate unbiased GRW algorithm (see [23, 25, 27]).

Figures 3.32 and 3.33 show results of a two-dimensional BGRW simulation, performed with the Matlab code given in Appendix A.3.3.1, of the same advection–diffusion process as that considered for the simulation presented in Figs. 3.28 and 3.29.

3.3.3.1 Boundary Conditions

The GRW and BGRW presented above were performed on lattices larger than the maximum spatial extension of the system of particles, therefore no boundary conditions were necessary. Dirichlet and Neumann boundary conditions are formulated through particle representations of the continuous field, solution of the Fokker–Planck equation [31]. The same in case of internal boundaries and discontinuities. For instance, in case of an interface at $y = 0$ and different coefficients, D_1 for $y \leq 0$ and D_2 for $y > 0$, $D_2 = 2D_1$, the continuity of the mass flux at $y = 0$ is expressed

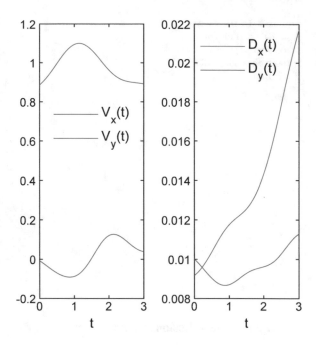

Fig. 3.33 Velocity of the center of mass components and effective diffusion coefficients corresponding to Fig 3.32

by the equality of the number of particles jumping to the right and to the left across the interface, $r_1 n_1 = r_2 n_2$ [21]. This relation, together with the conservation of the number of particles at $y = 0$, $n = n_1 + n_2$, gives the number of particles jumping from $(y \leq 0)$ to $(y > 0)$,

$$n_1 = \frac{r_2}{r_1 + r_2} n = \frac{D_2}{D_1 + D_2} n,$$

and the number of particles jumping from $(y > 0)$ to $(y \leq 0)$,

$$n_2 = \frac{r_1}{r_1 + r_2} n = \frac{D_1}{D_1 + D_2} n.$$

Figures 3.34 and 3.35 compare BGRW solutions for constant and discontinuous diffusion and drift coefficients.

3.3.3.2   Coupled MFEM-GRW Simulations

Flow and transport problems associated with Eqs. (1.2) and (1.4) can be efficiently solved by coupling MFEM flow solutions [2, 19] and GRW solutions of the advection–diffusion problem. The coupled MFEM-GRW approach benefits of accurate velocity fields while avoiding the drawback of numerical diffusion in MFEM methods.

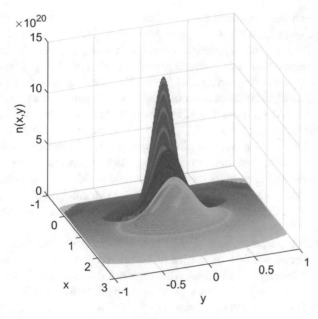

Fig. 3.34 BGRW solution for constant diffusion and drift coefficients

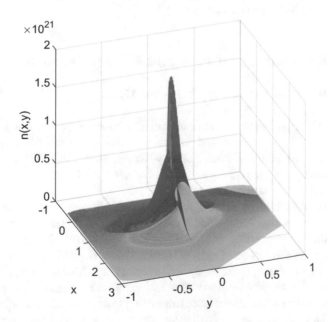

Fig. 3.35 BGRW solution for discontinuous diffusion and drift coefficients at the interface $y = 0$: $D(y > 0) = 2D(y < 0)$, and $V(y > 0) = 2V(y < 0)$

The approach was illustrated for two-dimensional transport of passive scalars in random velocity fields with short-range correlation [30]. A log-normal hydraulic conductivity field K with exponential correlation was generated by the Kraichnan method as a superposition of random periodic modes, by using a field generator similar to that given in Appendix C.3.1.1. The Kraichnan generator ensures reliable simulations of transport if the number of modes is of the order of the total simulation time [7, 26]. MFEM solutions to the incompressible flow problem associated with (1.2), for given samples of the K field, were computed in a rectangular domain and GRW simulations of the advection–diffusion process, with the algorithm described in Sect. 3.3.2.5 above, were performed in a smaller region inside the flow domain, to avoid boundary effects. Velocity values defined on MFEM elements were then interpolated to the nodes of the GRW lattice. A good resolution of the velocity field and small overshooting errors were ensured with a mean Courant number of 2/3 (see Sect. 3.3.2.1).

For validation purposes, full MFEM solutions to both flow and transport for a small grid-Peclet number of 0.1 were compared with coupled MFEM-GRW solutions. The Lagrangian velocity and the effective diffusion coefficients estimated by the two methods were in a very good agreement, with a small overestimation of the longitudinal diffusion coefficient by the full MFEM solution, the source of which could be the inherent numerical diffusion of the MFEM scheme [19]. The coupled MFEM-GRW achieved a speed-up of the computation by a factor of ten as compared to the full MFEM solution [30].

3.3.4 GRW Solutions for Flow and Reactive Transport

3.3.4.1 GRW Solutions for Flow in Porous Media

A random walk approach to flow in porous media, based on continuous time random walk (CTRW), has been proposed in the literature [4] as an effective way for accounting for the effect of heterogeneity. Rather than using particles, as other CTRW approaches [8], the flow solution is obtained by solving a Laplace transform Fokker–Planck equation with memory [3].

The GRW algorithm provides a new random walk approach to compute flow solutions under heterogeneous conditions by moving particles on regular lattices. To derive the corresponding Fokker–Planck equation, the Stratonovich second order operator in the left-hand side of Eq. (1.2) can be written as sum of an Itô diffusion term and a drift term, as shown in Sect. 1.2.2. Alternatively, a GRW flow solver can be constructed similarly to a staggered FD scheme. The procedure is illustrated below for a one-dimensional algorithm to approximate the solution of the flow equation (1.2) with scalar hydraulic conductivity K for incompressible flow ($f = 0$). The stationary solution of (1.2) is obtained as limit in time of the solution of the non-steady equation

$$\frac{\partial h}{\partial t} - \frac{\partial}{\partial x}\left(K\frac{\partial h}{\partial x}\right) = 0, \tag{3.45}$$

obtained with the staggered FD scheme

$$\frac{\partial h}{\partial t} \approx \frac{1}{\delta t}[h(i, k+1) - h(i, k)]$$

$$= \frac{1}{\delta x^2}\{[K(i+1/2, k)(h(i+1, k) - h(i, k))] - [K(i-1/2, k)(h(i, k) - h(i-1, k))]\},$$

for constant Dirichlet boundary conditions $h(-1) = 1$, $h(1) = 0$.

With the hydraulic head approximated by $h(i\delta x, k\delta t) \approx n(i, k)/N$, where $n(i, k)$ is the number of particles at lattice sites and N is the total number of particles, and jump probability defined by $r = K\delta t/\delta x^2$, one obtains the staggered GRW scheme

$$n(i, k+1) = \{1 - [r(i-1/2, k) + r(i+1/2, k)]\}n(i, k)$$

$$+ r(i-1/2, k)n(i-1, k) + r(i+1/2, k)n(i+1, k). \quad (3.46)$$

The numerical solution (3.46) is a sum of terms produced by the GRW algorithm which moves particles from sites j to $i = j \mp 1$,

$$n(j, k) = \delta(j, j, k) + \delta(j-1, j, k) + \delta(j+1, j, k),$$

$$\overline{\delta(j, j, k)} = \{1 - [r(j-1/2) + r(j+1/2)]\}\overline{n(j, k)},$$

$$\overline{\delta(j \mp 1, j, k)} = r(j \mp 1/2)\overline{n(j)}.$$

The numbers $r(i \pm 1/2, k)n(i, k)$ of jumping particles are binomial random variables [22]. The bias $r(i+1/2, k) - r(i-1/2, k)$ in the jump probabilities accounts for the second term in (1.3), hence the procedure is a BGRW algorithm. The velocity given by Darcy's law $V = -K dh/dx$ is approximated by $V(i, k) = -r(i, k)\delta x[n(i+1/2, k) - n(i-1/2, k)]/(N\delta t)$ at the same time as the hydraulic head h, similarly to MFEM approaches.

The BGRW procedure is illustrated in Figs. 3.36, 3.37, 3.38, and 3.39, for the solution of Eq. (3.45) obtained with the Matlab code given in Appendix A.3.3.2 in

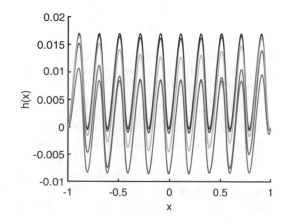

Fig. 3.36 Solution of the one-dimensional flow problem with periodic $K(x)$. The black line represents oscillation of the analytical solution about the linear trend imposed by the Dirichlet boundary condition. Colored lines show the convergence of the non-steady GRW solution towards the stationary solution

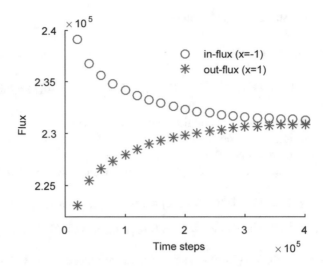

Fig. 3.37 Approach to the steady state indicated by the behavior in time of in- and out-flux; when they balance, a steady state solution is reached in the computational domain

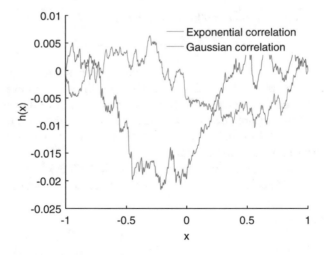

Fig. 3.38 Steady state solution to the same one-dimensional flow problem, for coefficients $K(x)$ given by realizations of a log-normal random function with different correlation shapes; correlation length equal to the spatial step δx

case of coefficient $K(x)$ given by a sin function as well as by realizations of random functions with exponential and Gaussian correlations, generated with the Matlab functions given in Appendices C.3.1.1 and C.3.1.2, respectively.

Giving up the particle indivisibility and representing n by real numbers, one obtains a deterministic version of the BGRW algorithm (BGRWD), where the numbers of particles jumping left/right are exactly given by $r(i \pm 1/2, k)n(i, k)$.

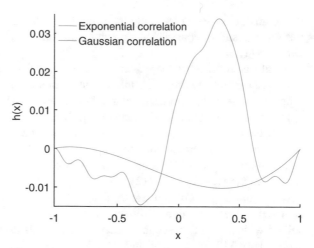

Fig. 3.39 Steady state solution for one-dimensional flow problem, for coefficients $K(x)$ given by realizations of a log-normal random function with different correlation shapes; correlation length equal to half the length of the computational domain

Since Eq. (3.45) does not contain drift terms, the BGRWD algorithm coincides with the staggered FD scheme. Both the BGRW and BGRWD schemes have the same convergence rate of the order $(\delta x)^2$ and can be used to evaluate flow solutions obtained by the faster, but less accurate, unbiased GRW algorithms.

Since the GRW schemes are non-steady and a transient time is needed to reach the stationary state, the computation time is in general larger than for classical approaches (e.g., the solution shown in Fig. 3.36 is obtained in about 10 s in case of BGRW and 5 s in case of BGRWD, a spectral Chebyshev collocation solution requires cca 3 s, and a FEM solution cca 1 s). However, GRW flow solutions are unconditionally stable, which is not the case of FEM solutions, and they may be computed on the same lattice as that used in GRW transport simulations. Such an integrated GRW flow and transport solution avoids interpolation errors, inherent when FEM flow solutions are used in GRW transport simulations. The integrated approach is also free of numerical diffusion, which is unavoidable in FEM solutions for advection-dominated transport in groundwater [19]. Moreover, since the time needed to import the FEM velocity solution into the GRW lattice is two orders of magnitude larger than the time of the GRW transport simulation itself [30, p. 35], the integrated approach could lead to important saving of computing time, even if GRW flow solutions can be ten times more expensive than spectral or FEM solutions.

3.3.4.2   Particle Solutions for Reactive Transport

Particle approaches to reactive transport were developed with sequential PT [1], or CTRW [8] schemes, as well as with global procedures like GRW algorithms [11, 17] or random walkers cellular automata (CA) [12, 13].

Within the sequential PT and CTRW approaches a reaction takes place when two particles representing different molecular species are close enough to interact. The decision is made by using the "interaction radius," an empirical parameter related to the grain size [8], or by using the probability that the two particles will occupy the same position, constructed as a convolution of the two Gaussian probability densities [1].

In global approaches there is no need to search for nearby groups of particles which can interact. A reaction takes place when a sufficient amount of particles representing reacting species meet at a lattice site [11, 12]. For instance, in the CA approach, assuming that local chemical equilibrium is attained on a time scale much shorter than advection and diffusion scales, both the reaction probabilities P_r and the rate of reaction production are proportional to the physical rate constant R and can be straightforwardly implemented as CA rules, e.g., $P_r \sim R$ and $n_c \sim Rn_a n_b^2$, for the reaction $a + 2b \rightarrow c$ [12].

While existing GRW approaches [11, 17] do not incorporate advection, the CA approach solves advection–diffusion–reaction problems but only reaction steps are solved with a global procedure, advection and diffusion steps being performed sequentially [13, Eqs. (2.2) and (2.3)]. A GRW-CA algorithm for advection–diffusion–reaction is a natural extension for both approaches which goes beyond their limitations. In addition, because in a reduced fluctuations GRW procedure [22] the number of particles can be as large as the number of molecules involved in chemical reactions, the difficulties related to the adjustment of the scaling factor relating concentrations and particle densities [13] will be avoided.

References

1. Benson, D.A., Meerschaert, M.M.: Simulation of chemical reaction via particle tracking: diffusion-limited versus thermodynamic rate-limited regimes. Water Resour. Res. **44**(12), W12201 (2008)
2. Brunner, F., Radu, F.A., Bause, M., Knabner, P.: Optimal order convergence of a modified BDM1 mixed finite element scheme for reactive transport in porous media. Adv. Water Resour. **35**, 163–171 (2012)
3. Cortis, A., Berkowitz, B.: Computing "anomalous" contaminant transport in porous media: the CTRW MATLAB toolbox. Ground Water **43**(6), 947–950 (2005)
4. Cortis, A., Knudby, C.: A continuous time random walk approach to transient flow in heterogeneous porous media. Water Resour. Res. **42**(10), W10201 (2006)
5. Dagan, G.: Theory of solute transport by groundwater. Water Resour. Res. **19**, 183–215 (1987)
6. Doob, J.L.: Stochastic Processes. Wiley, New York (1990)
7. Eberhard, J.P., Suciu, N., Vamos, C.: On the self-averaging of dispersion for transport in quasi-periodic random media. J. Phys. A: Math. Gen. **40**(4), 597–610 (2007)
8. Edery, Y., Scher, H., Berkowitz, B.: Particle tracking model of bimolecular reactive transport in porous media. Water Resour. Res. **46**(7), W07524 (2010)
9. El Haddad, R., Lécot, C., Venkiteswaran, G.: Diffusion in a nonhomogeneous medium: quasi-random walk on a lattice. Monte Carlo Methods Appl. **16**, 211–230 (2010)
10. Gardiner, C.W.: Stochastic Methods. Springer, Berlin (2009)
11. Izsák, F., Lagzi, I.: Models of Liesegang pattern formation. In: Lagzi, I. (ed.) Precipitation Patterns in Reaction-Diffusion Systems, pp. 207–217. Research Signpost, Kerala (2010)
12. Karapiperis, T.: Cellular automaton model of precipitation/dissolution coupled with solute transport. J. Stat. Phys. **81**(1–2), 165–180 (1995)

13. Karapiperis, T., Blankleider, B.: Cellular automaton model of reaction-transport processes. Physica D **78**, 30–64 (1994)
14. Kloeden, P.E., Platen, E.: Numerical Solutions of Stochastic Differential Equations. Springer, Berlin (1999)
15. Kozma, G., Tóth, B.: Central limit theorem for random walks in divergence-free random drift field: $\mathscr{H}_{-1}$ suffices. Ann. Probab. **45**(6b), 4307–4347 (2017)
16. Lécot, C., Coulibaly, I.: A particle method for some parabolic equations, J. Comput. Appl. Math. **90**, 25–44 (1998)
17. Nagy, N., Izsák, F.: Stability of reaction fronts in random walk simulations. Appl. Math. Res. eXpress **2012**(1), 114–126 (2011)
18. Papoulis, A., Pillai, S.U.: Probability, Random Variables and Stochastic Processes. McGraw-Hill, Singapore (2002)
19. Radu, F.A., Suciu, N., Hoffmann, J., Vogel, A., Kolditz, O., Park, C.-H., Attinger, S.: Accuracy of numerical simulations of contaminant transport in heterogeneous aquifers: a comparative study. Adv. Water Resour. **34**, 47–61 (2011)
20. Strikwerda, J.C.: Finite Difference Schemes and Partial Differential Equations. Wadsworth & Brooks, Pacific Grove (2004)
21. Suciu, N.: Global random walk algorithm for transport in media with discontinuous dispersion coefficients. Geophys. Res. Abstr. **15**, EGU2013-12751-1 (2013)
22. Suciu, N.: Diffusion in random velocity fields with applications to contaminant transport in groundwater. Adv. Water. Resour. **69**, 114–133 (2014)
23. Suciu, N., Vamoş, C.: Evaluation of overshooting errors in particle methods for diffusion by biased global random walk. Rev. Anal. Numer. Theor. Approx. **35**, 119–126 (2006)
24. Suciu, N., Vamoş, C., Vanderborght, J., Hardelauf, H., Vereecken, H.: Numerical modeling of large scale transport of contaminant solutes using the global random walk algorithm. Monte Carlo Methods Appl. **10**(2), 153–177 (2004)
25. Suciu, N., Vamoş, C., Knabner, P., Ruede, U.: Biased global random walk, a cellular automaton for diffusion. In: Hülsemann, F., Kowarschik, M., Rude, U. (eds.) Simulations technique, 18th Symposium in Erlangen, pp. 562–567. SCS Publishing House e. V., Erlangen (2005)
26. Suciu, N., Vamoş, C., Vanderborght, J., Hardelauf, H., Vereecken, H.: Numerical investigations on ergodicity of solute transport in heterogeneous aquifers. Water Resour. Res. **42**, W04409 (2006)
27. Suciu N., Vamoş, C., Eberhard, J.: Evaluation of the first-order approximations for transport in heterogeneous media. Water Resour. Res. **42**, W11504 (2006)
28. Suciu N., Vamos, C., Vereecken, H., Sabelfeld, K., Knabner, P.: Memory effects induced by dependence on initial conditions and ergodicity of transport in heterogeneous media. Water Resour. Res. **44**, W08501 (2008)
29. Suciu, N., Vamoş, C., Turcu, I., Pop, C.V.L., Ciortca, L.I.: Global random walk modeling of transport in complex systems. Comput. Vis. Sci. **12**, 77–85 (2009)
30. Suciu, N., Radu, F.A., Prechtel, A., Knabner, P.: A coupled finite element-global random walk approach to advection-dominated transport in porous media with random hydraulic conductivity. J. Comput. Appl. Math. **246**, 27–37 (2013)
31. Vamoş, C., Suciu, N., Vereecken, H.: Generalized random walk algorithm for the numerical modeling of complex diffusion processes. J. Comput. Phys. **186**(2), 527–544 (2003)
32. Vamoş, C., Şoltuz, Ş., Crǎciun, M.: (2007). arXiv:079.2963vl [physics.data-an]

Chapter 4
Diffusion in Random Velocity Fields

Abstract Classical stochastic theories for transport in subsurface are revisited and transport models are formulated as stochastic processes. The process of diffusion with space variable drift coefficients is proposed as a general frame for stochastic modeling in subsurface hydrology. Stochastic homogeneity properties, first order approximations, and the occurrence of anomalous diffusion, ergodic, and self-averaging properties are presented.

4.1 Classical Stochastic Theories Revisited

4.1.1 Taylor's Theory of Diffusion by Continuous Movements

4.1.1.1 Movements in Random Fields

In a paper published in 1921, Taylor [81] derived a process with asymptotic diffusive behavior, induced by a space random velocity field, as model for turbulent diffusion in the atmosphere. This is a pioneering work in the theory of stochastic processes, published a decade before the Kolmogorov's work on the mathematical basis of modern theory of diffusion processes [48]. Taylor's paper was the starting point for a vast literature on turbulent diffusion, transport in porous media, and plasma physics. A minimal mathematical frame for Taylor's theory of diffusion by continuous movements, based on definitions and notions presented in Chap. 2, can be formulated as follows.

Let $\vartheta : \Omega \longmapsto Y_V^\Lambda$, $Y_V \subseteq \mathbb{R}^3$, $\Lambda \subseteq \mathbb{R}^3$, be a random field, with a range of parameters in the physical space. We make the following hypotheses about the properties of the random field:

(H1)   *For every fixed $\omega \in \Omega$ the sample $V^{(\omega)}$, $V^{(\omega)} : \Lambda \longmapsto Y_V$, is a nonsingular vector field.*

(H2)   *For all the samples $\omega \in \Omega$ the functions $V^{(\omega)}(x)$ and their derivatives are continuous with respect to the space variable x.*

© Springer Nature Switzerland AG 2019

N. Suciu, *Diffusion in Random Fields*, Geosystems Mathematics,
https://doi.org/10.1007/978-3-030-15081-5_4

The nonsingularity of the samples *(H1)* means that the field cannot vanish. The continuity hypothesis *(H2)* ensures the existence of a function of x and ω defined by the samples of the field, $V(x, \omega) = V^{(\omega)}(x)$, measurable with respect to (x, ω) which is an equivalent representation of the random field ϑ (see Sect. 2.1.2.1). The measurability ensures the commutation between the space integral and the stochastic average

$$\int_\Lambda M_\Omega[V(x, \omega)]dx = M_\Omega \left[\int_\Lambda V(x, \omega)dx \right].$$

This property will be used in the next section to derive the connection between Lagrangian and Eulerian correlations.

Under the hypotheses *(H1)* and *(H2)*, the realizations ω of the random field ϑ generate differentiable dynamical systems $\{S_t^{(\omega)}\}_{t \in \mathbb{R}}$, with trajectories in $Y_x \subseteq \mathbb{R}^3$, $X^{(\omega)}(t; x_0, t_0) = X(t, \omega, x_0, t_0) = S_{t-t_0}^{(\omega)}(x_0)$ [2]. (Note that the space Y_x on which the trajectories are defined and the space of parameters Λ of the velocity field coincide.) The trajectories $x^{(\omega)}$ are solutions of the system of ordinary differential equations

$$\frac{dX^{(\omega)}}{dt} = V^{(\omega)}(X^{(\omega)}). \tag{4.1}$$

Equation (4.1) describes the movement of a "fluid particle," as in continuum mechanics (see, e.g., [65]).

The stochastic model of a fluid particle moving in the random velocity field is the statistical ensemble consisting of the infinite set of dynamical systems generated by the realizations $V^{(\omega)}$ of the random field. More precisely, the random field induces the process $\chi : \Omega \times Y_x \longmapsto Y_x^{\mathbb{R}}$, where, for fixed $\omega \in \Omega$ and $x_0 \in Y_x$, $\chi(\omega, x_0) = X(t, \omega, x_0, t_0)$ is a trajectory of the dynamical system $\{S_t^{(\omega)}\}_{t \in \mathbb{R}}$. Let $(\Omega, \mathscr{A}, P)$ be the probability space of the velocity field. The space of the initial positions Y_x can also be organized as a probability space $(Y_x, \mathscr{B}, P_x)$, with the σ-algebra Borel $\mathscr{B}$ and the probability measure P_x defined by

$$P_x(B) = \int_B c(x_0, t_0)dx_0, \text{ for all sets } B \in \mathscr{B},$$

where the density $c(x_0, t_0)$ is the normalized initial concentration of the fluid particles. Then, one defines the measure on the probability space $\Omega \times Y_x$ (space of elementary events for the process χ) as a normalized product measure, $P P_x$. The stochastic average (2.3) becomes

$$M_{\Omega \times Y_x}(f) = \int_\Omega P(d\omega) \int_{Y_x} f(\omega, x_0)c(x_0, t_0)dx_0 = M_\Omega \left[\int_{Y_x} f(\omega, x_0)c(x_0, t_0)dx_0 \right].$$
$$\tag{4.2}$$

As seen in Sect. 2.2.1, the dynamical system is a deterministic process in $Y_x \subseteq \mathbb{R}^3$, with a "degenerate" transition probability

$$p^{(\omega)}(x, t \mid x_0, t_0; \omega) = \delta(x - X(t, \omega, x_0, t_0)), \qquad (4.3)$$

which is a solution of the Liouville equation

$$\partial_t p^{(\omega)} + \nabla_x (V^{(\omega)} p^{(\omega)}) = 0, \qquad (4.4)$$

with initial condition $p^{(\omega)}(x, 0 \mid x_0, 0; \omega) = \delta(x - x_0)$ [36, Sect. 4.3.4]. Given the probability density of the initial positions $c(x_0, t_0)$, the normalized concentration at $t > t_0$ is the one-dimensional density (2.25) for fixed realization ω of the velocity field,

$$c^{(\omega)}(x, t) = \int_{Y_x} p^{(\omega)}(x, t \mid x_0, t_0; \omega) c(x_0, t_0) dx_0, \qquad (4.5)$$

which is also a solution of (4.4). Relation (4.5) is used to define the concentration in a single realization of the Darcy velocity field into the "Lagrangian framework" proposed in [19]. Since (4.5) is a solution of the Liouville equation (4.4), in a single realization of the velocity field the solute is not "diluted": the initial concentration $c(x_0, t_0)$ is advected with the velocity $V^{(\omega)}$, without any deformation of the concentration profile.

The normalized concentration field corresponding to the ensemble of fluid particles described by the process χ is the one-dimensional density obtained, according to (4.2) and (2.18), as an average over the probability space $\Omega \times Y_x$,

$$
\begin{aligned}
c(x, t) &= M_{\Omega \times Y_x}[\delta(x - X(t, \omega, x_0, t_0))] \\
&= \int_{Y_r} M_\Omega[\delta(x - X(t, \omega, x_0, t_0))] c(x_0, t_0) dx_0 \\
&= \int_{Y_x} p(x, t \mid x_0, t_0) c(x_0, t_0) dx_0, \qquad (4.6)
\end{aligned}
$$

where

$$p(x, t \mid x_0, t_0) = M_\Omega[p^{(\omega)}(x, t \mid x_0, t_0; \omega)] \qquad (4.7)$$

is a transition probability of the process χ. Hence, as follows from (4.5) and (4.7), relation (4.6) represents the average over Ω of the concentration in a single realization, $c(x, t) = M_\Omega[c^{(\omega)}(x, t)]$.

Because for given realizations ω of the random field the fluid particles move along the trajectories of a differentiable dynamical systems $\{S_t^{(\omega)}\}_{t \in \mathbb{R}}$, the process

χ is continuous, i.e., it verifies the property (i) of the diffusion processes (see Sect. 2.2.2.1). The coefficients ((2.28)–(2.29)) of the Fokker–Planck equation (equivalent with diffusion properties (ii) and (iii)) can be computed within the Lagrangian statistics as averages over trajectories of the process by using the conditional averages of the form (2.42).

For fixed ω, the values of the function $V(x, \omega)$ at points x lying on the trajectory of the process χ which starts from the initial position x_0 define the *Lagrangian velocity* $V(t, \omega, x_0, t_0) = V(X(t, \omega, x_0, t_0), \omega)$. Using it, the integral representation of (4.1) is given by

$$X(t, \omega, x_0, t_0) - x_0 = \int_{t_0}^{t} V(s, \omega, x_0, t_0)ds. \tag{4.8}$$

The drift coefficient (2.28), conditioned by a state (x_0, t_0) is obtained by using (4.8) and the conditional average (2.42) as

$$A(x_0, t_0) = \frac{d}{dt} M_{\Omega}[(X(t, \omega, x_0, t_0) - x_0)]|_{t=t_0}$$

$$= \frac{d}{dt} \int_{t_0}^{t} M_{\Omega}[V(s, \omega, x_0, t_0)]ds|_{t=t_0}.$$

Thus, the local mean of the Lagrangian velocity $\overline{V}(x_0, t_0) = M_{\Omega}[V(t_0, \omega, x_0, t_0)]$ defines the drift coefficient of the Fokker–Planck equation,

$$A(x_0, t_0) = \overline{V}(x_0, t_0). \tag{4.9}$$

The diffusion coefficients can also be computed by using (4.8) and the Lagrangian definition (2.45). The tensor of displacement variance conditioned by the state (x_0, t_0) is given by

$$\tilde{\sigma}^2(t; x_0, t_0) = \int_{t_0}^{t} ds \int_{t_0}^{t} \{M_{\Omega}[V(s, \omega, x_0, t_0)V(s', \omega, x_0, t_0)]$$

$$- M_{\Omega}[V(s, \omega, x_0, t_0)]M_{\Omega}[V(s', \omega, x_0, t_0)]\}ds', \tag{4.10}$$

and from (2.45) one obtains the diffusion tensor

$$\tilde{B}(x_0, t_0) = \frac{1}{2}\frac{d}{dt}\tilde{\sigma}^2(t; x_0, t_0)|_{t=t_0} = \tilde{0}. \tag{4.11}$$

The coefficients (4.9) and (4.11) have the same form at all points (x, t), and the Fokker–Planck equation takes the form (2.33a) of the Liouville equation,

$$\partial_t c(x, t) + \nabla_x (\overline{V}(x, t) c(x, t)) = 0. \tag{4.12}$$

Equation (4.12) is just the average over velocity realizations of the Liouville equation (4.4) associated with dynamical systems in each realization of the field. Since the coefficients $\tilde{B}$ (4.11) vanish, there is no local diffusive behavior for the fluid particle moving in a random velocity field.

4.1.1.2 Asymptotic Diffusive Behavior

When the asymptotic diffusive behavior property (2.65) holds, one expects that at large scales the process χ, described locally by the Liouville equation (4.12), could be approximated by a diffusion equation. One expects also that the transport behavior in single realizations is similar with the average over realizations of the random field. Only in these conditions, the stochastic model predicts the evolution of the concentration field for transport in heterogeneous media. Nevertheless, while the asymptotic diffusive behavior can be proved in some circumstances, the existence of an effective diffusion equation and the asymptotic diffusive behavior in all realizations of the field still remains an open problem.

To investigate the non-local behavior of transport, one defines the *Lagrangian correlation* tensor function as a complete average over the entire probability space $\Omega \times Y_x$, of the process χ,

$$\tilde{R}_L(s, s') = M_{\Omega \times Y_x}[V(s, \omega, x_0, t_0) V(s', \omega, x_0, t_0)]$$
$$- M_{\Omega \times Y_x}[V(s, \omega, x_0, t_0)] M_{\Omega \times Y_x}[V(s', \omega, x_0, t_0)]. \tag{4.13}$$

Similarly to the local variance (4.10) conditioned on initial state (x_0, t_0), the unconditional variance is expressed by using (4.13) as

$$\tilde{\sigma}^2(t) = \int_{t_0}^{t} ds \int_{t_0}^{t} \tilde{R}_L(s, s') ds'. \tag{4.14}$$

In particular, if the Lagrangian correlation is a *homogeneous* time function, $\tilde{R}_L(s, s') = \tilde{R}_L(\tau)$, where $\tau = s - s'$, and $\tilde{R}_L$ is an *even function* of τ, $\tilde{R}_L(-\tau) = \tilde{R}_L(\tau)$, (4.14) can be written as

$$\tilde{\sigma}^2(t) = 2 \int_0^t (t - \tau) \tilde{R}_L(\tau) d\tau. \tag{4.15}$$

This relation, first derived by Taylor [81] is known in the literature as "Taylor formula" [54, Eq. (9.30')].

The asymptotic diffusive behavior (2.65) requires the order relation $\sigma^2 = \mathcal{O}(t)$ for $t \longrightarrow \infty$. From (4.14) one finds the necessary condition

$$\lim_{t \to \infty} \int_{t_0}^{t} \tilde{R}_L(s, s')ds' < \infty,$$

equivalent with the condition of finite correlation time (2.73) for the process presented in Sect. 2.2.3.6. Condition (2.65) defines the upscaled diffusion coefficient by

$$\tilde{D}^* = \lim_{t \to \infty} \frac{\tilde{\sigma}^2(t)}{2t} = \lim_{t \to \infty} \frac{1}{2} \frac{d}{dt} \tilde{\sigma}^2(t) = \lim_{t \to \infty} \int_{t_0}^{t} \tilde{R}_L(t, s')ds', \qquad (4.16)$$

which has the form of a Green–Kubo relation (see Eq. (2.72)).

The existence of the diffusion limit (4.16) was inferred by Taylor [81] from the assumption that in turbulent flows the displacements (4.8) become uncorrelated at successive times and their sum can be approximated by a Gaussian process. This has the form of CLT discussed in Sect. 3.1.2.2. The randomness of the pore–grain geometry in natural porous media led to stochastic representations, as, for instance, the model of "complete disorder"[65], for which CLTs are assumed to hold. It is therefore believed that at space scales that are large as compared to the pore scale, by virtue of CLT the sum of displacements (4.8) can be approximated by a Gaussian process and the transport behaves asymptotically diffusive [82].

If the limit (4.16) exists, *one asserts* that it is possible to approximate asymptotically the advection equation (4.12) with the advection–diffusion equation

$$\partial_t c(x, t) + V^* \nabla c(x, t) = \tilde{D}^* \nabla^2 c(x, t), \qquad (4.17)$$

where $V^* = M_{Y_x}[\overline{V}(x_0, t_0)] = M_{\Omega \times Y_x}[V(t_0, \omega, x_0, t_0)]$ is the mean velocity. This is the general framework of theories on "Lagrangian passive transport in turbulent fields" and of the Green–Kubo type approaches from statistical mechanics [6, 42, 54, 81]. One remarks that while the existence of derivatives ((2.28)–(2.29)) determines the Fokker–Planck equation, the existence of the limit (4.16) alone does not ensure the existence of the advection–diffusion equation as an asymptotic approximation of the exact Liouville equation (4.12). The weak convergence of the probability distribution of the process χ (and implicitly of the mean concentration (4.6)) to an upscaled Gaussian diffusion process described by (4.17) was proved by Kesten and Papanicolaou [45] under the supplementary assumptions that the velocity field has small fluctuations around a constant mean and possess some "strong mixing" property (see Sect. 2.2.1), characterized by a suitable fast decay of the correlation function.

It should be noted that, due to the absence of the local diffusion, looking for the asymptotic diffusive behavior of single realizations of the process χ makes no sense. Indeed, (4.10) without average over realizations implies that $\sigma^2 = 0$ at all times.

4.1.1.3 Corsin's Conjecture

In experiments, the velocity field is measured at fixed space points, which corresponds to an Eulerian description. To pass to an Eulerian description we write the Lagrangian correlation (4.13) as

$$\tilde{R}_L(s, s') =$$

$$\int_{Y_x} \int_{Y_x} M_{\Omega \times Y_x}[\delta(x - X(s, \omega, x_0, t_0))\delta(x' - X(s', \omega, x_0, t_0))V(x, \omega)V(x', \omega)]dxdx'$$

$$- \int_{Y_x} M_{\Omega \times Y_x}[\delta(x - X(s, \omega, x_0, t_0))V(x, \omega)]dx$$

$$\times \int_{Y_x} M_{\Omega \times Y_x}[\delta(x' - X(s', \omega, x_0, t_0))V(x', \omega)]dx'. \tag{4.18}$$

In (4.18) one considers the extensions to $\mathbb{R}^3$ of the functions $V(x, \omega)$, defined for every fixed ω as $V(x, \omega) : \Lambda \longmapsto Y_V$, $V(x, \omega) = 0$ for $x \notin Y_x$ (remember that Λ and Y_x coincide), one uses the commutation between the space integral and the stochastic average ensured by *(H2)*, and one rewrites the Lagrangian velocity with the aid of the Dirac function as

$$V(s, \omega, x_0, t_0) = \int_{\mathbb{R}^3} \delta(x - X(s, \omega, x_0, t_0))V(x, \omega)dx$$

$$= \int_{Y_x} \delta(x - X(s, \omega, x_0, t_0))V(x, \omega)dx.$$

The *Corsin's conjecture* (in the form used by Saffman [64] in problems of turbulent diffusion) says that the averages over $\Omega \times Y_x$ factorize as follows:

$$M_{\Omega \times Y_x}[\delta(x - X(s, \omega, x_0, t_0))\delta(x' - X(s', \omega, x_0, t_0))V(x, \omega)V(x', \omega)] =$$

$$M_{\Omega \times Y_x}[\delta(x - X(s, \omega, x_0, t_0))\delta(x' - X(s', \omega, x_0, t_0))]$$

$$\times M_{\Omega \times Y_x}[V(x, \omega)V(x', \omega)], \tag{4.19}$$

and

$$M_{\Omega \times Y_x}[\delta(x - X(s, \omega, x_0, t_0))V(x, \omega)] =$$
$$M_{\Omega \times Y_x}[\delta(x - X(s, \omega, x_0, t_0))]M_{\Omega \times Y_x}[V(x, \omega)],$$
$$M_{\Omega \times Y_x}[\delta(x' - X(s', \omega, x_0, t_0))V(x', \omega)] =$$
$$M_{\Omega \times Y_x}[\delta(x' - X(s', \omega, x_0, t_0))]M_{\Omega \times Y_x}[V(x', \omega)]. \tag{4.20}$$

According to (2.17), the first factor in (4.19) defines the two-dimensional density of the process χ,

$$p(x, s; x', s') = M_{\Omega \times Y_x}[\delta(x - X(s, \omega, x_0, t_0))\delta(x' - X(s', \omega, x_0, t_0))], \tag{4.21}$$

and the first factor in each relation (4.20) defines one-dimensional densities,

$$p(x, s) = M_{\Omega \times Y_x}[\delta(x - X(s, \omega, x_0, t_0))],$$
$$p(x', s) = M_{\Omega \times Y_x}[\delta(x' - X(s', \omega, x_0, t_0))]. \tag{4.22}$$

The second factor in (4.19) can be written as

$$M_{\Omega \times Y_x}[V(x, \omega)V(x', \omega)] =$$

$$M_{\Omega}[V(x, \omega)V(x', \omega)] \int_{Y_x} c(x_0, t_0)dx_0 = M_{\Omega}[V(x, \omega)V(x', \omega)]$$

$$= \int_{Y_V} \int_{Y_V} vv' M_{\Omega}[\delta(v - V(x, \omega))\delta(v' - V(x', \omega))]dvdv',$$

where one uses the commutation property *(H2)* and the Dirac function, similarly to the derivation of (4.18). Using the two-dimensional density of the field ϑ, defined by (2.17),

$$p_v(v, x; v', x') = M_{\Omega}[\delta(v - V(x, \omega))\delta(v' - V(x', \omega))],$$

one obtains

$$M_{\Omega \times Y_x}[V(x, \omega)V(x', \omega)] = \int_{Y_V} \int_{Y_V} vv' p_v(v, x; v', x')dvdv', \tag{4.23}$$

and, similarly, for the second factors of (4.20),

$$M_{\Omega \times Y_x}[V(x, \omega)] = \int_{Y_V} v p_v(v, x) dv = \overline{V}(x),$$

$$M_{\Omega \times Y_x}[V(x', \omega)] = \int_{Y_V} v' p_v(v', x') dv' = \overline{V}(x'). \qquad (4.24)$$

Using (4.19)–(4.24) in (4.18), one obtains the final relation

$$\tilde{R}_L(s, s') = \int_{Y_x} \int_{Y_x} p(x, s; x', s') \tilde{R}_E(x, x') dx dx', \qquad (4.25)$$

where $\tilde{R}_E$ is the *Eulerian correlation* tensor

$$\tilde{R}_E(x, x') = \int_{Y_V} \int_{Y_V} [vv' - \overline{V}(x) \overline{V}(x')] p_v(v, x; v', x') dv dv'. \qquad (4.26)$$

To establish the connection between the Lagrangian and Eulerian descriptions by means of relations of the form (4.25) is one of the main goals in modeling turbulent diffusion and transport processes in heterogeneous porous media [41, 60].

4.1.2   Advection–Dispersion Models

4.1.2.1   Diffusion in Random Fields and Macrodispersion

Sometimes at *local scales* the transport processes in natural porous media are well described by advection–diffusion equations but at larger, *global scales,* the diffusion coefficients experimentally derived during the calibration of the model show an apparent increase as compared to the local ones. This is the so-called scale effect [35, 64]. The main question is: May the transport be globally described by an effective diffusion equation? This could be the case when the variability of the Darcy velocity produces a hydrodynamic dispersion of the solute which, at a suitable large scale, behaves similarly to a diffusion process. This phenomenon is sometimes called "macrodispersion" and the corresponding effective diffusion coefficient is referred to as "macrodispersion coefficient" [19]. The macrodispersion problem requires a two-scale description: at Darcy scale where the process of local diffusion is important and at the macrodispersion scale, where the solute transport is dominated by the variability of the advection velocity. A stochastic two-scale model can be formulated as a diffusion process in a random velocity field.

Let the velocity field be described by the random function $\vartheta : \Omega_V \longmapsto Y_V^\Lambda$, $Y_V \subseteq \mathbb{R}^3$, $\Lambda \subseteq \mathbb{R}^3$, and let us consider the Wiener process $W : \Omega_W \longmapsto Y_W^{[0,\infty]}$, $Y_W \subseteq \mathbb{R}^3$. For every fixed $\omega_V \in \Omega_V$ one considers the Itô equation

$$dX^{(\omega_V)}(t) = V(X^{(\omega_V)}(t), \omega_V)dt + \left(\tfrac{2}{3}D\right)^{\frac{1}{2}} dW(t, \omega_W). \tag{4.27}$$

A transport model of form (4.27) for homogeneous porous media was introduced, on the basis of a rigorous proof of a CLT, by Bhattacharya and Gupta [10].

Let us assume that the sufficient conditions for the existence of unique solutions of Itô equation, A1–A3 of Sect. 3.2.1, are fulfilled. Note that the Lipschitz condition A2 implies the continuity condition H2 which ensures the permutation of averages and integrals. Then, for every trajectory of the Wiener process W and sample of the random velocity field ϑ, and for a deterministic initial condition $X^{(\omega_V)}(t_0) = x_0$ there is a unique solution of Eq. (4.27), $X^{(\omega_V, \omega_W)}(t) = X(t, \omega_W, \omega_V, x_0, t_0)$. Under these conditions, a stochastic process χ can be defined in the corresponding product probability space, $\chi : \Omega_V \times \Omega_W \times Y_x \longmapsto Y_x^{[0,\infty]}$, $Y_x \subseteq \mathbb{R}^3$, with trajectories given by $X^{(\omega_V, \omega_W)}(t)$.

The mean concentration is given by the one-dimensional density of the process χ,

$$c(x, t) = M_{\Omega_V \times \Omega_W \times Y_x}[\delta(x - X(t, \omega_W, \omega_V, x_0, t_0))], \tag{4.28}$$

which generalizes the definition of the mean concentration (4.6) for the model of advective transport presented in the previous section.

To simplify the presentation, in the following one supposes $M_{\Omega_V}[V(x, \omega_V)] = 0$. Using the integral form of (4.27) and the Itô calculus (see Sect. 2.2.3.2) one obtains the variance of the process χ, as average over the probability space $\Omega_V \times \Omega_W \times Y_x$,

$$\tilde{\sigma}^2(t) = 2D(t - t_0) +$$

$$\int_{t_0}^{t} ds \int_{t_0}^{t} M_{\Omega_V \times \Omega_W \times Y_x}[V(s, \omega_W, \omega_V, x_0, t_0)V(s', \omega_W, \omega_V, x_0, t_0)]ds'. \tag{4.29}$$

Note that in derivation of (4.29) the cross correlation between advective and diffusive displacements cancels because $M_{\Omega_V}(V) = 0$ and because, by virtue of H2, the successive averages may be performed in any order. More generally, the ensemble average cross correlation also vanishes in case of statistically homogeneous velocity fields, because then the ensemble mean velocity $M_{\Omega_V}[V(X^{(\omega_V)})]$ is constant, thus independent of the Wiener process.

At finite times the second term from (4.29) behaves as $(t - t_0)^2$, which is an explanation of the "scale effect": the effective diffusion coefficients given by $\tilde{D}^* = \tilde{\sigma}^2(t)/(2t) \sim t$ increase in time.

The Lagrangian correlation under the integral in (4.29) is an average over the product probability space $\Omega_V \times \Omega_W \times Y_x$. Using the Dirac function, one obtains

$$\tilde{R}_L(s, s') =$$

$$\int\limits_{Y_x} \int\limits_{Y_x} M_{\Omega_V \times \Omega_W \times Y_x}[V(x, \omega_V)V(x', \omega_V)\delta(x - X(s, \omega_W, \omega_V, x_0, t_0))$$

$$\times \delta(x' - X(s', \omega_W, \omega_V, x_0, t_0))]dxdx' = \int\limits_{Y_x} \int\limits_{Y_x} M_{\Omega_V}\{V(x, \omega_V)V(x', \omega_V)$$

$$\times M_{\Omega_W \times Y_x}[\delta(x - X(s, \omega_W, \omega_V, x_0, t_0))\delta(x' - X(s', \omega_W, \omega_V, x_0, t_0)]\}dxdx'.$$

$$(4.30)$$

When Corsin's conjecture is used in the average over ω_V, the second factor from (4.30) defines the two-dimensional probability density of the process χ:

$$p(x, s; x', s') =$$
$$M_{\Omega_V \times \Omega_W \times Y_x}[\delta(x - X(s, \omega_W, \omega_V, x_0, t_0))\delta(x' - X(s', \omega_W, \omega_V, x_0, t_0))].$$

$$(4.31)$$

The first factor from (4.30) gives the Eulerian correlation (4.26), and, using (4.31), relation (4.30) takes the form

$$\tilde{R}_L(s, s') = \int\limits_{Y_x} \int\limits_{Y_x} p(x, s; x', s')\tilde{R}_E(x, x')dxdx', \qquad (4.32)$$

which gives the Lagrangian correlation as a space average of the Eulerian correlation weighted by the two-dimensional density of the process χ. Relation (4.32) has the same form as (4.25), with the difference that now the two-dimensional density is given by (4.31).

If the Lagrangian correlation $\tilde{R}_L$ decreases to zero fast enough as $t \longrightarrow \infty$, then the process χ behaves asymptotically diffusive with effective diffusion coefficient given by the variance (4.29) and the definition (2.65),

$$\tilde{D}^* = D\tilde{1} + \lim_{t \to \infty} \int\limits_{t_0}^{t} \tilde{R}_L(t, s')ds'. \qquad (4.33)$$

If the coefficients (4.33) exist, one usually asserts that for large times the concentration (4.28) is described by the diffusion equation

$$\partial_t c(x, t) = \tilde{D}^* \nabla^2 c(x, t).$$

The Lagrangian correlation (4.30) is an average over Ω_W of the velocity correlation on a single trajectory of the local dispersion. Thus, the effective diffusion coefficients for diffusion in random fields are not the sum between the local coefficient D and the effective coefficient (4.16) computed in the case without local diffusion. The local diffusion causes a faster decay of $\tilde{R}_L$ and enhances the diffusive behavior.

In a single realization of the velocity field, by dropping the average over ω_V in (4.30), one obtains

$$\tilde{R}_L^{(\omega_V)}(s, s') = \int\limits_{Y_x} \int\limits_{Y_x} p^{(\omega_V)}(x, s; x', s') V(x, \omega_V) V(x', \omega_V) dx dx',$$

where

$$p^{(\omega_V)}(x, s; x', s') = M_{\Omega_W \times Y_x}[\delta(x - X(s, \omega_W, \omega_V, x_0, t_0))\delta(x' - X(s', \omega_W, \omega_V, x_0, t_0))]$$

is the two-dimensional density of the diffusion process in the realization ω_V of the random field. Thus, the Lagrangian correlation $\tilde{R}_L^{(\omega_V)}$ is given by a space average over the solute plume (i.e., weighted by the probability density $p^{(\omega_V)}$ describing the diffusion process in a fixed realization of the velocity field). It is straightforward that the correlation (4.30) is the average over realizations of the correlations in individual realizations,

$$\tilde{R}_L(s, s') = M_{\Omega_V}[\tilde{R}_L^{(\omega_V)}(s, s')].$$

The effective diffusion coefficient in a single realization is

$$\tilde{D}^{*(\omega_V)} = D\tilde{1} + \lim_{t \to \infty} \int\limits_{t_0}^{t} \tilde{R}_L^{(\omega_V)}(t, s') ds'.$$

Comparing with (4.33) we find the relation

$$\tilde{D}^* = M_{\Omega_V}[\tilde{D}^{*(\omega_V)}]. \tag{4.34}$$

Due to (4.34), if all realizations ω_V have the self-averaging property (2.65) (i.e., all $\tilde{D}^{*(\omega_V)}$ are finite), then the process χ describing diffusion in random fields is self-averaging as well. If moreover

$$\lim_{t \to \infty} \int\limits_{t_0}^{t} \left(\tilde{R}_L^{(\omega_V)}(t, s') - \tilde{R}_L(t, s') \right) ds' = 0, \quad \text{for all } \omega_V \in \Omega_V,$$

then the effective diffusion coefficients have the same value $\tilde{D}^*$ in all realizations.

4.1.2.2 The Model of Matheron and de Marsily

An example where a Lagrangian description and the connection with the Eulerian statistics can be rigorously derived is the stratified aquifer model of Matheron and de Marsily [53]. The model describes the movement in the plan $\mathbb{R}^2$ of a particle, due to a two-dimensional diffusion process with longitudinal diffusion coefficient D_L and transverse diffusion coefficient D_T, on which one superposes a horizontal random field which depends only on the transverse coordinate, $V(z, \omega_V)$:

$$dX^{(\omega_V)}(t) = V(Z(t), \omega_V)dt + D_L dw(t),$$

$$dZ(t) = D_T dw(t). \tag{4.35}$$

In this case, the tensor of Lagrangian correlations (4.30) reduces to the component corresponding to horizontal velocities,

$$R_L(s, s') = \int_{\mathbb{R}} \int_{\mathbb{R}} M_{\Omega_V \times \Omega_W \times \mathbb{R}^2}[\delta(z - Z(s, \omega_W, z_0, t_0))$$

$$\times \delta(z' - Z(s', \omega_W, z_0, t_0))V(z, \omega_V)V(z', \omega_V)]dz dz' =$$

$$\int_{\mathbb{R}} \int_{\mathbb{R}} M_{\Omega_W \times \mathbb{R}^2}[\delta(z - Z(s, \omega_W, z_0, t_0))$$

$$\times \delta(z' - Z(s', \omega_W, z_0, t_0))]M_{\Omega_V}[V(z, \omega_V)V(z', \omega_V)]dz dz'. \tag{4.36}$$

Because the transverse diffusion in (4.35) is not influenced by the horizontal random field, the vertical component of the trajectory $Z(s, \omega_W, z_0, t_0)$ in (4.36) does not depend on ω_V. In this case, the average over Ω_V factorizes and the connection between Lagrangian and Eulerian correlations (4.32) holds without the hypothesis of Corsin's factorization. The variance of the longitudinal displacements becomes

$$\sigma_x^2(t) = 2D_L(t - t_0)$$

$$+ \int_{t_0}^{t} ds \int_{t_0}^{t} ds' \int_{\mathbb{R}} \int_{\mathbb{R}} p(z, s; z', s'; D_T)R_E(z, z')dz dz', \tag{4.37}$$

where

$$p(z, s; z', s'; D_T) = M_{\Omega_W \times \mathbb{R}^2}[\delta(z - Z(s, \omega_W, z_0, t_0))\delta(z' - Z(s', \omega_W, z_0, t_0))]$$

is the Gaussian two-dimensional density of the transverse diffusion process with coefficient D_T. This expression (4.37) is identical, with slightly different notations, with equation (6) of Matheron and de Marsily [53, Appendix 1].

The main feature of the model is that for Gaussian correlated fields, $R_E \sim e^{-z^2}$, the variance behaves as $\sigma_x^2 \sim t^{3/2}$. From (2.65) it follows that the behavior is superdiffusive at all times. This property is often used to check the validity of the numerical models of diffusion in random fields [6, 40]. In [53] it is proved that if the velocity V also has a vertical component the behavior becomes diffusive.

4.1.3 The Eulerian Statistics of the Travel-Time

Another case where, starting with a Lagrangian description, the final results can be exactly expressed by means of Eulerian averages is the simplified model of transport in soils presented in the following. By neglecting the pore scale dispersion, the Lagrangian description of the solute transport in soils can be obtained with an equation of form (4.8),

$$
Z(t, \omega) = \int_0^t V(Z(s, \omega))ds, \tag{4.38}
$$

where $Z(t, \omega)$ is the trajectory starting at $t_0 = 0$ from $z_0 = 0$. The velocity V is defined by $V(z, \omega) = V(Z(t, \omega))$ as a realization of a random field sampled at points z on the trajectory $Z(t, \omega)$ of the process. In the literature, it is often assumed that the trajectories (4.38) of the solute particles are vertical [83].

One of the quantities which describe the process is the travel-time $\tau(z)$ of a solute particle to reach a certain depth z and its statistics (average and variance). The expression of the depth z given by the trajectory equation (4.38),

$$
z = \int_0^\tau V(Z(s, \omega))ds,
$$

can be written as a function F, which implicitly defines τ as function of z,

$$
F(z, \tau(z)) = z - \int_0^\tau V(Z(s, \omega))ds \equiv 0. \tag{4.39}
$$

The total derivative of F with respect to z is

$$
\frac{dF}{dz} = \frac{\partial F}{\partial z} + \frac{\partial F}{\partial \tau}\frac{\partial \tau}{\partial z} = 0,
$$

and, using the Leibniz–Newton formula to write the derivative with respect to τ of the integral in (4.39) as $\partial F/\partial \tau = -V(Z(\tau, \omega)) = -V(z, \omega)$, we formally obtain the derivative of $\tau(z)$ with respect to z:

$$\frac{\partial \tau}{\partial z} = -\frac{\partial F}{\partial z} \Big/ \frac{\partial F}{\partial \tau} = 1/V(z, \omega). \tag{4.40}$$

Assuming $V > 0$, F becomes a monotonous function of τ and, together with the continuity hypothesis $H2$, ensures the existence of the function $\tau(z)$ on the entire range of z by the implicit function theorem from differential calculus. Integrating (4.40), one obtains

$$\tau(z) = \int_0^z \frac{\partial \tau}{\partial z} dz = \int_0^z \frac{1}{V(z, \omega)} dz. \tag{4.41}$$

Unlike the usual Lagrangian representations (see, e.g., [83, Eq. (23)], (4.41) is an Eulerian representation of the travel-time, because the velocity is function of the fixed depth z.

The mean travel-time is obtained as average of (4.41) over realizations of the velocity field. Using the property of commutation of stochastic average and space integral (under the hypothesis $H2$), the definition (2.18) of the one-dimensional density $p(v, z)$, and the Dirac function representation, one obtains

$$M_\Omega[\tau(z)] = \int_0^z dz \int_{-\infty}^\infty \frac{1}{v} M_\Omega[\delta(v - V(z, \omega)]dv = \int_0^z dz \int_{-\infty}^\infty \frac{1}{v} p(v, z)dv.$$

If one defines the harmonic average $v_H = 1/\overline{(1/v)}$, where $\overline{1/v} = \int_{-\infty}^\infty 1/v \, p(v, z)dv$ is the Eulerian average of $1/v$, the average travel-time is finally expressed as

$$M_\Omega[\tau(z)] = \frac{z}{v_H}. \tag{4.42}$$

The variance of the travel-time (4.41) can be also computed as an Eulerian average with respect to a two-dimensional density defined by (2.17),

$$\sigma^2[\tau(z)] = M_\Omega[(\tau(z) - M_\Omega[\tau(z)])^2] = M_\Omega[(\tau(z))^2] - (M_\Omega[\tau(z)])^2$$

$$= \int_0^z \int_0^z dz dz' \int_{-\infty}^\infty \int_{-\infty}^\infty \frac{1}{v}\frac{1}{v'} M_\Omega[\delta(v - V(z, \omega)\delta(v' - V(z', \omega)]dvdv' - \left(\frac{z}{v_H}\right)^2$$

$$= \int_0^z \int_0^z dz dz' \int_{-\infty}^\infty \int_{-\infty}^\infty \left(\frac{1}{v}\frac{1}{v'} - \left(\frac{1}{v_H}\right)^2\right) p(v, z; v', z')dvdv'.$$

The stochastic average in this expression defines the covariance of inverse velocities

$$cov_{\frac{1}{v},\frac{1}{v'}}(z, z') = \int\limits_{-\infty}^{\infty} \int\limits_{-\infty}^{\infty} \left(\frac{1}{v}\frac{1}{v'} - \left(\frac{1}{v_H}\right)^2 \right) p(v, z; v', z')dvdv'.$$

With the hypothesis that the random field is statistically homogeneous, $p(v, z; v', z') = p(v, z - z'; v')$ and $cov_{\frac{1}{v},\frac{1}{v'}}(z, z') = cov_{\frac{1}{v},\frac{1}{v'}}(z - z')$. Using the normalized covariance notation $\rho(\Delta z) = cov_{\frac{1}{v},\frac{1}{v'}}(z - z')/\sigma^2[\frac{1}{v}]$, where $\sigma^2[\frac{1}{v}] = cov_{\frac{1}{v},\frac{1}{v'}}(0)$, the variance of the travel-time can be written as

$$\sigma^2[\tau(z)] = 2\sigma^2 \left[\frac{1}{v}\right] \int\limits_0^z (z - \Delta z)\rho(\Delta z)dz. \qquad (4.43)$$

The average (4.42) and the variance (4.43) are identical with relations (13)–(14) of [68]. In the present approach the results are obtained from the Eulerian representation (4.41) of the travel-time, rigorously derived using the implicit function theorem under the differentiability hypothesis *(H2)* and assuming $V > 0$.

The travel-time is an observable that is easily accessible to measurements, as it is related to "breakthrough curves" (concentrations as functions of time for fixed points in space), the measurement of which is less expensive than the full sampling of the concentration over the experimental field necessary to estimate concentration moments. Under an implicit ergodicity assumption the mean and variance of the travel-time are inferred from the time moments of the measured breakthrough curve. Such estimations are based on the interpretation of the rate at which solute mass crosses the reference plane at which the breakthrough curve is measured as a probability density of the travel-time. With this interpretation one derives relations between concentrations C averaged over reference planes at the inlet and outlet boundaries of the transport model of the form

$$C_{out}(t) = \int_0^t g(t - t'|t')C_{in}(t')dt'. \qquad (4.44)$$

Relation (4.44) is similar to the evolution equation for the probability density of the Markov processes (2.25) and, in fact, it expresses the Markovian character of the travel-time random variable. Equation (4.44) is referred to as the *transfer function equation* [44] and is used as an alternative to the one-dimensional advection–dispersion equation for modeling transport in soils. However the predictions by the two models can be different depending on the assumed Gaussian or log-normal probability density of the travel time [44]. In this respect, estimations of form (4.42)–(4.43) which relate the statistics of the travel-time to that of the velocity field could be used to calibrate the models for solute transport in heterogeneous soils.

4.1.4   The Local Dispersivity Tensor

In Sects. 4.1.1 and 4.1.2 a constant local dispersion coefficient was considered for the purpose of illustration. But even in isotropic porous media the dispersion process may be anisotropic [66] and, in case of directed flow, the spreading caused by dispersion in the direction of the mean flow is greater than in the transverse direction [9]. The term *dispersion* has been introduced to distinguish from *diffusion*, which is a phenomenon at the molecular scale caused by the disordered movements of the molecules. Dispersion means that, due to the tortuosity of the flow paths, tracer particles injected at the same point in a porous medium arrive at different points after a given time interval. In statistics, dispersion is sometimes used as a synonym for variance. The latter provides a mathematical quantification of spreading caused by both molecular diffusion and dispersion in porous media. Mathematically they are both modeled as diffusion processes.

Simplified hypothetical models of porous media consider either a bundle of capillary tubes, with somewhat disordered distribution of diameters and orientations, or directly a stochastic model treating the flow path through porous media as a random walk [9]. Thus, more or less explicitly one assumes that at the local scale the transport is a simple Gaussian diffusion. This seems to be also indicated by experimental results [66]. For a Gaussian concentration distribution with constant parameters (see, e.g., (3.32)), the variance is given by

$$\sigma^2 = 2Dt = 2D\frac{x}{v},$$

where $x = vt$ is the displacement with the constant velocity v (see [9, Eq. (25)] or [66, Eqs. (4)–(5)]). If the latter is identified with the mean flow velocity through porous media, one arrives at the expression of the dispersion coefficient D given by

$$D = \frac{\sigma^2}{2x}v = av. \tag{4.45}$$

The constant a, called *dispersivity*, should describe the property of the porous medium to disperse a fluid as a consequence of the intricate geometry of its pore–channel system [9, Eq. (5)]. Relation (4.45) can however be considered only as a rough approximation of the dispersion coefficient and dispersivity constant. By studying the variance after an arbitrary number of dispersion episodes with different constant velocities, Bear [9] concluded that the dispersivity should be a fourth rank tensorial quantity. Scheidegger [66] arrived at the same conclusion from the observation that the velocity is a vector and the dispersion coefficient is a second rank tensor. The starting point of its analysis was the relation

$$D_{ik} = a_{iklm}\frac{v_l v_m}{|v|}. \tag{4.46}$$

The dispersivity tensor a_{iklm} introduced by (4.46) is a fourth rank tensor with 81 components for three-dimensional transport problems (16, in two-dimensional case) in anisotropic media. a_{iklm} possesses the obvious symmetry $a_{iklm} = a_{ikml}$, and, by assumption of thermodynamic equilibrium and Onsager's principle of microscopic reversibility, $a_{iklm} = a_{kilm}$. This reduces the number of independent components to 36. Bear [9] also considered the symmetry with respect to the groups of first two and last two indices, $a_{iklm} = a_{lmik}$, which reduces the number of independent components to 21. For an isotropic porous medium one requires the symmetry with respect to rotations about the three spatial directions. The only non-vanishing components which remain [66, Eqs. (27)–(30)] are

$$a_{1111} = a_{2222} = a_{3333} = a_L$$

$$a_{1122} = a_{1133} = a_{2233} = a_{2211} = a_{3311} = a_T$$

$$a_{1212} = a_{1313} = a_{2323} = a_{2121} =$$

$$a_{3131} = a_{3232} = a_{1221} = a_{1331} =$$

$$a_{2332} = a_{2112} = a_{3113} = a_{3223} = \frac{1}{2}(a_L - a_T), \tag{4.47}$$

where a_L and a_T are the longitudinal and transverse dispersivities. For instance, if the velocity has the components $(v, 0, 0)$ the transport equation becomes

$$\partial_t c + v\partial_{x_1} c = a_L v\partial_{x_1}^2 c + a_T v\partial_{x_2}^2 c + a_T v\partial_{x_3}^2 c. \tag{4.48}$$

For a two-dimensional problem the last term of (4.48) drops out. Thus, in the simplest cases of transport in isotropic porous media, in general different longitudinal and transverse dispersivities have to be considered. Nevertheless, when the local dispersion is an isotropic process but the velocity field is strongly heterogeneous and is modeled as a sample of a random field, the local longitudinal dispersion has a small impact on the time behavior of the first two moments of the concentration. In such situations, letting $a_L = a_T$ or even $a_L = 0$ practically does not change the simulation results. Moreover, for fluctuations of the velocity of the order of the local dispersion coefficients it was found that the approximation (4.45), $D = a_T \bar{v}$, where $\bar{v}$ is the mean flow velocity, is accurate enough for the purpose of large-scale simulations of transport [73].

4.1.5   First Order Approximations for Small Variance of the Log-Hydraulic Conductivity

Many results concerning macrodispersion are obtained using first order linear approximations of the velocity field and particle displacements. We consider a velocity field given by Darcy's law for an isotropic medium, stationary in time and divergence free,

$$V_i = -\frac{K}{n}\frac{\partial H}{\partial x_i}, \quad \sum_{i=1}^{3}\frac{\partial V_i}{\partial x_i} = 0. \tag{4.49}$$

where $i = 1, 2, 3$, K is the isotropic hydraulic conductivity, n is the porosity supposed to be constant, and H is the hydraulic head. The solution of (4.49) for appropriate boundary conditions provides the components V_i of the incompressible flow in isotropic porous media. The hydraulic conductivity is conveniently described by its logarithm, $Y(\mathbf{x}) = \ln K(\mathbf{x})$, for which statistical models can be inferred from measurements. On the basis of field-scale experimental data, it is often assumed that Y is a random function, statistically homogeneous in space, normally distributed, with constant mean $Y_0 = M_{\Omega_Y}(Y)$ and covariance function of the fluctuations $y = Y - Y_0$ of the form

$$C_{yy}(\mathbf{r}) = \sigma_y^2 \exp\left(-\frac{|\mathbf{r}|^\alpha}{\lambda_y^\alpha}\right), \tag{4.50}$$

where σ_y^2 is the variance of y, $\mathbf{r}$ is the separation vector, and λ_y is the isotropic correlation length [16, 38, 62, 63]. The most used are the exponential ($\alpha = 1$) and the Gaussian ($\alpha = 2$) shapes of the covariance function. With these assumptions, the velocity field governed by Eq. (4.49) becomes a random function with properties depending on Y.

The first order approximation theory assumes that the fluctuations of Y are small with respect to the constant mean value Y_0 and considers the expansion of Y of in powers of σ_y,

$$Y(\mathbf{x}) = Y_0 + y(\mathbf{x})\sigma_y + \mathcal{O}(\sigma_y^2), \text{ for } \sigma_y \longrightarrow 0. \tag{4.51}$$

The hydraulic conductivity becomes $K(\mathbf{x}) = K_0 \exp(y(\mathbf{x})\sigma_y + \mathcal{O}(\sigma_y^2)) \cong K_0(1 + y(\mathbf{x})\sigma_y) + \mathcal{O}(\sigma_y^2)$, where $K_0 = \exp(Y_0) = \exp[M_{\Omega_Y}(\ln K)]$ is the geometric mean of the hydraulic conductivity. With a formal expansion of H and V_i, $H(\mathbf{x}) = H_0(\mathbf{x}) + h(\mathbf{x})\sigma_y + \mathcal{O}(\sigma_y^2)$ and $V_i(\mathbf{x}) = V_{0i}(\mathbf{x}) + v_i(\mathbf{x})\sigma_y + \mathcal{O}(\sigma_y^2)$, and using the first equation (4.49) one obtains the approximations of order $\mathcal{O}(1)$ and $\mathcal{O}(\sigma_y)$ of the velocity,

$$V_{0i}(\mathbf{x}) = -\frac{K_0}{n}\frac{\partial H_0(\mathbf{x})}{\partial x_i}, \quad v_i(\mathbf{x}) = -\frac{K_0}{n}\left[y(\mathbf{x})\frac{\partial H_0(\mathbf{x})}{\partial x_i} + \frac{\partial h(\mathbf{x})}{\partial x_i}\right]. \tag{4.52}$$

Substituting the first equation (4.49) into the second and using the homogeneity property $\partial Y_0/\partial x_i = 0$, one obtains the asymptotic approximations of the equation for the hydraulic head of the orders $\mathcal{O}(1)$ and $\mathcal{O}(\sigma_y)$, respectively,

$$\sum_{i=1}^{3} \frac{\partial^2 H_0(\mathbf{x})}{\partial x_i^2} = 0, \quad \sum_{i=1}^{3} \left[\frac{\partial^2 h(\mathbf{x})}{\partial x_i^2} + \frac{\partial y(\mathbf{x})}{\partial x_i} \frac{\partial H_0(\mathbf{x})}{\partial x_i} \right] = 0. \tag{4.53}$$

In the following one assumes that $H_0(\mathbf{x}) = H_0(x_1)$ and one considers as boundary condition a constant head gradient $\partial H_0/\partial x_1(L_1) = \partial H_0/\partial x_1(L_2) = -J$, where L_1 and L_2 are the limits of the domain in the x_1 direction. The first equation (4.53) becomes $\partial^2 H_0(x_1)/\partial x_1^2 = 0$ and its solution is $H_0(x_1) = -Jx_1$. Then the first order approximation (4.52) becomes

$$V_{01}(\mathbf{x}) = K_0 J/n, \quad V_{0i}(\mathbf{x}) = 0, \quad i = 2, 3,$$

$$v_1(\mathbf{x}) = \frac{K_0}{n} \left[Jy(\mathbf{x}) - \frac{\partial h(\mathbf{x})}{\partial x_1} \right],$$

$$v_i(\mathbf{x}) = -\frac{K_0}{n} \frac{\partial h(\mathbf{x})}{\partial x_i}, \quad i = 2, 3. \tag{4.54}$$

These relations, well known in the literature [19, 37, 38], were derived without any assumption on shape of the covariance C_{yy} and on the magnitude of the variance σ_y^2. However, the fluctuating components $v_i(\mathbf{x})$ of the velocity depend on the statistics of the fluctuating part $y(\mathbf{x})$ of the log-normal hydraulic conductivity field. Spectral representations of (4.54) and statistical models of the random field $y(\mathbf{x})$ are used to generate numerically realizations of random velocity fields in the first order approximation (see the randomization representation (C.4) used in velocity field generation codes presented in Appendix C.3.2).

Note that the random velocity is defined on the space of events Ω_Y and the velocity fluctuations v_i are of the order $\mathcal{O}(\sigma_y)$ of the truncated asymptotic expansion (4.51) of the log-hydraulic conductivity Y. The leading terms V_{0i} in (4.54) are defined by the geometric mean K_0 of the hydraulic conductivity and by the constant head gradient J. Assuming the statistical homogeneity of the head fluctuations $h(\mathbf{x})$ and using the equations in first order approximation (4.52) to evaluate terms of order σ_y^2, Gelhar [37] found that in the isotropic two-dimensional case the mean velocity is given by $U = V_{0i} = K_0 J/n$. Thus, in this case the mean hydraulic conductivity coincides with the geometric mean, $M_{\Omega_Y}(K) = Un/J = K_0$. Similarly, for the isotropic three-dimensional case Gelhar and Axness [38] found that the two means no longer coincide and $U = K_0 J/n(1 + \sigma_y^2/6)$. In anisotropic three-dimensional cases, second order corrections (of order σ_y^4) should be considered but they are unimportant as far as $\sigma_y^2 \ll 1$ [22]. For larger variances σ_y^2, the mean hydraulic conductivity in the isotropic case can be estimated as follows: $M_{\Omega_Y}(K) = M_{\Omega_Y}[\exp(Y_0 + y)] \approx K_g(1 + \sigma_y^2/2)$. Further, considering the terms in the parenthesis as the first two terms of a Taylor expansion for the exponential function, one obtains $M_{\Omega_Y}(K) \approx K_g \exp(\sigma_y^2/2)$ (also see Appendix C.3.1 for application to numerical simulations).

The contribution of the velocity field to the upscaled dispersion coefficient (4.33) is given by the Lagrangian velocity correlation $\tilde{R}_L$. The latter is the correlation of the velocity sampled on the trajectories of the Itô equation (4.27). In their turn, the trajectories depend on velocity statistics. The nonlinearity is removed by successive iterations of the Itô equation.

With the transformations $X = \hat{X}\lambda_y$, $t = \hat{t}\lambda_y/U$, $V = U + \sigma_y \hat{u}U$, and $W = \hat{W}\sqrt{D\lambda_y/U}$ the Itô equation (4.27) can be written in the integral non-dimensional form

$$\hat{X}(\hat{t}) = \hat{t} + \sigma_y \int_0^{\hat{t}} \hat{u}(\hat{X}(\hat{t}'))d\hat{t}' + Pe^{-1/2} \int_0^{\hat{t}} d\hat{W}(\hat{t}'),$$

where $Pe = U\lambda_y/D$ is the Péclet number. For advection dominated transport problems, the order of magnitude relation $\sigma_y = \mathcal{O}(Pe^{-1/2})$ can be considered. In a consistent expansion of $\hat{X}$ for $\sigma_y \to 0$, the leading term cannot contain contributions of order higher than σ_y^0. Thus, consistent first approximations can be obtained by a first iteration of the Itô equation around the mean flow trajectory Ut [74]. This is tantamount to replace the arguments of the velocity in (4.30) by Ut', where t' are suitable time moments, and leads to a result even simpler than (4.32) based on Corsin's conjecture. Now the Lagrangian correlation is simply the Eulerian correlation evaluated on the mean flow trajectory

$$R_{L,ii}(t, t') = R_{E,ii}(Ut, Ut') = M_{\Omega_Y}[v_i(Ut)v_i(Ut')]. \tag{4.55}$$

With the change of variable $x_1 = Ut$, and assuming $D_I = D_L = D$, the coefficients D_{ii}^* (4.33) become

$$D_{ii}^* = D + \frac{1}{U} \lim_{x_1 \to \infty} \int_0^{x_1/U} R_{E,ii}(x_1, x_1')dx_1' = \frac{\sigma_{v_i}^2 \lambda_{u_i}}{U}, \tag{4.56}$$

where

$$\lambda_{u_i} = \frac{1}{\sigma_{v_i}^2} \lim_{x_1 \to \infty} \int_0^{x_1/U} R_{E,ii}(x_1, x_1')dx_1' \tag{4.57}$$

is the correlation length of v_i and $\sigma_{v_i}^2$ is the corresponding variance. The correlation $R_{E,ii}$ can be computed using the linear approximation (4.54). It was shown [16, 62] that at large distances only the term $y(\mathbf{x})$ from (4.54) has a non-vanishing contribution, while $\partial h(\mathbf{x})/\partial x_i$ gives no contribution to the first order. Then, the correlation length of the velocity in the mean flow direction becomes

$$\lambda_{u_1} = \frac{1}{\sigma_{v_1}^2} \lim_{x_1 \to \infty} \int_0^{x_1/U} R_{E,ii}(x_1, x_1') dx_1' = \frac{U^2}{\sigma_{v_1}^2} \lim_{x_1 \to \infty} \int_0^{x_1/U} C_{yy}(|x_1 - x_1'|) dx_1' = \frac{U^2}{\sigma_{u_i}^2} \sigma_y^2 \lambda_y,$$

and for the transverse directions $\lambda_{u_2} = \lambda_{u_3} = 0$. Using these results in the first order approximation (4.56) one obtains the upscaled dispersion coefficients

$$D_{11}^* = D + U\sigma_y^2 \lambda_y, \quad D_{22}^* = D_{33}^* = D. \tag{4.58}$$

The coefficients (4.58) are well known results of Dagan's theory of solute transport in groundwater [16, 17, 19]. Gelhar and Axness [38], using a linearization of the advection–diffusion equation and an Eulerian spectral perturbation approach, obtained the longitudinal dispersion coefficient $D + U\sigma_y^2 \lambda_y/(1 + \sigma_y^2/6)^2$ and transverse dispersion coefficients greater than D.

4.2   Diffusion with Space Variable Drift

The process of diffusion in random velocity fields is intensively used to describe phenomena which are not reproducible experimentally under macroscopically identical conditions, e.g., turbulent flows [58, 64, 81], plasmas [7], or ionized gases [8]. The same process is used to model transport processes for which the incomplete knowledge of the physical parameters is compensated by stochastic parameterizations, as in subsurface hydrology [19, 37]. Common issues in these modeling approaches are the *scale effect*, i.e., increase of apparent diffusion coefficients with the scale of observation, and *memory effects*, e.g., persistent influence of initial conditions. A primary mechanism responsible for such effects is the spatial variability of the drift coefficients of the Fokker–Planck and Itô equations [70].

4.2.1   Fokker–Planck Equation with Variable Drift

Scale and memory effects are already present in case of diffusion processes with deterministic coefficients if the drift coefficients vary in space.

The density of the transition probability $g(\mathbf{x}, t \mid \mathbf{x}_0, t_0)$ of a real diffusion process $\{X_i(t), t \geq 0, X_i(t) \in \mathbb{R}, i = 1, 2, 3\}$ with drift coefficients V_i and diffusion coefficients D_{ij} is the solution of the Cauchy problem for the Fokker–Planck equation

$$\frac{\partial g}{\partial t} + \frac{\partial}{\partial x_i}(V_i g) = \frac{\partial^2}{\partial x_i \partial x_j}(D_{ij} g) \tag{4.59}$$

with the initial condition $g(\mathbf{x}, t_0 \mid \mathbf{x}_0, t_0) = \delta(\mathbf{x} - \mathbf{x}_0)$. Here and in the following summation over repeated indices is implied.

The transition probability g governs the evolution of the concentration,

$$c(\mathbf{x}, t) = \int g(\mathbf{x}, t \mid \mathbf{x}_0, t_0) c(\mathbf{x}_0, t_0) d\mathbf{x}_0, \tag{4.60}$$

where $c(\mathbf{x}_0, t_0)$ is the initial concentration. If $c(\mathbf{x}_0, t_0)$ is normalized to unity, so is $c(\mathbf{x}, t)$, and both can be interpreted as one-point probability densities.

By virtue of (4.60), the concentration $c(\mathbf{x}, t)$ verifies Eq. (4.59) with the initial condition $c(\mathbf{x}, t_0) = c(\mathbf{x}_0, t_0)$. The equivalence with the Stratonovich form of the concentration balance equation, consistent with Fick's law of diffusion,

$$\frac{\partial c}{\partial t} + \frac{\partial}{\partial x_i}(V_i^* c) = \frac{\partial}{\partial x_i}(D_{ij} \frac{\partial c}{\partial x_j}), \tag{4.61}$$

is established by the relation $V_i^* = V_i - \partial D_{ij}/\partial x_j$ between velocity components V_i^* and drift coefficients V_i (see Sect. 2.2.3.3).

A diffusion process, defined by conditions (i)–(iii) in Sect. 2.2.2.1, satisfies uniformly in $\mathbf{x}$ and t, for all $\epsilon > 0$ the conditions

$$\lim_{\Delta t \to 0} \frac{1}{\Delta t} \int_{|\mathbf{x}'-\mathbf{x}| \geq \epsilon} g(\mathbf{x}', t + \Delta t \mid \mathbf{x}, t) d\mathbf{x}' = 0, \tag{4.62}$$

$$V_i(\mathbf{x}, t) = \lim_{\Delta t \to 0} \frac{1}{\Delta t} \int_{|\mathbf{x}'-\mathbf{x}| < \epsilon} (x_i' - x_i) g(\mathbf{x}', t + \Delta t \mid \mathbf{x}, t) d\mathbf{x}', \tag{4.63}$$

$$D_{ij}(\mathbf{x}, t) = \frac{1}{2} \lim_{\Delta t \to 0} \frac{1}{\Delta t} \int_{|\mathbf{x}'-\mathbf{x}| < \epsilon} (x_i' - x_i)(x_j' - x_j) g(\mathbf{x}', t + \Delta t \mid \mathbf{x}, t) d\mathbf{x}'. \tag{4.64}$$

Condition (4.62) prevents instantaneous jumps and ensures the almost sure continuity of the sample paths $X(t)$, (4.63) defines the drift coefficients, and (4.64) the diffusion coefficients [47].

In applications to transport in groundwater, the class of diffusion processes is restricted by imposing conditions for finite first and second spatial moments of g at finite times [46, 77]:

$$\lim_{\Delta t \to 0} \frac{1}{\Delta t} \int_{|\mathbf{x}'-\mathbf{x}| \geq \epsilon} x_i' \, g(\mathbf{x}', t + \Delta t \mid \mathbf{x}, t) d\mathbf{x}' = 0, \tag{4.65}$$

$$\lim_{\Delta t \to 0} \frac{1}{\Delta t} \int_{|\mathbf{x}'-\mathbf{x}| \geq \epsilon} x_i' x_j' \, g(\mathbf{x}', t + \Delta t \mid \mathbf{x}, t) d\mathbf{x}' = 0. \tag{4.66}$$

With (4.65) and (4.66), the integrals in (4.63) and (4.64) extend over the entire $\mathbb{R}^3$. In fact, the local averages, over spheres of radius ϵ, are used in (4.62)–(4.64) to avoid the hypothesis that the first two moments exist [27, p. 276]. However, the latter is always true when the drift coefficients are samples of random velocity fields with finite-range correlations, as well as for samples of fractional Gaussian noise velocity fields, the two situations considered in the following.

The conditions (4.65) and (4.66) are fulfilled, for instance, by the one-dimensional Gaussian diffusion with affine mean and linear variance,

$$\mu(t) = \int xc(x, t)dx = x_0 + V(t - t_0),$$

$$s(t) = \int [x - \mu(t)]^2 c(x, t)dx = 2D(t - t_0),$$

constant drift and diffusion coefficients, $V = \frac{d}{dt}\mu(t)$, $D = \frac{1}{2}\frac{d}{dt}s(t)$, and transition probability density invariant to spatial and temporal translations,

$$g(x, t \mid x_0, t_0) = (4\pi Dt)^{-1/2} \exp\left(-(x - x_0 - V(t - t_0))^2/4D(t - t_0)\right).$$
$$\tag{4.67}$$

4.2.2 Dispersion and Memory Terms

For diffusion processes with variable coefficients, the linear time behavior of the first two moments no longer holds. General relations between moments and coefficients are derived in Appendix D. For a general diffusion process satisfying (4.62)–(4.64) and the conditions for finite moments (4.65) and (4.66), the components of the first moment, μ_i, and of the covariance, s_{ij}, $i, j = 1, 2, 3$, are given by (Appendix D.1):

$$\mu_i(t, t_0) = \int x_i c(\mathbf{x}, t)d\mathbf{x} = \mu_i(t_0) + \int_{t_0}^t \overline{V_i}(t')dt', \tag{4.68}$$

$$s_{ij}(t, t_0) = \int (x_i - \mu_i(t))(x_j - \mu_j(t))c(\mathbf{x}, t)d\mathbf{x}$$

$$= s_{ij}(t_0) + 2\int_{t_0}^t dt' \int D_{ij}(\mathbf{x}, t')c(\mathbf{x}, t')d\mathbf{x} + s_{u,ij}(t, t_0) + m_{ij}(t, t_0), \tag{4.69}$$

$$s_{u,ij}(t, t_0) = \int_{t_0}^t dt' \int_{t_0}^{t'} dt'' \int c(\mathbf{x}_0, t_0)d\mathbf{x}_0 \int \int [u_i(\mathbf{x}', t'')u_j(\mathbf{x}, t')$$

$$+ u_j(\mathbf{x}', t'')u_i(\mathbf{x}, t')]g(\mathbf{x}, t' \mid \mathbf{x}', t'')g(\mathbf{x}', t'' \mid \mathbf{x}_0, t_0)d\mathbf{x}d\mathbf{x}', \tag{4.70}$$

$$m_{ij}(t, t_0) = \int_{t_0}^{t} dt' \int c(\mathbf{x}_0, t_0) d\mathbf{x}_0 \int [(x_{0j} - \mu_j(t_0)) u_i(\mathbf{x}, t')$$

$$+ (x_{0i} - \mu_i(t_0)) u_j(\mathbf{x}, t')] g(\mathbf{x}, t' \mid \mathbf{x}_0, t_0) d\mathbf{x} , \qquad (4.71)$$

where $\overline{V}_i(t) = \int V_i(\mathbf{x}, t) c(\mathbf{x}, t) d\mathbf{x}$ and $u_i(\mathbf{x}, t) = V_i(\mathbf{x}, t) - \overline{V}_i(t)$.

According to (4.69), the covariance s_{ij} is decomposed into the sum between the initial covariance, a diffusion term produced by the diffusion coefficient D_{ij}, a dispersion term $s_{u,ij}$, positive definite, expressed through correlations of the drift coefficients, (4.70), and a *memory term*, consisting of correlations between drift coefficients and initial positions (4.71), which are no longer positive definite (see numerical results presented in [77, Fig. 1]). The decomposition (4.69) can be viewed as an extension of an earlier result of Kitanidis [46]. Using Eq. (D.2) given in Appendix D.1 one can readily check that the cross-correlation position-velocity in the relation derived by Kitanidis [46, Eq. (25)] is precisely the sum between the dispersion term (4.70) and the memory term (4.71).

Modeling motions in random environments often requires relations between spatial moments of probability densities and the statistics of the process trajectories (e.g., [51, p. 287]). A systematical derivation of such relations can be achieved by using the suitably defined, consistent finite dimensional probability distributions (2.17) introduced in Chap. 2. This is illustrated below for the diagonal components of the displacement covariance.

To proceed, let us consider $\Omega = \mathbb{R}^3 \times \Omega_D$, where Ω_D is the space of events of the diffusion process starting from a fixed initial position. Let $\langle \cdot \rangle = \langle \langle \cdot \rangle_D \rangle_{X_0} = \langle \cdot \rangle_{DX_0}$ be the expectation defined as average over Ω_D and over the initial positions $\mathbf{X}(0) \in \mathbb{R}^3$. Further, consider a constant diffusion coefficient D and the time-stationary drift coefficients $V_i(\mathbf{x})$. Since D is a constant, the second term of (4.69) becomes $2D(t - t_0)$. For $i = j$, the average with respect to the joint probability density $g(\mathbf{x}, t' \mid \mathbf{x}', t'') g(\mathbf{x}', t'' \mid \mathbf{x}_0, t_0) c(\mathbf{x}_0, t_0) = p(\mathbf{x}, t'; \mathbf{x}', t''; \mathbf{x}_0, t_0)$ in (4.70) can be computed, according to (2.17), as average with respect to the probability density $\langle \delta(x - X_{t'}(\omega)) \delta(x' - X_{t''}(\omega)) \delta(x_0 - X_{t_0(\omega)}) \rangle_{DX_0}$. Similarly, the average with respect to $g(\mathbf{x}, t' \mid \mathbf{x}_0, t_0) c(\mathbf{x}_0, t_0) = p(\mathbf{x}, t'; \mathbf{x}_0, t_0)$ in (4.71) becomes an average with respect to the probability density $\langle \delta(x - X_{t'}(\omega)) \delta(x_0 - X_{t_0(\omega)}) \rangle_{DX_0}$. With these, the decomposition (4.69) for the diagonal components of the variance of the process governed by the Fokker–Planck equation (4.59) can be recast in terms of trajectories as follows:

$$s_{ii}(t, t_0) = s_{ii}(t_0) + 2D(t - t_0)$$

$$+ 2 \int_{t_0}^{t} dt' \int_{t_0}^{t'} \langle u_i(\mathbf{X}(t')) u_i(\mathbf{X}(t'')) \rangle_{DX_0} dt''$$

$$+ 2 \int_{t_0}^{t} \langle [X_i(t_0) - \langle X_i(t_0) \rangle_{DX_0}] u_i(\mathbf{X}(t')) \rangle_{DX_0} dt'. \qquad (4.72)$$

For given coefficients V_i and D of the Fokker–Planck equation (4.59) it is possible to construct a process satisfying the Itô equation (e.g., [47, p. 144])

$$X_i(t) = X_{0i} + \int_{t_0}^{t} V_i(\mathbf{X}(t'))dt' + W_i(t - t_0), \qquad (4.73)$$

where W_i is any Wiener process with mean $E(W_i) = 0$ and variance $E(W_i^2) = 2D(t - t_0)$. Equation (4.73) describes the diffusion process in a weak sense, that is, only the coefficients are specified but not the Wiener process [47]. If sufficient conditions for the existence of weak solutions of (4.73) are fulfilled, the process has the same probability distribution as the diffusion process governed by the corresponding Fokker–Planck equation. Doob [27, Chap. 6, Sec. 3] also proved the equivalence of Itô and Fokker–Planck representations in a strong sense. That means, under more restrictive conditions, the pathwise unique solutions of the Itô equation are diffusion processes satisfying (4.62)–(4.64) and, given a diffusion process with transition probability solving the Fokker–Planck equation, the associated Itô equation admits pathwise unique solutions [47, Theorems 4.6.1, and 4.7.1].

If the variance of the process (4.73) is computed for a fixed Wiener process, e.g., by using the Itô formula and Itô–Taylor expansions, one obtains, in addition to the terms of (4.72), a term consisting of correlations between the Wiener process and the velocity fluctuations (Appendix D.2). This term is canceled, for instance, when a weak solution to (4.73) is constructed by successive approximations with independent Wiener processes in each iteration (Appendix D.3). However, the representation (4.72) is equivalent with the decomposition (4.69), for either strong or weak solutions of the Itô equation, provided that the diffusion process satisfies the supplementary conditions (4.65) and (4.66). As shown in Appendix B, weak solutions obtained with GRW algorithms described in Sect. 3.3 provide numerical approximations for such diffusion processes.

To emphasize the role of the initial conditions, it is useful to rewrite the variance (4.72) as

$$s_{ii}(t, t_0) = s_{ii}(t_0) + \widetilde{s}_{ii}(t, t_0) + m_{ii}(t, t_0), \qquad (4.74)$$

where the sum of the second and third term of (4.72) is expressed in terms of displacements $\widetilde{X}_i(t) = X_i(t) - X_i(t_0)$,

$$\widetilde{s}_{ii}(t, t_0) = \langle (\widetilde{X}_i(t) - \langle \widetilde{X}_i(t) \rangle_{DX_0})^2 \rangle_{DX_0}.$$

The term $\widetilde{s}_{ii}$ describes an enhanced dispersion with respect to the local dispersion $2D(t - t_0)$ in (4.72), which explains the scale effect in modeling subsurface transport processes (e.g., [35]). The last term in (4.74),

$$m_{ii}(t, t_0) = 2\langle (X_i(t_0) - \langle X_i(t_0) \rangle_{DX_0})(\widetilde{X}_i(t) - \langle \widetilde{X}_i(t) \rangle_{DX_0}) \rangle_{DX_0}, \qquad (4.75)$$

quantifies the memory effects.

4.2.3 Memory Effects and Transition Probabilities

Similarly to (4.74), for any three successive times, $t_1 < t_2 < t_3$, the variances of the increments of the process are related by

$$\widetilde{s}_{ii}(t_3, t_1) = \widetilde{s}_{ii}(t_2, t_1) + \widetilde{s}_{ii}(t_3, t_2) + m_{ii}(t_3, t_2, t_1), \qquad (4.76)$$

where

$$\widetilde{s}_{ii}(t_3, t_1) = \text{var}\{X_i(t_3) - X_i(t_1)\},$$

$$\widetilde{s}_{ii}(t_2, t_1) = \text{var}\{X_i(t_2) - X_i(t_1)\},$$

$$\widetilde{s}_{ii}(t_3, t_2) = \text{var}\{X_i(t_3) - X_i(t_2)\},$$

$$m_{ii}(t_3, t_2, t_1) = 2\text{cov}\{(X_i(t_2) - X_i(t_1)), (X_i(t_3) - X_i(t_2))\},$$

var$\{\cdot\}$ denotes the variance, and cov$\{\cdot\}$ the covariance. In fact, (4.76) is the general binomial rule saying that the variance of the sum of two random variables is the sum of variances plus two times their covariance (e.g., [57, p. 213]) and as such it is valid for any stochastic process $X(t)$. Relation (4.74) is retrieved for the increments of the process (4.73) with the expectation defined by $\langle \cdot \rangle_{DX_0}$.

The condition of vanishing memory terms, $m_{ii}(t_1, t_2, t_3) = 0$, leads to

$$E\{(X_i(t_2) - X_i(t_1))(X_i(t_3) - X_i(t_2))\} = E\{X_i(t_2) - X_i(t_1)\}E\{X_i(t_3) - X_i(t_2)\},$$

which expresses the uncorrelatedness of the increments of the process. As follows from (4.76), vanishing memory terms is equivalent with the additivity of the variance of the increments $\widetilde{\Sigma}_{ii}$ with respect to nonoverlapping time intervals [69, p. 5]. In the particular case of diffusion with constant coefficients the increments are uncorrelated, $m_{ii}(t_1, t_2, t_3) = 0$, and (4.76) expresses the linearity in time of the variance $\widetilde{s}_{ii}(t_n, t_m) = 2D(t_n - t_m)$, $t_m < t_n$. It is also easy to see that the uncorrelatedness of the increments is a consequence of the translation invariance of the transition probability density (4.68). Moreover, it has been shown that the only Itô-diffusion process with space-homogeneous transition probabilities are Gaussian diffusion processes with constant drift and diffusion coefficients [1].

In general, according to Theorem II.3.2 of Doob [27, p. 74], for any real process with $\langle |X(t)|^2 \rangle < \infty$ and uncorrelated increments there exists a wide sense version of $X(t)$ (i.e., with the same first two moments) which is a Gaussian process with independent increments [27, p. 100]. By the definition of the statistical independence, this process has space-homogeneous transition probabilities. Since the converse is clearly true, i.e., processes with homogeneous transition probabilities have uncorrelated increments, we have the following corollary [69, p. 5]:

If the memory terms of a real process with finite first and second moments vanish for arbitrary successive time increments, then, the transport process is a wide-sense version of a Gaussian processes with spatially homogeneous transition probabilities.

Thus, memory-free processes have homogeneous transition probabilities. Inhomogeneous transition probabilities as memory effects were also identified in case of rare and extreme events (where the memory-free limit corresponds to independent identically distributed variables) and for non-Markovian processes (generalized Langevin equation, diffusion equations with memory, or fractional diffusion) [69]. It is worth noting that the memory effects quantified by (4.75) are not necessarily an indication of non-Markovian behavior of the processes. They characterize all Itô processes such as (4.73), which are Markovian processes [47, Chap. 4], as far as their increments are correlated.

4.3 Diffusion in Random Fields Model of Passive Transport

4.3.1 Ensemble-, Effective-, and Center of Mass-Dispersion

Models of transport in highly heterogeneous media such as atmosphere, plasmas, industrial devices, or groundwater are based on stochastic partial differential equations of parabolic type [3–5, 7, 11, 15, 38, 51]. In case of transport in saturated aquifers, essential features of the transport, such as the scale dependence, may be described by the simple advection–diffusion equation without sources and with constant diffusion coefficients (e.g., [19, 24, 29, 46]),

$$\partial_t c + \mathbf{V} \cdot \nabla c = D\nabla^2 c, \tag{4.77}$$

where $c(\mathbf{x}, t)$ is the concentration field, D is a local dispersion coefficient, and $\mathbf{V}(\mathbf{x})$ is a sample of a random velocity field. The latter is a solution of continuity and Darcy equations

$$\nabla \cdot \mathbf{V} = 0, \ \mathbf{V} = -K\nabla\psi, \tag{4.78}$$

where ψ is the piezometric head and K is the hydraulic conductivity, which is a sample of a space random function. This model reflects the specificity of transport in groundwater, where, unlike in the case of turbulence, the flow is laminar and randomness is introduced by a stochastic parameterization of the flow equations (4.78).

Equation (4.77) is the particular case of the concentration balance equation (4.61) for divergence-free velocity field $\mathbf{V}$ and constant diffusion coefficient D. Since for constant D the velocity components coincide with the drift coefficients, Eq. (4.77) is a Fokker–Planck equation (compare (4.59) and (4.61)). Thus, the concentration $c(\mathbf{x}, t)$ normalized to unity may be interpreted as the probability density function of the diffusion process described by the Itô equation (4.73).

For a given realization of the velocity field, corresponding, via (4.78), to a realization of the hydraulic conductivity, Eq. (4.77) describes the diffusion with space variable drift analyzed in Sect. 4.2.1. The space of events of this process of diffusion in random velocity fields is the Cartesian product $\Omega = \mathbb{R}^3 \times \Omega_D \times \Omega_V$, where Ω_V is the space of realizations of the random velocity field. Accordingly, the expectation will be formally written as $\langle \cdot \rangle = \langle\langle\langle \cdot \rangle_D\rangle_{X_0}\rangle_V = \langle \cdot \rangle_{DX_0V}$. With these, we define three centered processes of mean zero, $X_i^{eff}(t)$, $X_i^{ens}(t)$, $X_i^{cm}(t)$, $i = 1, 2, 3$, so that their variance describes the "effective" and the "ensemble" dispersion, $S_{ii}(t)$ and $\Sigma_{ii}(t)_{ii}$, and the fluctuations, or dispersion, of the center of mass, $R_{ii}(t)$ [78, 79]:

$$X_i^{eff}(t) = X_i(t) - \langle X_i(t)\rangle_{DX_0}, \quad S_{ii}(t) = \langle(X_i^{eff}(t))^2\rangle$$

$$X_i^{ens}(t) = X_i(t) - \langle X_i(t)\rangle_{DX_0V}, \quad \Sigma_{ii}(t) = \langle(X_i^{ens}(t))^2\rangle$$

$$X_i^{cm}(t) = \langle X_i(t)\rangle_{DX_0} - \langle X_i(t)\rangle_{DX_0V}, \quad R_{ii}(t) = \langle(X_i^{cm}(t))^2\rangle. \quad (4.79)$$

Using the relations between spatial moments of probability densities and trajectory's statistics (2.17) (also see Sect. 4.2.2), it can easily be seen that the variances $S_{ii}(t)$, $\Sigma_{ii}(t)$, and $R_{ii}(t)$ of the three processes defined in (4.79) correspond, respectively, to the expectation (average over velocity realizations) of the second spatial moment of the single-realization concentration $c(\mathbf{x}, t)$, to the second moment of the ensemble average concentration $\langle c(\mathbf{x}, t)\rangle_V$, and to the variance of the first spatial moment of $c(\mathbf{x}, t)$. These quantities are related by

$$S_{ii} = \Sigma_{ii} - R_{ii}. \quad (4.80)$$

This identity was used by Le Doussal and Machta [49], in the context of measurements methods for diffusion coefficients, to define "quenched" and "annealed" coefficients, $S_{ii}/(2t)$ and $\Sigma_{ii}/(2t)$, respectively. Attinger et al. [3] considered the same quantities to define effective and ensemble dispersion coefficients, $\frac{1}{2}dS_{ii}/dt$ and $\frac{1}{2}d\Sigma_{ii}/dt$, respectively. Kitanidis [46] also obtained the identity (4.80) after computing $S_{ii}(t)$ and $\Sigma_{ii}(t)$ by averaging an advection–diffusion equation with random coefficients.

If the necessary joint measurability conditions which allow permutations of averages are fulfilled (see, e.g., [85]) the second moment of the mean concentration can be expressed as (see [75] and, for the case $D = 0$, [71])

$$\Sigma_{ii} = S_{ii}(0) + \langle X_{ii}\rangle_{X_0} + M_{ii} + Q_{ii}, \quad (4.81)$$

where $S_{ii}(0) = \langle(X_{0i} - \langle X_{0i}\rangle_{X_0})^2\rangle_{X_0}$, $X_{ii} = \langle(\widetilde{X}_i - \langle\widetilde{X}_i\rangle_{DV})^2\rangle_{DV}$ is the "one-particle dispersion" (defined by averaging with respect to D and V for a fixed initial position), $M_{ii} = \langle m_{ii}\rangle_V$ is the ensemble mean of the memory term (4.75), and

$Q_{ii} = \langle ((\widetilde{X}_i)_{DV} - (\widetilde{X}_i)_{DX_0V})^2 \rangle_{X_0}$ is the spatial variance of the one-particle center of mass $\langle \widetilde{X}_i \rangle_{DV}$, computed by averages over X_0.

As follows from (4.73), the trajectory $\mathbf{X}(t)$ depends on the Lagrangian velocity field $V_i(\mathbf{X}(t))$, which consists of observations at random locations on the trajectory of the random Eulerian velocity (which is defined in a fixed reference frame) [85]. If the Lagrangian velocity field is statistically homogeneous the one-particle center of mass $\langle \widetilde{X}_i \rangle_{DV}$ and dispersion X_{ii} are independent of X_0. Then M_{ii} and Q_{ii} vanish and from (4.80) and (4.81) one obtains

$$S_{ii} = S_{ii}(0) + X_{ii} - R_{ii}. \tag{4.82}$$

The validity of the Lagrangian homogeneity hypothesis and of the relation (4.82), first derived by Dagan [20], is crucial for the interpretation of the field measurements and for the inference of the upscaled diffusion coefficients.

4.3.2 Statistical Homogeneity Properties

Lagrangian homogeneity implies the independence on the deterministic initial position $X_i(t_0)$ of the increment $\widetilde{X}_i(t) = X_i(t) - X_i(t_0)$. Since the transition density $g(\mathbf{x}, t | \mathbf{x}_0, t_0)$ is the probability density of $\widetilde{X}_i$, the statistical homogeneity of $\widetilde{X}_i$ is equivalent to the invariance to space translations of the ensemble averaged transition density $\langle g \rangle_V$.

The usual setup for statistical homogeneity is as follows. Let V be a homogeneous random function defined on the canonical probability space $(\Omega, \mathscr{A}, P)$, usually denoted by $V(\omega, \mathbf{x}) = \omega(\mathbf{x})$. Measure-preserving shifts on Ω are defined through $(\tau_{\mathbf{x}_0} \omega)(\mathbf{x}) = \omega(\mathbf{x} + \mathbf{x}_0)$, $P \circ \tau_{\mathbf{x}_0}^{-1} = P$. A composed function $F(V)$ is also homogeneous if it depends on ω and $\mathbf{x}_0$ only through measure-preserving shifts $F = F(\tau_{\mathbf{x}_0} \omega)$ [85].

Let us consider the transition density solving (4.77) in a translated reference system, $\widetilde{\mathbf{x}} = \mathbf{x} - \mathbf{x}_0$,

$$\frac{\partial g(\widetilde{\mathbf{x}}, t | \mathbf{0}, t_0)}{\partial_t} + \nabla \cdot (\mathbf{V}(\widetilde{\mathbf{x}} + \mathbf{x}_0) g(\widetilde{\mathbf{x}}, t | \mathbf{0}, t_0)) = D \nabla^2 g(\widetilde{\mathbf{x}}, t | \mathbf{0}, t_0). \tag{4.83}$$

As follows from (4.83), g depends on velocity statistics only through $\tau_{\mathbf{x}_0} \omega = \mathbf{V}(\widetilde{\mathbf{x}} + \mathbf{x}_0)$; hence, it is statistically homogeneous if the Eulerian velocity field $\mathbf{V}(\mathbf{x}, \omega)$ is homogeneous. If, in addition, the solutions of (4.83) are unique in a classical sense, then $g(\mathbf{x}, t | \mathbf{x}_0, t_0, \omega) = g(\mathbf{x} - \mathbf{x}_0, t | \mathbf{0}, t_0, \tau_{\mathbf{x}_0} \omega)$ and the measure-preserving property of $\tau_{\mathbf{x}_0}$ implies the translation invariance of the mean transition density, that is, $\langle g \rangle_V (\mathbf{x}, t | \mathbf{x}_0, t_0) = \langle g \rangle_V (\mathbf{x} - \mathbf{x}_0, t | \mathbf{0}, t_0)$.

According to (4.73), to the Fokker–Plank equation (4.83) one associates an Itô equation solving for displacements $\widetilde{X}_i(t) = X_i(t) - x_{0i}$ from the deterministic initial position $X_i(t_0) = x_{0i}$,

$$\widetilde{X}_i(t) = \int_0^t V_i(\widetilde{\mathbf{X}}(t') + \mathbf{x}_0) dt' + W_i(t). \tag{4.84}$$

If the solutions of (4.84) are pathwise unique, then the displacement field $\widetilde{\mathbf{X}}(t; \mathbf{x}_0, \omega) = \widetilde{\mathbf{X}}(t; \mathbf{0}, \tau_{\mathbf{x}_0}\omega)$ is homogeneous [85, Remark 6.7 and Proposition 6.1]. The homogeneity of $\widetilde{\mathbf{X}}(t)$ implies the homogeneity of the Lagrangian velocity field, $V_i^L(\mathbf{x}_0, t) = V_i(\mathbf{X}(t)) = V_i(\widetilde{\mathbf{X}}(t) + \mathbf{x}_0)$, which depends on the statistics of $\widetilde{\mathbf{X}}$ through measure-preserving shifts. Conversely, assuming the homogeneity of $V_i(\mathbf{x}_0, t)$, Eq. (4.84) implies the homogeneity of $\widetilde{\mathbf{X}}$ (see also [69, p. 2]).

We have thus the following result which summarizes the homogeneity properties of the process of diffusion in random velocity fields:

If (1) the Eulerian velocity field $\mathbf{V}(\mathbf{x}, \boldsymbol{\omega})$ is statistically homogeneous and (2) the Fokker–Plank equation (4.77) admits unique classical solutions for (3) deterministic initial conditions, then

(a) *The ensemble averaged transition density $\langle g \rangle_V$ is invariant to spatial translations, $\langle g \rangle_V(\mathbf{x}, t | \mathbf{x}_0, t_0) = \langle g \rangle_V(\mathbf{x} - \mathbf{x}_0, t | \mathbf{0}, t_0)$.*

   If, in addition, (4) the associated Itô equation (4.84) admits pathwise unique solutions, then

(b) *the following statements are equivalent:*

   (b1) *The displacement field $\widetilde{\mathbf{X}} = X_i(t) - x_{0i}$ is statistically homogeneous.*
   (b2) *The Lagrangian velocity field $V_i^L(\mathbf{x}_0, t)$ is statistically homogeneous.*
   (b3) *The ensemble averaged transition density $\langle g \rangle_V$ is translation-invariant.*

The statements (b1) and (b2) can be proved independently, without requiring the existence of a density g for the transition probability. The first proof of homogeneity property (b2) and of the equality between Lagrangian and Eulerian means was given by Lumley [50] for purely advective transport, under the implicit assumption of analytical velocity realizations. Port and Stone [59] extended this result by considering diffusion in random advection fields and provided a rigorous proof for the equality of the Lagrangian and Eulerian one-point probability distributions under milder conditions, i.e., the continuity of the first order spatial derivatives of the velocity samples. Zirbel [85] extended the previous homogeneity results to statistical stationarity in case of space–time velocity fields and generalized the results of Port and Stone by replacing the Wiener process by a family of martingales which allow including the diffusion in the random environment.

The results obtained so far do not go beyond the equality of the one-dimensional probability distributions of the Lagrangian and Eulerian velocity fields. Since the higher order distributions do not coincide, the probability laws of the Lagrangian and Eulerian fields are in general different [85]. The invariance to spatial translations of the mean density $\langle g \rangle_V$ (statement (a)) is also a one-point statistical property, which implies the homogeneity of $\langle \widetilde{X}_i \rangle_{DV}$ and X_{ii}, cancels the mean of the memory terms (4.75), and ensures the validity of expression (4.82) for the second moments.

These homogeneity properties hold true for unique solutions of the transport equations. The irregularity of the velocity samples for the exponentially correlated $\ln K$ field used in MC simulations (see [69, pp. 2–3 and Fig. 1]), which do not meet the conditions for uniqueness of the solutions, may explain the non-vanishing mean memory terms indicated by Fig. 2 of [69].

Also essential is the assumption of deterministic initial conditions. In case of random initial conditions, the velocity with the spatial argument translated by $\mathbf{x}_0(\omega)$ in equations (4.83) and (4.84) is not a measure-preserving shift and we are no longer in the frame of the usual homogeneity setup. Since, as follows from (4.73) and (4.76) the mean memory terms are time integrals of the Lagrangian velocity covariance (see [77, Eq. (25)]), they are non-vanishing as long as the velocity is correlated. This implies that translation invariance of the mean transition probabilities associated with successive increments of the process may be expected only in case of uncorrelated velocity fields or asymptotically in the long time limit, for velocity fields with finite correlation scales. In such cases, by the corollary formulated at the end of Sect. 4.2.3, the transport process is a wide-sense version of a Gaussian diffusion.

4.3.3 Anomalous Diffusion, Ergodicity, and Self-averaging

Anomalous diffusion behavior is commonly characterized by the time dependence of the variance. If it grows like t^α with $\alpha < 1, \alpha = 1$, or $\alpha > 1$ one considers that the process is subdiffusive, diffusive, or superdiffusive, respectively. There are however cases where this classification scheme might be misleading, as, for instance, the apparently diffusive behavior with $\alpha = 1$ resulted from the competition between subdiffusion and Lévi flights [28]. To overcome such situations, O'Malley and Cushman proposed a renormalization-group classification of diffusive processes, which generalizes the notion of self-similarity [55, 56]. The new scheme performs better in case of processes with infinite variance and/or nonstationary increments and yields the same classification as the variance scheme for processes with stationary increments, as those considered in the following. Another special case is that of the process made up of the sum of a subdiffusion and a normal diffusion, with linear long time behavior of the variance, which is, however, not a normal diffusion because of the indefinite persistence of the memory terms [69]. Though it cannot classify different types of diffusive behavior, the criterion of vanishing memory terms allows unambiguous identification of normal diffusion processes (see Sect. 4.2.3).

The ergodicity of the center of mass process $X_i^{cm}(t)$ (see (4.79)) indicated by the decay to zero at long times of its variance shown by numerical simulations [75, Fig. 1] has been associated with that of the space random fields with finite correlation range and with the normal diffusive behavior of the process at large times [20]. In the more general case of space–time random fields, arguments have been put forward that temporally ergodic flows satisfy the diffusion limit criterion, consisting of convergence of the integral of Lagrangian correlation, and that the violation of this criterion may lead to anomalous diffusion [31]. For time-independent fields and small velocity fluctuations, some relations between the ergodicity of the random fields, the ergodicity of the dispersion coefficients, and the type of diffusive behavior are readily available in the frame of the consistent first order approximation.

A consistent first order approximation is obtained by the first iteration (D.11) of the Itô equation starting with a particular solution $\mathbf{X}^{(0)}(t)$. If we consider the mean flow trajectory in dimensionless variables, $\mathbf{X}^{(0)}(t) = (t, 0, 0)$, the first order approximation of the Itô equation (4.73), for $X_{0i} = 0$ at $t_0 = 0$, is given by

$$X_i(t) = \int_0^t V_i(t')dt' + \widetilde{W}_i(t),$$

where $\widetilde{W}_i$ is the dimensionless Wiener process (see Sect. 4.1.5). Since V_i is independent of the Wiener process, the latter does not influence the ergodic properties of the process X_i, determined by those of the velocity field, and can be disregarded. As follows from (4.79), $X_i^{eff}(t) \equiv 0$ and $X_i^{cm}(t) = X_i^{ens}(t) = X_i(t) - \langle X_i(t) \rangle_V$. The process X_i^{ens} has the explicit form

$$X_i^{ens}(t) = \int_0^t u_i(t')dt', \tag{4.85}$$

where $u_i = V_i - \langle V_i \rangle_V$, and its variance becomes

$$\Sigma_{ii}(t) = \int_0^t \int_0^t \langle u_i(t')u_i(t'') \rangle_V dt'dt''. \tag{4.86}$$

Next, similarly to the definition (2.29) of the local diffusion coefficient D, we define an apparent, ensemble diffusion coefficient,

$$D_{ii}^{ens}(t) = \frac{1}{2} \frac{d\Sigma_{ii}(t)}{dt} = \int_0^t \langle u_i(t)u_i(t') \rangle_V dt'. \tag{4.87}$$

Consider first velocity fields with short-range correlations. In such cases the integral range is finite and the space random velocity is necessarily ergodic [13]. Typical examples are exponential and Gaussian short-range correlations, for which the first order approximation of the Lagrangian covariance behaves like $\mathrm{cov}_\mathbf{u}(t) \sim e^{-t}$ and $\mathrm{cov}_\mathbf{u}(t) \sim e^{-t^2}$, respectively.

An ergodic estimator in the first order of approximation of the dispersion coefficient for the process X_i^{ens} can be obtained by replacing in (4.87) the Lagrangian covariance $\text{cov}_\mathbf{u}(t - t') = \langle u_i(t)u_i(t')\rangle_V$ by the time average of the product $u_i(t)u_i(t')$,

$$D_{ii}^*(t) = \int_0^t dt' \frac{1}{T-t} \int_0^{T-t} u_i(t+s)u_i(t'+s)ds = \int_0^t \text{cov}_\mathbf{u}^T(t, t')dt'. \quad (4.88)$$

Since to the first order the velocity fields are Gaussian of mean zero (e.g., [20, 38]), the fourth moments are completely determined by the correlation function (e.g., [84, Eq. (3.29)]). The limit, in the mean-square sense, $\langle[\text{cov}_\mathbf{u}^T(t, t') - \text{cov}_\mathbf{u}(t-t')]^2\rangle_V \to 0$ as $T \to \infty$ exists if and only if the condition of Slutsky's theorem for ergodic variance is fulfilled [84, p. 234], i.e.,

$$\frac{1}{T} \int_0^T (\text{cov}_\mathbf{u}(s))^2 ds \xrightarrow[T\to\infty]{} 0. \quad (4.89)$$

The validity of (4.89) for short-range correlation fields implies the mean-square convergence of the estimator D_{ii}^* given by (4.88) towards the ensemble coefficient (4.87).

Power-law velocity correlations also may occur in hydrological modeling if the analysis of the field data shows a dependence on the observation scale of the log-hydraulic conductivity. A discussion on this topic and explicit relations of the longitudinal dispersion coefficients for two important classes of long-range correlated $\ln K$ fields, namely fractional Gaussian noise and fractional Brownian motion correlation types, can be found in [33].

Consider in the following *fractional Gaussian noise* velocities with power-law correlations $\text{cov}_\mathbf{u}(t) \sim t^{-\beta}, 0 < \beta < 2$. According to (4.86), the variance of the process X^{ens} behaves like $\Sigma_{ii} = \langle Y_i^2 \rangle \sim t^{2-\beta}$. If $\beta \neq 1$, X^{ens} is a *fractional Brownian motion* with *Hurst exponent* $H = 1 - \beta/2, 0 < H < 1, H \neq 1/2$ (*superdiffusion* if $0 < \beta < 1$ and *subdiffusion* if $1 < \beta < 2$). If $\beta = 1$ (the case of "$1/x$" noise), then $\Sigma_{ii} \sim t \ln t - t$. In all these cases of anomalous diffusion, the increments of the process are correlated and the memory terms $M_{ii} \sim t^{2-\beta}$ persist indefinitely [69].

For the case of fractional Brownian motion, Deng and Barkai [21] proved the ergodicity of the variance by a direct analytical computation of the variance of its ergodic estimator. Since condition (4.89) holds for all $\beta > 0$, ergodicity is also a corollary of Slutsky's theorem.

Summarizing, we have the following result: *Ensemble dispersion coefficients for diffusion in velocity fields with short-range correlations as well as in fractional Gaussian noise random fields with correlations* $\text{cov}_\mathbf{u} \sim t^{-\beta}, 0 < \beta < 2, \beta \neq 1$, *are ergodic within the precision of the consistent first order approximation.*

The normal diffusion is retrieved as a particular case of fractional Brownian motion, together with the super- and subdiffusive cases, if one considers correlations of fractional Gaussian noise of the form $\text{cov}_\mathbf{u} \sim c_1(1 - \beta)t^{-\beta} + c_2 t^{1-\beta}\delta(t)$, where δ is the Dirac function (see, e.g., [43]). With Hurst exponent defined as above by $H = 1 - \beta/2$, normal diffusion corresponds to $\beta = 1$ and $H = 1/2$.

It is worth noting that anomalous diffusion is generated, through (4.87), by the power-law correlation of the Lagrangian velocity without necessarily requiring a power-law correlation of the Eulerian velocity field. This is illustrated by the model of Matheron and de Marsily presented in Sect. 4.1.2.2, where the interplay between the local dispersion and the longitudinal Eulerian velocity with finite correlation length along the transverse direction produces a Lagrangian velocity with longitudinal correlation scaling as $\sim t^{-1/2}$ [12, 53]. This results in superdiffusive behavior with Hurst exponent $H = 3/4$ in two dimensions [69]. The same model also shows subdiffusive behavior in three dimensions, and a $\sim t \ln t - t$ behavior when extended to higher dimensions [52].

To check whether the variance of the process X^{ens} is also self-averaging, consider the single-trajectory quantity similar to the Taylor formula (4.15),

$$\Sigma_{ii}^*(t) = 2 \int_0^t (t - \tau) d\tau \left(\frac{1}{t - \tau} \int_0^{t-\tau} u_i(s) u_i(s + \tau) ds \right).$$

The expression in the brackets is an ergodic estimator of the velocity covariance $\text{cov}_{\mathbf{u}}(t)$. If this estimator is accurate enough at finite times, then

$$\Sigma_{ii}^*(t) = 2 \int_0^t d\tau \int_0^{t-\tau} u_i(s) u_i(s + \tau) ds \approx 2 \int_0^t (t - \tau) \text{cov}_{\mathbf{u}}(\tau) d\tau, \qquad (4.90)$$

that is, Σ_{ii}^* is a self-averaging estimator of Σ_{ii}. The self-averaging property (4.90) has been demonstrated numerically and used to estimate ensemble dispersion coefficients $D_{ii}^{ens} \approx \Sigma_{ii}^*/(2t)$ on a single trajectory of *diffusion in short-range correlated velocity fields* [72]. Such self-averaging estimates have been used as input parameters in models for the evolution of the probability density of the random concentration [80].

Consistent first order approximations of effective dispersion coefficients $S_{ii}/(2t)$ are computed according to (4.80) by subtracting from the ensemble coefficients the coefficients of the center of mass $R_{ii}/(2t)$ evaluated by spatial integrals of two-particle velocity covariances over the support of the initial plume [20, 76]. For singular point-like sources, the consistent first order approximation yields $\Sigma_{ii} - R_{ii} = 2Dt$ and the effective coefficients equal the local dispersion coefficient D [74]. Useful first order approximations for point sources can be obtained in an Eulerian frame by computing small perturbations around the process of diffusion in the constant mean velocity field [3, 24]. Even though analytical results for two-dimensional transport in velocity fields with short-range correlations and point sources may suggest non-vanishing variance of the longitudinal effective coefficients in the long time limit [23, Fig. 1], direct numerical estimations, using ensembles of single-realization effective coefficients computed by the same perturbation approach [29], clearly show the decay of the variance at large times [74, Fig. 3]. This self-averaging property, demonstrated numerically for both two- and three-dimensional cases [30], is also consistent with the results on asymptotic ergodicity presented in the next chapter. One the other side, in case of anomalous diffusion with $H = 3/4$, generated by the model of Matheron and de Marsily, the sample-to-sample fluctuations calculated analytically do not converge to zero and the longitudinal effective dispersion coefficient is not self-averaging [14, Fig. 4].

4.4   First Order Approximations

4.4.1   Lagrangian and Eulerian Representations of Diffusion in Random Velocity Fields

Let $\{X_{i,t} = X_i(t), t \geq 0, X_i(t) \in \mathbf{R}, i = 1, \cdots, d\}$ be an advection–dispersion process in d spatial dimensions, with a space variable drift $\mathbf{V}(\mathbf{x})$, sample of a statistically homogeneous random velocity field, and a local dispersion coefficient D, assumed constant, described by the Itô equation

$$X_{i,t} = x_{i,0} + \int_0^t V_i(\mathbf{X}_{t'})dt' + W_{i,t}, \qquad (4.91)$$

where $x_{i,0}$ is a deterministic initial position and W_i are the components of a Wiener process of mean zero and variance $\langle W_{i,t}^2 \rangle = 2Dt$.

The solution g of the Fokker–Planck equation

$$\partial_t g + \mathbf{u}\nabla g = D\nabla^2 g, \qquad (4.92)$$

which corresponds to Green's function in Eulerian approaches to dispersion in random fields, is the density $g(\mathbf{x}, t | \mathbf{x}_0)$ of the transition probability of the Itô process (4.91). Equation (4.92) is the starting point in Eulerian perturbation approaches [3, 24, 25].

The half derivative of the variances defined in (4.79) defines, respectively, the *effective*, the *ensemble*, and the *center of mass* dispersion coefficients,

$$D_{ii}^{eff}(t) = \frac{1}{2}\frac{dS_{ii}(t)}{dt}, \quad D_{ii}^{ens}(t) = \frac{1}{2}\frac{d\Sigma_{ii}(t)}{dt}, \quad D_{ii}^{cm}(t) = \frac{1}{2}\frac{dR_{ii}(t)}{dt}, \qquad (4.93)$$

related by $D_{ii}^{eff}(t) = D_{ii}^{ens}(t) - D_{ii}^{cm}(t)$.

According to (4.91), the process $X_{i,t}^{ens}$ starting from $x_{i,0} = 0$ verifies

$$X_{i,t}^{ens} = \int_0^t u_i(\mathbf{X}_{t'})dt' + W_{i,t}, \qquad (4.94)$$

where $u_i = V_i - \overline{\langle V_i \rangle}$ (to simplify the notation, hereafter in this section angular brackets denote averages over realizations of the Wiener process and overbars averages over realizations of the velocity or of the random conductivity field). The variance of the process (4.94) is obtained as follows:

$$\Sigma_{ii}(t) = \int_0^t \int_0^t \overline{\langle u_i(\mathbf{X}_{t'})u_i(\mathbf{X}_{t''})\rangle} dt' dt'' + 2\int_0^t \overline{\langle u_i(\mathbf{X}_{t'})W_{i,t}\rangle} dt' + \left\langle W_{i,t}^2 \right\rangle$$

$$= 2Dt + 2\int_0^t dt' \int_0^{t'} \overline{\langle u_i(\mathbf{X}_{t'})u_i(\mathbf{X}_{t''})\rangle} dt'',$$

where the cross-correlation term is canceled as a consequence of statistical homogeneity of the velocity field, $\overline{\langle u_i\rangle} = 0$, and of the independence of the Wiener process on the velocity statistics. The process $X_{i,t}^{cm}$ verifies (4.94) with $u_i = \langle V_i \rangle - \overline{\langle V_i \rangle}$ and the second term set to zero and has the variance

$$R_{ii}(t) = 2\int_0^t \int_0^t \overline{\langle u_i(\mathbf{X}_{t'})\rangle \langle u_i(\mathbf{X}_{t''})\rangle} dt' dt''.$$

Further, the definitions (4.93) yield

$$D_{ii}^{ens}(t) = D + \int_0^t \overline{\langle u_i(\mathbf{X}_t)u_i(\mathbf{X}_{t'})\rangle} dt', \qquad (4.95)$$

$$D_{ii}^{cm}(t) = \int_0^t \overline{\langle u_i(\mathbf{X}_t)\rangle \langle u_i(\mathbf{X}_{t'})\rangle} dt', \qquad (4.96)$$

$$D_{ii}^{eff}(t) = D_{i,i}^{ens}(t) - D_{i,i}^{cm}(t). \qquad (4.97)$$

The consistent first order approximation presented in Sect. 4.1.5 is obtained by iterating once the Itô equation (4.91) about the trajectory of the mean flow, $X_{i,t}^{(0)} = \delta_{j,1}Ut$. In this approximation the arguments of u_i are replaced by $X_i^{(0)}$ and from (4.95)–(4.97) it follows that the effective dispersion coefficient is not influenced by the statistics of the random velocity field, $D_{ii}^{ens}(t) = D$. Nontrivial effective dispersion coefficients are obtained by including in $X_i^{(0)}$ the local dispersion process. This approach yields an approximation which is no longer a consistent asymptotic expansion in σ_Y^2 but accounts, up to the first order, for the enhanced effective dispersion produced by velocity fluctuations.

The first iteration (D.11) of (4.94) about the unperturbed problem consisting of diffusion with constant coefficient D in the mean flow field $U = \overline{\langle V_1 \rangle}$ (oriented along the longitudinal axis), with trajectory $X_{i,t}^{(0)} = \delta_{j,1}Ut + W_{i,t}$, leads to replacing $\mathbf{X}_t$ by $\mathbf{X}_t^{(0)}$ in (4.95) and (4.96). Then, (4.95) gives the following first order approximation of the ensemble dispersion coefficient:

$$D_{ii}^{ens}(t) - D = \int_0^t \overline{\langle u_i(\mathbf{X}_t^{(0)})u_i(\mathbf{X}_{t'}^{(0)})\rangle} dt'$$

$$= \int_0^t dt' \int \int \overline{u_i(\mathbf{x})u_i(\mathbf{x}')} p(\mathbf{x}, t; \mathbf{x}', t') d\mathbf{x} d\mathbf{x}', \qquad (4.98)$$

where $p(\mathbf{x}, t; \mathbf{x}', t') = \langle \delta(\mathbf{x} - \mathbf{X}_t^{(0)}) \delta(\mathbf{x}' - \mathbf{X}_{t'}^{(0)}) \rangle$ is the Gaussian joint probability density of the process $X_{i,t}^{(0)}$ expressed with the aid of the Dirac δ-function (see (2.17)). Similarly, with $p(\mathbf{x}, t) = \langle \delta(\mathbf{x} - \mathbf{X}_t^{(0)}) \rangle$, (4.96) gives the first order approximation of the center of mass coefficient

$$D_{ii}^{cm} = \int_0^t dt' \int \int \overline{u_i(\mathbf{x}) u_i(\mathbf{x}')} p(\mathbf{x}, t) p(\mathbf{x}', t') d\mathbf{x} d\mathbf{x}'. \qquad (4.99)$$

Finally, $D_{ii}^{eff}(t)$ is obtained from (4.97). Together, (4.98–4.99) give the (stochastic) Lagrangian formulation of the first order approximation of the dispersion coefficients, which is equivalent to that obtained in the Eulerian perturbation approach [3, 24, 25].

Remark 4.1 As follows from (4.98)–(4.99), for a superposition $u_i(\mathbf{x}) = u_i^1(\mathbf{x}) + \cdots + u_i^n(\mathbf{x})$ of statistically independent velocity fields the dispersion coefficients estimated to the first order are given by the sum of terms determined by the corresponding correlations $\overline{u_i^1(\mathbf{x}) u_i^1(\mathbf{x}')}, \cdots, \overline{u_i^n(\mathbf{x}) u_i^n(\mathbf{x}')}$.

Dispersion coefficients associated with memory terms (4.75) can also be derived as follows. The variance of the process (4.94) written as sum of increments,

$$X_{i,t}^{ens} = (X_{i,\tau}^{ens} - X_{i,0}^{ens}) + (X_{i,t}^{ens} - X_{i,\tau}^{ens}),$$

is given, according to (4.76), by

$$\Sigma_{t,0} = \Sigma_{\tau,0} + \Sigma_{t,\tau} + M_{t,\tau,0}, \qquad (4.100)$$

and the ensemble dispersion coefficient (4.93) defined by half derivative of (4.100) with respect to t becomes

$$D_{ii}^{ens}(t, 0) = D_{ii}^{ens}(t, \tau) + M D_{ii}^{ens}(t, \tau, 0), \qquad (4.101)$$

where $M D_{ii}^{ens}(t, \tau, 0) = \frac{1}{2} \frac{d M_{ii}(t, \tau, 0)}{dt}$ is a memory dispersion coefficient.

From (4.95) and (4.101) one obtains the Lagrangian expression

$$M D_{ii}^{ens}(t, \tau, 0) = D_{ii}^{ens}(t, 0) - D_{ii}^{ens}(t, \tau) = \int_0^\tau \overline{\langle u_i(\mathbf{X}_t) u_i(\mathbf{X}_{t'}) \rangle} dt'. \qquad (4.102)$$

With (4.98), the first order approximation of (4.102) is given by

$$M D_{ii}^{ens}(t, \tau, 0) = \int_0^\tau dt' \int \int \overline{u_i(\mathbf{x}) u_i(\mathbf{x}')} p(\mathbf{x}, t; \mathbf{x}', t') d\mathbf{x} d\mathbf{x}'. \qquad (4.103)$$

Memory coefficients associated with $X_{i,t}^{eff}$ and $X_{i,t}^{cm}$ are defined in a similar way.

4.4.2 Explicit First Order Results for Transport in Aquifers

Approximations of the flow and transport equations for small variance of the logarithm of the hydraulic conductivity K of the aquifer system lead to explicit functional dependencies of the dispersion coefficients on the covariance C_Y of the random field $Y = \ln K$ [38].

With the common convention for the Fourier transform

$$\hat{u}_j(\mathbf{k}) = \int u_j(\mathbf{x}) \exp(-i\mathbf{k} \cdot \mathbf{x}) d\mathbf{x}, \tag{4.104}$$

$$u_j(\mathbf{x}) = \frac{1}{(2\pi)^d} \int \hat{u}_j(\mathbf{k}) \exp(i\mathbf{k} \cdot \mathbf{x}) d\mathbf{k}, \quad j = 1 \cdots d, \; d = 2, 3, \tag{4.105}$$

the first order approximation in σ_Y^2 yields

$$\overline{\hat{u}_j(\mathbf{k})\hat{u}_j(\mathbf{k}')} = (2\pi)^d U^2 \delta(\mathbf{k} + \mathbf{k}') p_j^2(\mathbf{k}) \hat{C}_Y(\mathbf{k}), \tag{4.106}$$

where $p_j(\mathbf{k}) = (\delta_{j,1} - k_1 k_j / k^2)$ are projectors which ensure the incompressibility of the flow (see, e.g., [24, 25, 38, 67]).

Remark 4.2 The Dirac function $\delta(\mathbf{k} + \mathbf{k}')$ which ensures the statistical homogeneity in (4.106) has to be changed into $\delta(\mathbf{k} - \mathbf{k}')$ to obtain a similar first order relation for $\hat{u}_j(\mathbf{k})\hat{u}_j^*(\mathbf{k}')$, where $\hat{u}_j^*$ denotes the complex conjugate [17, 18, 67]. Even though it leads to the same results, this relation renders the computation more complicated.

Using the inverse Fourier transform (4.105) and relation (4.106), the Eulerian velocity correlation from (4.98) can be computed as follows:

$$\overline{u_j(\mathbf{x})u_j(\mathbf{x}')} = \frac{1}{(2\pi)^{2d}} \int \int \overline{\hat{u}_j(\mathbf{k})\hat{u}_j(\mathbf{k}')} \exp(i\mathbf{k} \cdot \mathbf{x}) \exp(i\mathbf{k}' \cdot \mathbf{x}') d\mathbf{k} d\mathbf{k}'$$

$$= \frac{U^2}{(2\pi)^d} \int p_j^2(\mathbf{k}) \hat{C}_Y(\mathbf{k}) \exp(i\mathbf{k} \cdot \mathbf{x}) d\mathbf{k} \int \delta(\mathbf{k} + \mathbf{k}') \exp(i\mathbf{k}' \cdot \mathbf{x}') d\mathbf{k}'$$

$$= \frac{U^2}{(2\pi)^d} \int p_j^2(\mathbf{k}) \hat{C}_Y(\mathbf{k}) \exp[i\mathbf{k} \cdot (\mathbf{x} - \mathbf{x}')] d\mathbf{k}. \tag{4.107}$$

With (4.107), the integrand of the time integral in (4.98) becomes

$$\int \int \overline{u_j(\mathbf{x})u_j(\mathbf{x}')} p(\mathbf{x}, t; \mathbf{x}', t') d\mathbf{x} d\mathbf{x}' =$$

$$= \frac{U^2}{(2\pi)^d} \int p_j^2(\mathbf{k}) \hat{C}_Y(\mathbf{k}) d\mathbf{k} \int \int \exp[i\mathbf{k} \cdot (\mathbf{x} - \mathbf{x}')] p(\mathbf{x}, t; \mathbf{x}', t')] d\mathbf{x} d\mathbf{x}'$$

$$= \frac{U^2}{(2\pi)^d} \int p_j^2(\mathbf{k}) \hat{C}_Y(\mathbf{k}) \exp[i k_1 U(t - t') - k^2 D(t - t')] d\mathbf{k}. \tag{4.108}$$

The passage to the last line in (4.108) was achieved by using the expression

$$\int \int \exp[i\mathbf{k} \cdot (\mathbf{x} - \mathbf{x}')] p(\mathbf{x}, t; \mathbf{x}', t') d\mathbf{x} d\mathbf{x}' = \exp[i k_1 U(t - t') - k^2 D(t - t')]$$

of the characteristic function of the increment $(\mathbf{X}_t^{(0)} - \mathbf{X}_{t'}^{(0)})$ of the Gaussian unperturbed process $X_{j,t}^{(0)} = \delta_{j,1} U t + W_{j,t}$ (see, e.g., [57]).

Remark 4.3 Changing the sign convention of the Fourier transform,

$$\hat{u}_j(\mathbf{k}) = \int u_j(\mathbf{x}) \exp(i\mathbf{k} \cdot \mathbf{x}) d\mathbf{x},$$

changes the sign of the first term of the characteristic function without altering the sign of the second term, $\exp[-i k_1 U(t - t') - k^2 D(t - t')]$ (see [57, Example 5–28, p. 153] for a detailed computation of characteristic function for Gaussian random variables). However, changing the sign of the imaginary exponent is equivalent to changing U into $-U$ (this also changes the sign of p_j (see [38, Eq. (28a)]), but (4.108) depends only on p_j^2). Since second moments and dispersion coefficients do not depend on the sign of the mean flow, expression (4.108) gives the same result irrespective of the sign of the imaginary exponent.

Inserting (4.108) into (4.98) one obtains the explicit expression of the advective contribution to the ensemble dispersion coefficient,

$$\delta\{D_{ii}^{ens}(t)\} = D_{ii}^{ens}(t) - D$$

$$= \frac{U^2}{(2\pi)^d} \int_0^t dt' \int p_j^2(\mathbf{k}) \hat{C}_Y(\mathbf{k}) \exp[i k_1 U(t - t') - k^2 D(t - t')] d\mathbf{k}. \quad (4.109)$$

With the change of variable $t'' = t - t'$, (4.109) becomes

$$\delta\{D_{ii}^{ens}(t)\} = -\frac{U^2}{(2\pi)^d} \int_t^0 dt'' \int p_j^2(\mathbf{k}) \hat{C}_Y(\mathbf{k}) \exp[i k_1 U t'' - k^2 D t''] d\mathbf{k}$$

$$= \frac{U^2}{(2\pi)^d} \int_0^t dt'' \int p_j^2(\mathbf{k}) \hat{C}_Y(\mathbf{k}) \exp[i k_1 U t'' - k^2 D t''] d\mathbf{k}. \quad (4.110)$$

Remark 4.4 The time derivative of (4.110) is identical with the result of Dagan ([17, Eq. (3.14)] and [18, Eq. (17)]), and Schwarze et al. [67, Eq. (13)]. Dagan [18] derived the result by using the changed sign in the homogeneity condition (Remark 4.2) and opposite Fourier transform convention (Remark 4.3).

Remark 4.5 The opposite Fourier transform convention and the same homogeneity condition as in (4.106) change $i k_1 U t''$ into $-i k_1 U t''$ and let $-k^2 D t''$ unchanged (see Remark 4.3). This is the case in the derivation of the variance Σ_{ii} by Fiori and Dagan [34, Eq. (14)], which is exactly the integral from (4.110) with $i k_1 U t''$

replaced by $-ik_1Ut''$. Dentz et al. [24] used the same conventions as Fiori and Dagan (i.e., equations (16) and (25) of [24]) and obtained the same result. The result given by their equation (40) with $\tilde{g}_0$ replaced by the characteristic function of the unperturbed problem [24, Eq. (25)] is just relation (4.110) from above, with exponent ik_1Ut'' replaced by $-ik_1Ut''$.

The corresponding memory coefficient is similarly obtained by inserting (4.108) into (4.103),

$$MD_{ii}^{ens}(t,\tau,0) = \frac{U^2}{(2\pi)^d} \int_0^\tau dt' \int p_j^2(\mathbf{k})\hat{C}_Y(\mathbf{k}) \exp[ik_1U(t-t') - k^2D(t-t')]d\mathbf{k}.$$

(4.111)

Making the change of variable $t'' = t - t'$ in (4.111), we see that the ensemble memory coefficients can be computed, after a simple change of the integration limits in the time integral, with the explicit expression (4.110) of the advective contributions $\delta\{D_{ii}^{ens}\}$ to the ensemble coefficients:

$$MD_{ii}^{ens}(t,\tau,0) = \frac{U^2}{(2\pi)^d} \int_{t-\tau}^t dt'' \int p_j^2(\mathbf{k})\hat{C}_Y(\mathbf{k}) \exp[ik_1Ut'' - k^2Dt'']d\mathbf{k}.$$

(4.112)

Remark 4.6 The memory coefficient (4.111) can also be obtained within the Eulerian approach of Dentz et al. by replacing in equation (18) of [25] the initial concentration $\hat{\rho}$ with the perturbation solution for point source at time τ given by equation (26) of [24], i.e., $\hat{\rho}(k,\tau) = g(k,\tau)$. Then, equation (18) of [25] gives $\hat{g}(k, t-\tau)$. Further, inserting $\hat{g}(k, t-\tau)$ in equation (14) of [25] and "expanding the logarithm consistently up to second order in the fluctuations" (as described at the beginning of Section 3.2 of [25]) should reproduce expression (4.111) of the memory coefficient $MD^{ens}(t,\tau,0)$.

To derive the explicit form of the effective coefficients (4.97) one needs an explicit expression of the center of mass coefficients (4.99). Similarly to (4.110), one obtains

$$\int\int \overline{u_j(\mathbf{x})u_j(\mathbf{x}')}p(\mathbf{x},t)p(\mathbf{x}',t')d\mathbf{x}d\mathbf{x}'$$

$$= \frac{U^2}{(2\pi)^d} \int p_j^2(\mathbf{k})\hat{C}_Y(\mathbf{k})d\mathbf{k} \int\int \exp[i\mathbf{k}\cdot(\mathbf{x}-\mathbf{x}')p(\mathbf{x},t)p(\mathbf{x}',t')]d\mathbf{x}d\mathbf{x}'$$

$$= \frac{U^2}{(2\pi)^d} \int p_j^2(\mathbf{k})\hat{C}_Y(\mathbf{k}) \exp[ik_1U(t-t') - k^2D(t+t')]d\mathbf{k}.$$

(4.113)

The passage to the last line in (4.113) is now achieved by using the expressions

$$\int\int \exp(i\mathbf{k}\cdot\mathbf{x})p(\mathbf{x},t)d\mathbf{x} = \exp[ik_1Ut - k^2Dt]$$

and

$$\int \int \exp(-i\mathbf{k} \cdot \mathbf{x}')p(\mathbf{x}', t')d\mathbf{x}' = \exp[-ik_1Ut' - k^2Dt']$$

of the characteristic functions of the Gaussian unperturbed process $X_{j,t}^{(0)} = \delta_{j,1}Ut + W_{j,t}$ [57] and the property pointed out in Remark 4.3. Inserting (4.113) into (4.99), one obtains the dispersion coefficient of the center of mass

$$D_{ii}^{cm}(t) = \frac{U^2}{(2\pi)^d} \int_0^t dt' \int p_j^2(\mathbf{k})\hat{C}_Y(\mathbf{k}) \exp[ik_1U(t - t') - k^2D(t + t')]d\mathbf{k}. \tag{4.114}$$

With the change of variable $t'' = t - t'$ (which implies $t + t' = 2t - t''$) expression (4.114) of the center of mass coefficients becomes

$$D_{ii}^{cm}(t) = -\frac{U^2}{(2\pi)^d} \int_t^0 dt'' \int \exp(-2k^2Dt) p_j^2(\mathbf{k})\hat{C}_Y(\mathbf{k}) \exp[ik_1Ut'' + k^2Dt'']d\mathbf{k}$$

$$= \frac{U^2}{(2\pi)^d} \int_0^t dt'' \int \exp(-2k^2Dt) p_j^2(\mathbf{k})\hat{C}_Y(\mathbf{k}) \exp[ik_1Ut'' + k^2Dt'']d\mathbf{k}. \tag{4.115}$$

The center of mass coefficients (4.115) and the advective contribution (4.110) to the ensemble dispersion coefficients determine the advective contributions to the effective coefficients according to (4.97):

$$\delta\{D_{ii}^{eff}(t)\} = D_{ii}^{ens}(t) - D_{i,i}^{cm}(t) - D = \delta\{D_{ii}^{ens}(t)\} - D_{i,i}^{cm}(t). \tag{4.116}$$

Remark 4.7 The ensemble contribution (4.110) and the center of mass coefficient (4.114) particularize to isotropic local diffusion coefficient the general relations M^- and M^+, respectively, derived by Dentz et al. [24, Eq. (47)]. Also note that (4.114) is the time derivative of relation (15) of Fiori and Dagan [34], which, together with Remarks 4.4 and 4.5, proves the equivalence between the Eulerian expressions for the ensemble and effective dispersion coefficients of Dentz et al. [24, 25] and the Lagrangian expressions derived by Fiori and Dagan [34] and by Vanderborght [83] for the corresponding variances. An advantage of the Eulerian approach is that it also provides first order approximations of the concentration (e.g., Eq. (26) of [24]).

The memory coefficient of the center of mass is readily obtained, similarly to (4.112), by a change of the integration limits in expression (4.115) of the center of mass dispersion coefficients:

$$MD_{ii}^{cm}(t, \tau, 0) = \frac{U^2}{(2\pi)^d} \int_{t-\tau}^t dt'' \int \exp(-2k^2Dt) p_j^2(\mathbf{k})\hat{C}_Y(\mathbf{k}) \exp[ik_1Ut'' + k^2Dt'']d\mathbf{k}. \tag{4.117}$$

Finally, the memory coefficient of the effective dispersion is obtained from relations (4.115), (4.116), and (4.112),

$$MD_{ii}^{eff}(t) = MD_{ii}^{ens}(t) - MD_{i,i}^{cm}(t).$$
(4.118)

4.4.3 First Order Results for Power-Law Correlated ln K Fields

The first order results for $\ln K$ fields with finite correlation lengths can be further used to develop approaches for a superposition of scales [26, 61]. To proceed, let us consider the particular case of an isotropic power-law covariance of $Y = \ln K$,

$$C_Y^{PL}(\mathbf{x}) = \sigma_Y^2 z^{-\beta}, \text{ where } z = \left(1 + \frac{|\mathbf{x}|^2}{L^2}\right)^{\frac{1}{2}}, \ |\mathbf{x}|^2 = \sum_{j=1}^{d} x_j^2, \ 0 \le \beta \le 2,$$
(4.119)

where L is some length scale associated with the spatial extension of the system. Power-law covariances can be constructed by integrating covariances with continuously increasing finite integral scales λ [39, p. 370, expression 3.478 (1.)] by the relation

$$\int_0^{\infty} \lambda^{\beta-1} \exp\left(-\mu\lambda^p\right) d\lambda = \frac{1}{p} \mu^{-\frac{\beta}{p}} \Gamma\left(\frac{\beta}{p}\right),$$
(4.120)

where $\mu > 0$, $\beta > 0$, and Γ is the gamma function. For $\mu = z^2$ and $p = 2$, (4.120) yields a power-law covariance as a superposition of Gaussian covariances:

$$C_Y^{PL}(\mathbf{x})/\sigma_Y^2 = z^{-\beta} = \frac{2}{\Gamma(\frac{\beta}{2})} \int_0^{\infty} \lambda^{\beta-1} \exp(-z^2\lambda^2) d\lambda$$

$$= \frac{2}{\Gamma(\frac{\beta}{2})} \int_0^{\infty} \lambda^{\beta-1} \exp\left[-\left(1 + \frac{|\mathbf{x}|^2}{L^2}\right)\lambda^2\right] d\lambda$$

$$= \int_0^{\infty} d\Lambda(\lambda) \exp\left(-\frac{1}{2}\frac{|\mathbf{x}|^2}{l(\lambda)^2}\right),$$
(4.121)

where

$$d\Lambda(\lambda) = \frac{2}{\Gamma(\frac{\beta}{2})} \lambda^{\beta-1} \exp(-\lambda^2) d\lambda \quad \text{and} \quad l(\lambda) = \frac{L}{\lambda\sqrt{2}}.$$
(4.122)

The integrand in (4.121) has the form of the Gaussian C_Y considered by Dentz et al. [24, 25].

Redefining z in (4.119) as $z = (1 + |\mathbf{x}|/L)$ and choosing $\mu = z$ and $p = 1$, one constructs a power-law covariance via (4.120) as a superposition of exponential covariances:

$$
C_Y^{PL}(\mathbf{x})/\sigma_Y^2 = z^{-\beta} = \frac{1}{\Gamma(\beta)} \int_0^\infty \lambda^{\beta-1} \exp(-z\lambda) d\lambda
$$

$$
= \frac{1}{\Gamma(\beta)} \int_0^\infty \lambda^{\beta-1} \exp\left[-\left(1 + \frac{|\mathbf{x}|}{L}\right)\lambda\right] d\lambda
$$

$$
= \int_0^\infty d\Lambda(\lambda) \exp\left(-\frac{|\mathbf{x}|}{l(\lambda)}\right), \tag{4.123}
$$

where

$$
d\Lambda(\lambda) = \frac{1}{\Gamma(\beta)} \lambda^{\beta-1} \exp(-\lambda) d\lambda \quad \text{and} \quad l(\lambda) = \frac{L}{\lambda}. \tag{4.124}
$$

The integrand in (4.123) is the isotropic version of the exponential C_Y considered by Fiori [32] and Vanderborght [83].

Remark 4.8 For $\beta = 1$ in either (4.121) or (4.123) one obtains the covariance of the "$1/z$ noise" as an integral of Gaussian or exponential covariances with continuous scale parameter λ. Further, by defining $z = |\mathbf{x}|/L$ the covariance $\sigma_Y^2 z^{-1}$ is simply given by integrating Gaussian covariances with respect to $d\Lambda(\lambda) = \frac{2}{\sqrt{\pi}} d\lambda$ in (4.121) or exponential covariances with respect to $d\Lambda(\lambda) = d\lambda$ in (4.123). This corresponds to the limit of infinite superposition of $\ln K$ fields with finite correlation scales.

By the linearity of the superposition relations (4.121), respectively (4.123), one obtains the general relations

$$
C_Y^{PL}(\mathbf{x}) = \int_0^\infty d\Lambda(\lambda) C_Y(\mathbf{x}, \lambda), \quad \widehat{C_Y^{PL}}(\mathbf{k}) = \int_0^\infty d\Lambda(\lambda) \widehat{C}_Y(\mathbf{k}, \lambda). \tag{4.125}
$$

By inserting (4.125) in (4.109) and in (4.114) and using (4.97), one finally obtains the superposition relations

$$\delta\{D_{ii}^{PL,ens}\}(t) = \int\limits_0^\infty d\Lambda(\lambda)\{D_{ii}^{ens}\}(t,\lambda)$$

$$\delta\{D_{ii}^{PL,eff}\}(t) = \int\limits_0^\infty d\Lambda(\lambda)\{D_{ii}^{eff}\}(t,\lambda). \tag{4.126}$$

According to (4.126), the memory coefficients (4.110), (4.117), and (4.118) are given by integrals with respect to $d\Lambda(\lambda)$ of the coefficients for given λ derived in the previous section.

References

1. Aït-Sahalia, Y.: Telling from discrete data whether the underlying continuous-time model is a diffusion. J. Financ. **57**, 2075–2112 (2002)
2. Arnold, V.I.: Ordinary Differential Equations. Springer, Berlin (1992)
3. Attinger, S., Dentz, M., Kinzelbach, H., Kinzelbach, W.: Temporal behavior of a solute cloud in a chemically heterogeneous porous medium. J. Fluid Mech. **386**, 77–104 (1999)
4. Avellaneda, M., Majda, M.: Stieltjes integral representation and effective diffusivity bounds for turbulent diffusion. Phys. Rev. Lett. **62**(7), 753–755 (1989)
5. Avellaneda, M., Majda, M.: Superdiffusion in nearly stratified flows. J. Stat. Phys. **69**(3/4), 689–729 (1992)
6. Avellaneda, M., Elliot, F. Jr., Apelian, C.: Trapping, percolation and anomalous diffusion of particles in a two-dimensional random field. J. Stat. Phys. **72**(5/6), 1227–1304 (1993)
7. Balescu, R.: Transport Processes in Plasmas. North-Holland, Amsterdam (1988)
8. Balescu, R., Wang, H-D., Misguich, J.H.: Langevin equation versus kinetic equation: subdiffusive behavior of charged particles in a stochastic magnetic field. Phys. Plasmas **1**(12), 3826–3842 (1994)
9. Bear, J.: On the tensor form of dispersion in porous media. J. Geophys. Res. **66**(4), 1185–1197 (1961)
10. Bhattacharya, R.N., Gupta, V.K.: A theoretical explanation of solute dispersion in saturated porous media at the Darcy scale. Water Resour. Res. **19**, 934–944 (1983)
11. Bouchaud, J.-P., Georges, A.: Anomalous diffusion in disordered media: statistical mechanisms, models and physical applications. Phys. Rep. **195**, 127–293 (1990)
12. Bouchaud, J.-P., Georges, A., Koplik, J., Provata, A., Redner, S.: Superdiffusion in random velocity fields. Phys. Rev. Lett. **64**, 2503–2506 (1990)
13. Chilès, J.P., Delfiner, P.: Geostatistics: Modeling Spatial Uncertainty. Wiley, New York (1999)
14. Clincy, M., Kinzelbach, H.: Stratified disordered media: exact solutions for transport parameters and their self-averaging properties. J. Phys. A. Math. Gen. **34**, 7142–7152 (2001)
15. Colucci, P.J., Jaberi, F.A., Givi, P.: Filtered density function for large eddy simulation of turbulent reacting flows. Phys. Fluids **10**(2), 499–515 (1998)
16. Dagan, G.: Solute transport in heterogeneous porous formations. J. Fluid Mech. **145**, 151–177 (1984)

17. Dagan, G.: Theory of solute transport by groundwater. Annu. Rev. Fluid Mech. **19**, 183–215 (1987)
18. Dagan, G.: Time-dependent macrodispersion for solute transport in anisotropic heterogeneous aquifers. Water Resour. Res. **24**, 1491–1500 (1988)
19. Dagan, G.: Flow and Transport in Porous Formations. Springer, Berlin (1989)
20. Dagan, G.: Transport in heterogeneous porous formations: spatial moments, ergodicity, and effective dispersion. Water Resour. Res. **26**(6), 1281–1290 (1990)
21. Deng, W., Barkai, E.: Ergodic properties of fractional Brownian-Langevin motion. Phys. Rev. E **79**, 011112 (2009)
22. Deng, F.W., Cushman, J.H.: On higher-order corrections to the flow velocity covariance tensor. Water Resour. Res. **31**(7), 1659–1672 (1995)
23. Dentz, M., de Barros, F.P.J.: Dispersion variance for transport in heterogeneous porous media. Water Resour. Res. **49**, 3443–3461 (2013)
24. Dentz, M., Kinzelbach, H., Attinger, S., Kinzelbach, W.: Temporal behavior of a solute cloud in a heterogeneous porous medium 1. Point-like injection. Water Resour. Res. **36**, 3591–3604 (2000)
25. Dentz, M., Kinzelbach, H., Attinger, S., Kinzelbach, W.: Temporal behavior of a solute cloud in a heterogeneous porous medium 2. Spatially extended injection. Water Resour. Res. **36**, 3605–3614 (2000)
26. Di Federico, V., Neuman, S.P.: Scaling of random fields by means of truncated power variograms and associated spectra. Water Resour. Res. **33**, 1075–1085 (1997)
27. Doob, J.L.: Stochastic Processes. Wiley, New York (1990)
28. Dybiec, B., Gudowska-Nowak, E.: Discriminating between normal and anomalous random walks. Phys. Rev. E **80**, 061122 (2009)
29. Eberhard, J.: Approximations for transport parameters and self-averaging properties for point-like injections in heterogeneous media. J. Phys. A. Math. Gen. **37**, 2549–2571 (2004)
30. Eberhard, J., Suciu, N., Vamos, C.: On the self-averaging of dispersion for transport in quasi-periodic random media. J. Phys. A: Math. Theor. **40**, 597–610 (2007)
31. Fannjiang, A., Komorowski, T.: Diffusive and nondiffusive limits of transport in nonmixing flows. SIAM J. Appl. Math. **62**, 909–923 (2002)
32. Fiori, A.: Finite Peclet extensions of Dagan's solutions to transport in anisotropic heterogeneous formations. Water Resour. Res. **32**, 193–198 (1996)
33. Fiori, A.: On the influence of local dispersion in solute transport through formations with evolving scales of heterogeneity. Water Resour. Res. **37**(2), 235–242 (2001)
34. Fiori, A., Dagan, G.: Concentration fluctuations in aquifer transport: a rigorous first-order solution and applications. J. Contam. Hydrol. **45**, 139–163 (2000)
35. Fried, J.J.: Groundwater Pollution. Elsevier, New York (1975)
36. Gardiner, C.W.: Stochastic Methods. Springer, Berlin (2009)
37. Gelhar, L.W.: Stochastic subsurface hydrology from theory to applications. Water Resour. Res. **22**(9S), 135S–145S (1986)
38. Gelhar, L.W., Axness, C.: Three-dimensional stochastic analysis of macrodispersion in aquifers. Water Resour. Res. **19**(1), 161–180 (1983)
39. Gradshteyn, I.S., Ryzhik, I.M.: Table of Integrals, Series, and Products. Elsevier, Amsterdam (2007)
40. Honkonen, G.: Stochastic processes with stable distributions in random environments. Phys. Rev. E **53**(1), 327–331 (1996)
41. Isichenko, M.B.: Percolation, Statistical Topography, and Transport in Random Media. Rev. Mod. Phys. **64**, 961 (1992)
42. Jaekel, U., Vereecken, H.: Renormalization group analysis of macrodispersion in a directed random flow. Water Resour. Res. **33**, 2287–2299 (1997)
43. Jeon, J.-H., Metzler, R.: Fractional Brownian motion and motion governed by the fractional Langevin equation in confined geometries. Phys. Rev. E **81**, 021103 (2010)
44. Jury, W.A., Sposito, G.: Field calibration and validation of solute transport models for the unsaturated zone. Soil Sci. Soc. Am. J. **49**, 1331–1241 (1985)

45. Kesten, H., Papanicolaou, G.C.: A limit theorem for turbulent diffusion. Commun. Math. Phys. **65**, 97–128 (1979)
46. Kitanidis, P.K.: Prediction by the method of moments of transport in a heterogeneous formation. J. Hydrol. **102**, 453–473 (1988)
47. Kloeden, P.E., Platen, E.: Numerical Solutions of Stochastic Differential Equations. Springer, Berlin (1999)
48. Kolmogorov, A.N.: Grundbegriffe der Warscheinlichkeitsrechnung. Springer, Berlin (1933)
49. Le Doussal, P., Machta, J.: Annealed versus quenched diffusion coefficient in random media. Phys. Rev. B **40**(12), 9427–9430 (1989)
50. Lumley, J.L.: The mathematical nature of the problem of relating Lagrangian and Eulerian statistical functions in turbulence. In: Mécanique de la Turbulence. Coll. Intern. du CNRS à Marseille (Ed. CNRS, Paris, 1962), pp. 17–26
51. Majda, A.J., Kramer, P.R.: Simplified models for turbulent diffusion: theory, numerical modelling, and physical phenomena. Phys. Rep. **14**, 237–574 (1999)
52. Majumdar, S.N.: Persistence of a particle in the Matheron–de Marsily velocity field. Phys. Rev. E **68**, 050101(R) (2003)
53. Matheron, G., de Marsily, G.: Is transport in porous media always diffusive? Water Resour. Res. **16**, 901–917 (1980)
54. Monin, A.S., Yaglom, A.M.: Statistical Fluid Mechanics, Volume II: Mechanics of Turbulence. MIT Press, Cambridge (1975)
55. O'Malley, D., Cushman, J.H.: A renormalization group classification of nonstationary and/or infinite second moment diffusive processes. J. Stat. Phys. **146**(5), 989–1000 (2012)
56. O'Malley, D., Cushman, J.H.: Two scale renormalization group classification of diffusive processes. Phys. Rev. E **86**(1), 011126 (2012)
57. Papoulis, A., Pillai, S. U.: Probability, Random Variables and Stochastic Processes. McGraw-Hill, Singapore (2009)
58. Pope, S.B.: PDF methods for turbulent reactive flows. Prog. Energy Combust. Sci. **11**(2), 119–192 (1885)
59. Port, S.C., Stone, C.J.: Random measures and their application to motion in an incompressible fluid. J. Appl. Prob. **13**, 498–506 (1976)
60. Reuss, J.-D., Misguish, J.H.: Low frequency percolation scaling for particle diffusion in electrostatic turbulence. Phys. Rev. E **54**(2), 1857–1869 (1996)
61. Ross, K., Attinger, S.: Temporal behaviour of a solute cloud in a fractal heterogeneous porous medium at different scales. Geophys. Res. Abstr. **12**, EGU2010-10921-2 (2010)
62. Russo, D.: On the velocity covariance and transport modeling in heterogeneous anisotropic porous formations. Water Resour. Res. **31**(1), 129–137 (1995)
63. Russo, D.: A note on ergodic transport of a passive solute in partially saturated porous formations. Water Resour. Res. **32**(12), 3623–3628 (1996)
64. Saffman, P.G.: Application of the Wiener-Hermite expansion to the diffusion of a passive scalar in a homogeneous turbulent flow. Phys. Fluids **12**(9), 1786–1798 (1969)
65. Scheidegger, A.E.: Statistical hydrodynamics in porous media. J. Appl. Phys. **25**(8), 994–1001 (1954)
66. Scheidegger, A.E.: General theory of dispersion in porous media. J. Geophys. Res. **66**(10), 3273–3278 (1961)
67. Schwarze, H., Jaekel, U., Vereecken, H.: Estimation of macrodispersivity by different approximation methods for flow and transport in randomly heterogeneous media. Transp. Porous Media **43**, 265–287 (2001)
68. Shapiro, A.M., Cvetkovic, V.D.: Stochastic analysis of solute travel time in heterogeneous Porous media. Water Resour. Res. **24**(10), 1711–1718 (1988)
69. Suciu, N.: Spatially inhomogeneous transition probabilities as memory effects for diffusion in statistically homogeneous random velocity fields. Phys. Rev. E **81**, 056301 (2010)
70. Suciu, N.: Diffusion in random velocity fields with applications to contaminant transport in groundwater. Adv. Water Resour. **69**, 114–133 (2014)

71. Suciu, N., Vamoş, C.: Comment on "Nonstationary flow and nonergodic transport in random porous media" by G. Darvini and P. Salandin. Water Resour. Res. **43**, W12601 (2007)
72. Suciu, N., Vamoş, C.: Ergodic estimations of upscaled coefficients for diffusion in random velocity fields. In: L'Ecuyér, P., Owen, A.B. (eds.) Monte Carlo and Quasi-Monte Carlo Methods 2008, pp. 617–626. Springer, Berlin (2009)
73. Suciu, N., Vamoş, C., Vanderborght, J., Hardelauf, H., Vereecken, H.: Numerical investigations on ergodicity of solute transport in heterogeneous aquifers. Water Resour. Res. **42**, W04409 (2006)
74. Suciu, N., Vamoş, C., Eberhard, J.: Evaluation of the first-order approximations for transport in heterogeneous media. Water Resour. Res. **42**, W11504 (2006)
75. Suciu, N., Vamos, C., Vereecken, H., Sabelfeld, K., Knabner, P.: Memory effects induced by dependence on initial conditions and ergodicity of transport in heterogeneous media. Water Resour. Res. **44**, W08501 (2008)
76. Suciu, N., Vamos, C., Vereecken, H., Sabelfeld, K., Knabner, P.: Ito equation model for dispersion of solutes in heterogeneous media. Rev. Anal. Num. Theor. Approx. **37**, 221–238 (2008)
77. Suciu, N., Vamos, C., Radu, F.A., Vereecken, H., Knabner, P.: Persistent memory of diffusing particles. Phys. Rev. E **80**, 061134 (2009)
78. Suciu, N., Attinger, S., Radu, F.A., Vamoş, C., Vanderborght, J., Vereecken, H., Knabner, P.: Solute transport in aquifers with evolving scale heterogeneity. Print No. 346, Mathematics Department—Friedrich-Alexander University Erlangen-Nuremberg (2011)
79. Suciu, N., Attinger, S., Radu, F.A., Vamoş, C., Vanderborght, J., Vereecken, H. Knabner, P.: Solute transport in aquifers with evolving scale heterogeneity. Analele Universitatii "Ovidius" Constanta-Seria Matematica **23**(3), 167–186 (2015)
80. Suciu, N., Schüler, L., Attinger, S., Knabner, P.: Towards a filtered density function approach for reactive transport in groundwater. Adv. Water Resour. **90**, 83–98 (2016)
81. Taylor, G.I.: Diffusion by continuous movements. Proc. Lond. Math. Soc. **2**(20), 196–212 (1921)
82. Trefry, M.G., Ruan, F.P., McLaughlin, D.: Numerical simulations of preasymptotic transport in heterogeneous porous media: departures from the Gaussian limit. Water Resour. Res. **39**(3), 1063 (2003)
83. Vanderborght, J.: Concentration variance and spatial covariance in second order stationary heterogeneous conductivity fields. Water Resour. Res. **37**(7), 1893–1912 (2001)
84. Yaglom, A.M.: Correlation Theory of Stationary and Related Random Functions, Volume I: Basic Results. Springer, New York (1987)
85. Zirbel, C.L.: Lagrangian observations of homogeneous random environments. Adv. Appl. Prob. **33**, 810–835 (2001)

Chapter 5
Monte Carlo GRW Simulations of Passive Transport in Groundwater

Abstract Monte Carlo GRW simulations of passive transport in groundwater are used to investigate ergodic properties, dependence on initial conditions, and the occurrence of anomalous diffusion. It is shown that memory effects produced by dependence on initial conditions are responsible for the lack of ergodicity of the transport, in the sense of approach to the theoretical upscaled process. Evolving scale heterogeneity of groundwater systems, consisting of a superposition of spatial scales, enhances the memory effects and may explain the occurrence of anomalous diffusion behavior.

5.1 Numerical Investigations on Memory Effects and Ergodicity

An ideal tracer experiment, consisting of passive transport of substance under different deterministic initial conditions, was simulated numerically with the unbiased GRW method presented in Sect. 3.3.2. Two-dimensional Cauchy problems for the advection–diffusion equation (4.77), with initial conditions given by uniform concentrations in rectangular slabs and by a point instantaneous injection, were solved by simultaneously tracking large collections of computational particles (ten billions in most cases, see, e.g., [18, 19, 21, 22, 24]) with the GRW algorithm (see implementation details in [21, Appendix A]). To mimic an infinite medium, the simulations were conducted in a two-dimensional computational grid larger than the largest extension of the plume.

For statistically homogeneous log-normal random hydraulic conductivity fields K with finite correlation lengths and small variance, first order approximations of the velocity field are obtained by a formal asymptotic expansion of the flow equations (4.78) [4, 11]. Velocity realizations are computed numerically by the Kraichnan method [13] as sums between a constant mean and a superposition of random periodic fluctuations, according to (C.4), with the algorithm presented in Appendix C.3.2.1. In this way one obtains fast estimations of samples of random velocity fields which allow the computation of thousands of transport simulations at moderate computational costs. For finite number of random periodic modes the

© Springer Nature Switzerland AG 2019
N. Suciu, *Diffusion in Random Fields*, Geosystems Mathematics,
https://doi.org/10.1007/978-3-030-15081-5_5

fluctuations of the estimated dispersion quantities may be affected by an artificial logarithmic increase [8, Appendix A]. An empirical recipe to avoid this artifact is to choose a number of modes of the order of the total computation time [8]. The convergence of the MC estimates for this problem is ensured by using several hundreds of simulations [20, Fig. 4]. The transport simulations presented in this chapter, as well as those presented in Chap. 6, were performed with 6400 random periodic modes and 10^{10} particles used in the GRW algorithm. Statistical ensembles of solutions were computed by running GRW codes written in C++ on parallel computers. The MC ensembles used to derive the results presented in this section consisted of 1024 realizations GRW solutions.

The problem is parameterized by an isotropic local dispersion coefficient $D = 0.01\,\mathrm{m^2/day}$, mean velocity with components $\langle V_1 \rangle = 1\,\mathrm{m/day}$ and $\langle V_2 \rangle = 0\,\mathrm{m/day}$, normal $\ln K$ field with exponential correlation, correlation length $\lambda = 1\,\mathrm{m}$, and a small variance $\sigma^2 = 0.1$, chosen to ensure the validity of the first order approximation of the velocity field. With these parameters, the first order approximation (4.58) of the ensemble dispersion coefficients in the limit of long time is given by $D_{11}^{ens} = D + \langle V_1 \rangle \sigma^2 \lambda = 0.11\,\mathrm{m^2/day}$ and $D_{22}^{ens} = 0$. The Peclet number $Pe = \langle V_1 \rangle \lambda / D = 100$ indicates an advection-dominated transport problem.

5.1.1   Ergodicity of the Center of Mass

Figure 5.1 shows the long-time decay of the variance R_{ii} of the center of mass process $X_i^{cm}(t)$. This implies, according to (4.80), that at large times the expectation

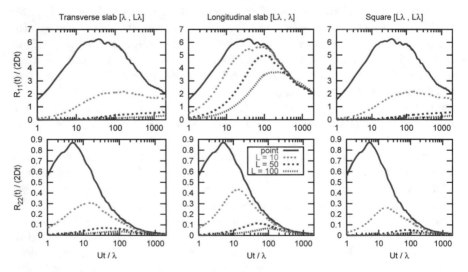

Fig. 5.1 The variance of the center of mass $R_{ii}/(2Dt)$, $i = 1, 2$, decreases uniformly with increasing source dimension and goes to zero for large times

of the second moment of the concentration may be approximated by the second moment of the mean concentration, $S_{ii} \approx \Sigma_{ii}$. Figure 5.1 also shows the decrease of R_{ii} with increasing supports of the initial concentration. One expects therefore that, for sufficiently large contaminant sources, the approximation $S_{ii} \approx \Sigma_{ii}$ also holds at finite times. Following the terminology introduced by Dagan [5], large plumes, for which the expected second moment can be approximated by the second moment of the ensemble mean concentration, are usually called "ergodic plumes" in the hydrogeological literature. One hypothesizes also that the mean second moment S_{ii} as well as the unaveraged moment s_{ii} of an ergodic plume can be approximated, according to (4.82), by the one-particle dispersion X_{ii} [18, 19].

Another ergodic property was formulated by Sposito et al. [16]. The transport in groundwater is called *asymptotically ergodic* if the solution to (4.78) for a given realization of the velocity field approaches that of the "macrodispersion" model, an upscaled advection–diffusion equation supposed to describe the transport in random velocity fields with finite correlation scales [4]. Various meanings of ergodicity in hydrogeological literature are particular cases of the general formulation proposed in [21]: an observable of the transport process is *ergodic with respect to a stochastic model* if the root mean-square distance from the model prediction is smaller than a given threshold. The squared distance can be decomposed as sum between the squared deviation of the ensemble mean of the observable from the reference stochastic model and the variance of the observable about its mean [21, definition 5]. The usual statistical inference for ergodic estimators of the mean [29] is retrieved in this formulation when the observable is an average over the parameter range (time or space) and the stochastic model is the ensemble mean of the random function (see also [19]). The *self-averaging property* from statistical physics [2] corresponds to the particular case when the observable is the (unaveraged) process itself and the stochastic model is the ensemble mean of the process. The self-averaging property is thus ensured by a vanishing variance in the long time limit, and, obviously, it is implied by the asymptotic ergodicity.

The decay with time of the variance R_{ii} indicates the self-averaging of the center of mass process $X_i^{cm}(t)$. The self-averaging of the center of mass corresponds to a self-averaging property of the mean Lagrangian velocity $\langle u_i(\mathbf{X}(t)) \rangle_{DX_0}$ [25, Fig. 2]. The variance of $(X_i^{cm})^2(t)$ also was found to decrease in time and with increasing source size [24, Fig. 1]. This implies the self-averaging of the quantity $(X_i^{cm})^2/(2t)$, which is the single-realization dispersion coefficient of the center of mass [23, Eq. (23)]. Since, according to Slutsky's theorem [29] a vanishing variance is a sufficient condition for ergodicity, the self-averaging implies the "usual" ergodicity, that is, the convergence of the time and space averages of the observables $X_i^{cm}(t))$, $\langle u_i(\mathbf{X}(t)) \rangle_{DX_0}$, and $(X_i^{cm})^2/(2t)$.

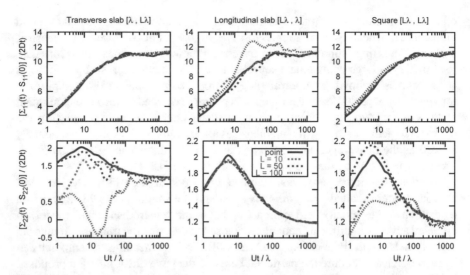

Fig. 5.2 For sources with large dimensions on the i-direction the ensemble dispersion Σ_{ii} depends on the initial conditions

5.1.2 Dependence on Initial Conditions

The variance of the process $X_i^{ens}(t)$, i.e., the second moment Σ_{ii} of the mean concentration, computed for different shapes and sizes of the source is shown in Fig. 5.2. Significant dependence on initial conditions of the ensemble dispersion corrected for the initial second moment, $\Sigma_{ii} - S_{ii}(0)$, manifests in case of asymmetric sources with large extension on the i-axis while the initial conditions have negligible influence for sources with direction of largest extension perpendicular to the i-axis. This behavior was attributed to the mean memory terms, which may be significantly large in the first case and negligible in the second case (according to relation (4.80), where the influence of the Q_{ii} term was found to be negligible, see [19, 24, 25]).

From Fig. 5.2 we can conclude that the second moments of the mean concentration depend on the size, geometry, and orientation of the source. They approximate the one-particle dispersion only in special cases of narrow sources with small extension on i-th direction and small memory terms (4.75). Otherwise, $\Sigma_{ii} \neq S_{ii}(0) + X_{ii}$. This indicates that *the Lagrangian stationarity, which would imply (4.82), fails even though the velocity field considered in simulations is statistically homogeneous.*

5.1.3 Non-ergodic Effective Dispersion at Finite Times

The variance (4.74) of the effective process $X_i^{eff}(t)$ computed for fixed realizations of the velocity field shows large sample to sample fluctuations in cases where the mean Σ_{ii} is also strongly influenced by the initial conditions (Fig. 5.3). The one-particle dispersion X_{ii} shown in Fig. 5.3 was approximated by $\Sigma_{ii} - S_{ii}(0)$ in ergodic situations consisting of large slab sources perpendicular to the i-axis. The deviation $S_{ii} - X_{ii}$ from the one-particle dispersion of the mean of s_{ii} is one to two orders of magnitude smaller than its standard deviation $SD(s_{ii})$. These are the two quantities which determine the deviation from ergodic behavior in the general formulation presented in Sect. 5.1.1. Thus, the results from Fig. 5.3 indicate that the single realization dispersion s_{ii} is in general non-ergodic with respect to the one-particle dispersion X_{ii} at finite times.

Ergodicity may be expected, within acceptable small root mean-square distances, for the longitudinal dispersion in case of large transverse slab sources and for transverse dispersion in case of longitudinal slab sources. However, the MC results contradict a common belief on ergodic plumes: *large transverse plumes do not necessarily imply the ergodicity of both longitudinal and transverse dispersion*. On the contrary, increasing the plume dimensions might result in dramatic non-ergodic behavior, mainly for the transverse dispersion.

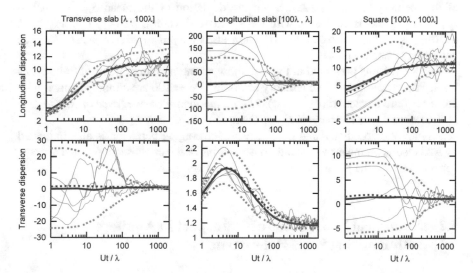

Fig. 5.3 Single-realization dispersion $(s_{ii} - S_{ii}(0))/(2Dt)$ (thin red lines), one-particle dispersion $X_{ii}/(2Dt)$ (blue dots), ensemble averages $(S_{ii} - S_{ii}(0))/(2Dt)$ (thick red lines), and $(S_{ii} - S_{ii}(0) \pm SD(s_{ii}))/(2Dt)$ (grey dots) [18]

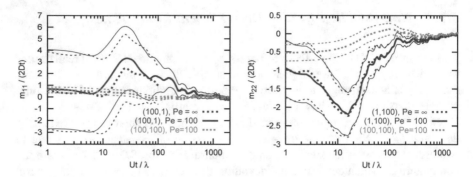

Fig. 5.4 Longitudinal (left) and transverse (right) memory terms in non-ergodic cases, for finite and infinite Peclet numbers. Solid lines correspond to M_{ii} and thin lines correspond to $(M_{ii} \pm SD(m_{ii})/R^{1/2})/(2Dt)$, where $R = 1024$ is the number of MC simulations [23]

5.1.4   Loss of Memory and Asymptotic Ergodicity

In ergodic cases shown in Fig. 5.3 (slab sources perpendicular to i-axis) the relation $s_{ii} - S_{ii}(0) \approx X_{ii}$ is an acceptable approximation of (4.74). In non-ergodic cases, the memory terms (4.75) are no longer negligible and (4.74) is approximated by $s_{ii} - S_{ii}(0) \approx X_{ii} + m_{ii}$, which allows estimations of means and standard deviations of the memory terms m_{ii} [25]. Since the deviation of the mean $S_{ii} - S_{ii}(0) - X_{ii}$ is negligible as compared with the standard deviation $SD(s_{ii})$ (see Fig. 5.3), $SD(m_{ii}) \approx SD(s_{ii})$ also quantifies the non-ergodicity of s_{ii} *with respect to* X_{ii} (see definition in Sect. 5.1.1).

The MC results presented in Fig. 5.4 show strong memory effects at finite times for asymmetric sources. The memory terms for $Pe = 100$ are almost identical with those for pure advection ($Pe = \infty$). *The mean-square convergence of m_{ii} to zero indicates the asymptotic ergodicity of the actual dispersion s_{ii}.*

Asymptotic ergodicity, and implicitly self-averaging behavior, is also indicated by MC results on means and standard deviations of the cross-section space average concentration at the plume center of mass [21, Figs. 1 and 2].

5.2   Numerical Simulations of Transport in Aquifers with Evolving Scale Heterogeneity

In Sect. 4.4.3 it was shown that power low correlated $\ln K$ fields can be obtained as superpositions of fields with short range correlations. Such fields could be useful in modeling flow and transport at an observation scale where aquifer heterogeneities have long-range correlations indicated by measurement data [14]. In other practical situations, the modeling should account for a hierarchy of observational scales with finite but increasing correlation lengths [10]. This kind of heterogeneity has been simulated by a superposition of statistically independent $\ln K$ fields with isotropic exponential correlation and constant variance $\sigma_Y^2 = 0.1$ [26, 27]. The integral scales (defined as integrals of the correlation function, not divided by the variance) of the $\ln K$ fields were increased with a constant step $\lambda = 1$ m, so that at the n-th step $\lambda_n = \lambda_{n-1} + n$, which results in $\lambda_n = (n+1)n/2$. For instance, the superposition of 2 fields has the variance 0.2 and the integral scale $\lambda_2 = 3$ m and the superposition of seven fields (the largest one investigated) has the variance 0.7 and $\lambda_7 = 28$ m (also see Remark 4.1 in Sect. 4.4.1). MC results obtained with this model of evolving scale heterogeneity will be used in the following to illustrate strengthening memory effects and the approach to anomalous diffusive behavior.

5.2.1   Quantifying Anomalous Diffusion by Memory Terms

In the following, let X_t be one of the processes $X_{i,t}^{eff}$, $X_{i,t}^{ens}$, or $X_{i,t}^{cm}$ defined in (4.79). Further, consider the uniform time partition $0 < \Delta < \cdots < k\Delta < \cdots < (K-1)\Delta < K\Delta$ and for $k > 0$, define the increments $X_{k,k-1} = X_{t_k} - X_{t_{k-1}}$ and note their square by $s_{k,k-1} = (X_{k,k-1})^2$. For a sum of two increments $X_{k,0} = X_{k-1,0} + X_{k,k-1}$ we have the relation

$$s_{k,0} = s_{k-1,0} + s_{k,k-1} + m_{k,k-1,0} \qquad (5.1)$$

where $m_{k,k-1,0} = 2X_{k-1,0}X_{k,k-1}$. By iterating the binomial formula (5.1) one obtains

$$s_{1,0} = s_{1,0}$$

$$s_{2,0} = s_{1,0} + s_{2,1} + m_{2,1,0}$$

$$\dots\dots\dots\dots\dots\dots\dots$$

$$s_{k,0} = s_{k-1,0} + s_{k,k-1} + m_{k,k-1,0}$$

$$\dots\dots\dots\dots\dots\dots\dots$$

$$s_{K-1,0} = s_{K-2,0} + s_{K-1,K-2} + m_{K-1,K-2,0}$$

$$s_{K,0} = s_{K-1,0} + s_{K,K-1} + m_{K,K-1,0}$$

The sum of these relations gives

$$s_{K,0} = \sum_{k=1}^{K} s_{k,k-1} + \sum_{k=1}^{K} m_{k,k-1,0}, \qquad (5.2)$$

which, using the relation

$$m_{k,k-1,0} = 2X_{k-1,0}X_{k,k-1} = 2X_{k,k-1} \sum_{l=1}^{k-1} X_{l,l-1}$$

can be further expressed as

$$s_{K,0} = \sum_{k=1}^{K} s_{k,k-1} + 2\sum_{k=1}^{K}\sum_{l=1}^{k-1} X_{k,k-1}X_{l,l-1}. \tag{5.3}$$

Relations (5.2) and (5.3) are discrete versions of (4.76) written for K terms. The expression (5.3) is in fact a discrete Taylor formula which can be utilized in numerical simulations to investigate the structure of the dispersion coefficients [28] (also see Appendix F.1).

For the "ensemble process" $X_t = X_t^{ens}$, the ensemble average of (5.1), $\Sigma_{k,0} = \langle s_{k,0} \rangle$, verifies

$$\Sigma_{k,0} = \Sigma_{k-1,0} + \Sigma_{k,k-1} + M_{k,k-1,0}, \tag{5.4}$$

where $M_{k,k-1,0} = \langle m_{k,k-1,0} \rangle$. For continuous time $t = k\Delta$, (5.4) becomes

$$\Sigma_{t,0} = \Sigma_{t-\Delta,0} + \Sigma_{t,t-\Delta} + M_{t,t-\Delta,0}. \tag{5.5}$$

This relation quantifies the departure from normal diffusive behavior with variance linear in time by a memory term M consisting of correlations of successive increments (see Sect. 4.2.3). Such correlations were used by Jeon and Metzler [12] to quantify the memory of two types of anomalous diffusion processes (fractional Brownian motion and processes governed by fractional Langevin equation). For instance, the increment correlation obtained by Jeon and Metzler [12, Eq. (4.4)] in case of fractional Brownian motion with variance $\Sigma_{t_2,t_1} = (t_2 - t_1)^{2H}$ and Hurst exponent $0 < H < 1$ is given by (5.5) written for two successive time intervals Δ,

$$M_\Delta = 2\langle X_{\Delta,0}X_{2\Delta,\Delta} \rangle = \Sigma_{2\Delta,0} - \Sigma_{\Delta,0} - \Sigma_{2\Delta,\Delta} = (2^{2H} - 2)\Delta^{2H}.$$

Similar relations hold for the "effective" and "center of mass" processes X_t^{eff} and X_t^{cm}. They can be used in numerical simulations to measure the deviation from normal diffusion of these processes over the last time step Δ by memory terms $M_{t,t-\Delta,0}$. The expectation of the second term of Eq. (5.2),

$$CM_t = \sum_{k=1}^{t/\Delta} \langle m_{k\Delta,(k-1)\Delta,0} \rangle, \tag{5.6}$$

is a cumulative memory term which, according to (5.3), contains a whole hierarchy of correlations between increments along paths of the transport process. Estimations of such cumulative memory terms by simulations of transport in media with evolving scale heterogeneity will be presented in the next section. It will be shown that cumulative memory terms of form (5.6) have the most important contribution to the ensemble, effective, and center of mass dispersion coefficients.

5.2.2 Ensemble and Memory Coefficients

The behavior of several transport observables was investigated by GRW simulations of a two-dimensional isotropic diffusion in random velocity fields, consisting of superpositions of statistical independent fields, with the numerical approach described in Sect. 5.1. MC estimates were obtained as averages over ensembles of 256 realizations of the GRW solution for the same initial condition given by an instantaneous point injection. Variances, dispersion coefficients, and memory terms were estimated for the three processes defined in (4.79).

The ensemble dispersion coefficients $\Sigma_{ii}(t)/(2t)$ and the one-step memory dispersion coefficients $M_{ii}(t, t - \Delta, 0)/(2t)$ are presented in Figs. 5.5, 5.6, 5.7, and 5.8. (See Ref. [26] for results on effective and center of mass coefficients.) The extinction of the memory coefficients (Figs. 5.7 and 5.8) corresponds to the approach to a normal diffusive behavior with constant ensemble dispersion coefficients, shown in Figs. 5.5 and 5.6, in agreement with Eq. (5.5).

Figures 5.9 and 5.10 show the cumulative memory coefficients $CM(t)/(2t)$, with CM given by Eq. (5.6), compared with the corresponding ensemble, effective, and center of mass coefficients, in case of the seven-scales velocity field. The cumulative memory coefficients are close to the corresponding dispersion coefficients, the small difference between the two coefficients being just the half slope of the average of the first term in (5.2). The latter is a dispersion coefficient describing the displacement of solute particles over one step Δ (in our case, the simulation time

Fig. 5.5 Longitudinal ensemble dispersion coefficients for $\ln K$ fields with increasing integral scales

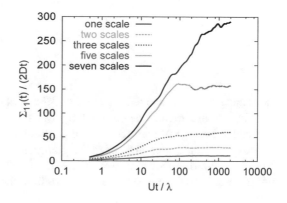

Fig. 5.6 Transverse
ensemble dispersion
coefficients for ln K fields
with increasing integral scales

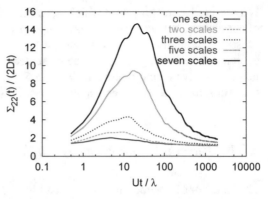

Fig. 5.7 Longitudinal
ensemble memory
coefficients for ln K fields
with increasing integral scales

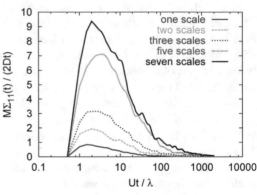

Fig. 5.8 Transverse
ensemble memory
coefficients for ln K fields
with increasing integral scales

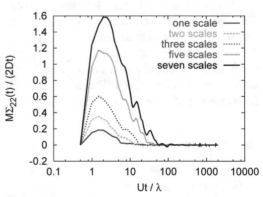

step) and corresponds to the linear part of the total dispersion. Thus, the dispersion
coefficients are mainly determined by the hierarchy of increment correlations
(ensemble average of the second term of the discrete Taylor formula (5.3)) which
approximates the correlations of the Lagrangian velocity. This makes the difference
from genuine diffusion processes (Brownian motions) where the increments are
uncorrelated and the one-step dispersion defines the constant diffusion coefficient.

Fig. 5.9 Cumulative longitudinal memory coefficients compared with the corresponding ensemble, effective, and center of mass dispersion coefficients ($\lambda_7 = 28$ m)

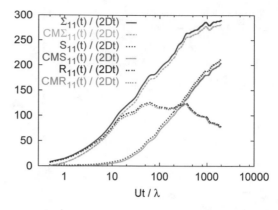

Fig. 5.10 Cumulative transverse memory coefficients compared with the corresponding ensemble, effective, and center of mass dispersion coefficients ($\lambda_7 = 28$ m)

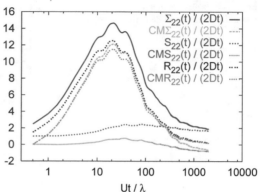

Figures 5.5 and 5.9 show that both the ensemble and the effective coefficients behave anomalously in time windows which increase with the number of superposed scales.

5.2.3 Breakthrough Curves and Cross-Section Concentrations

Anomalous diffusive behavior is also indicated by the asymmetry of the solute plume at 1000 days, presented in Figs. 5.11 and 5.12, as well as by the breakthrough curves $c(t)$ of the concentration spatially averaged over the vertical section of the simulation domain, recorded at increasing distances along the mean flow direction, shown in Figs. 5.13 and 5.14.

The asymmetry of the solute plume at 1000 days, presented in Fig. 5.11, is more pronounced than that recorded at 2000 days (Fig. 5.12). The asymmetry of the plume indicates that many solute particles move slower than the center of mass and remain behind the front of the plume. The explanation is, very likely, the persistence of negative velocity fluctuations, rather than the increase of the fluctuations (note that

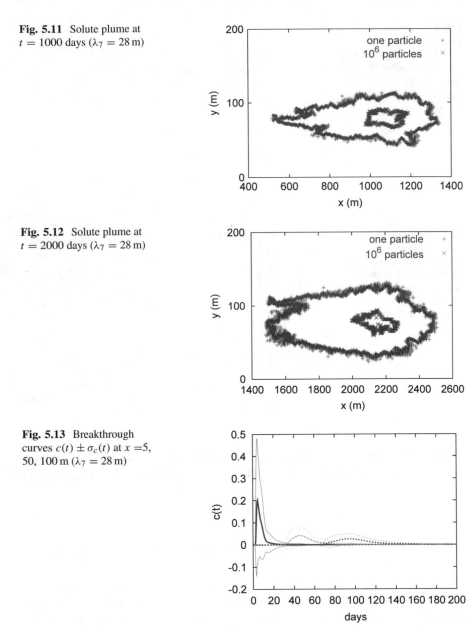

Fig. 5.11 Solute plume at $t = 1000$ days ($\lambda_7 = 28$ m)

Fig. 5.12 Solute plume at $t = 2000$ days ($\lambda_7 = 28$ m)

Fig. 5.13 Breakthrough curves $c(t) \pm \sigma_c(t)$ at $x = 5$, 50, 100 m ($\lambda_7 = 28$ m)

the variance of the ln K field is only 0.7). This agrees with the anomalous behavior of the dispersion coefficients over the first 1000 days (Figs. 5.5 and 5.6) and with expected approach to a Fickian regime in the limit of very large times, ensured by the finiteness of the integral scale [25].

Fig. 5.14 Breakthrough curves $c(t) \pm \sigma_c(t)$ at $x =$ 300, 500, 1000, and 1500 m ($\lambda_7 = 28$ m)

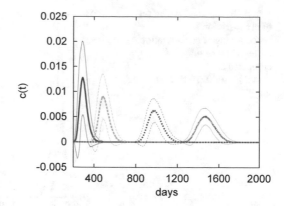

Fig. 5.15 Ensemble average of cross-section space-averaged concentration at the center of mass recorded at $x =$ 500, 1000, 1500, 2000 m

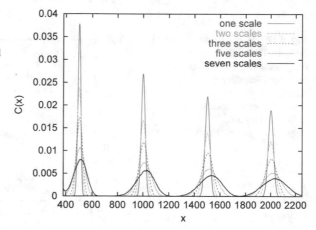

The breakthrough curves shown in Figs. 5.13 and 5.14 are also strongly asymmetric at early stages and become more and more symmetrical as the process approaches the Fickian regime. The ensembles of simulated breakthrough curves also provide estimations of the mean, the variance, and the probability distribution of the travel-time [26].

The concentration at the center of mass, integrated over the transverse dimension of the plume, is shown in Fig. 5.15. One remarks a small asymmetry of the concentration profiles with the increase of the integral scale. The effect of multiscale heterogeneity is better illustrated by the asymmetry of the corresponding standard deviation (Fig. 5.16), which is more pronounced and persists over larger distances as the integral scale increases.

Fig. 5.16 Standard deviation of cross-section space-averaged concentration at the center of mass computed at $x = 500, 1000, 1500, 2000$ m

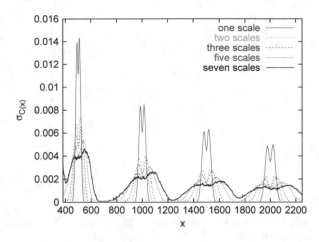

Fig. 5.17 Power-law fit between 10 and 500 days of longitudinal ensemble and effective coefficients; standard fit-errors for power-law exponents of 0.58% and 0.46%, respectively ($\lambda_7 = 28$ m)

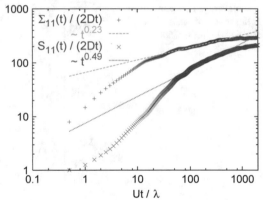

5.2.4   Anomalous Diffusion Behavior

The anomalous behavior induced by the multiscale structure of the velocity field was further investigated by fitting dispersion coefficients with power-law functions. Figures 5.17 and 5.18 indicate a power law behavior of the coefficients. One remarks that ensemble and effective coefficients and variances have different time behavior, with a larger slope for the effective quantities. This seems to be at variance with a theoretical result for power-law correlated fields obtained by Fiori [9].

As shown in Sect. 4.4.3, power-law covariances can be constructed by integrating covariances with continuously increasing integral scales. An infinite sum of equally weighted exponential correlations results in a $1/x$-type correlation (see Remark 4.8 of Sect. 4.4.3). Such correlations lead to the $\sim(t \ln t - t)$ behavior of the ensemble variance Σ_{ii} (see Sect. 4.3.3). Dispersion coefficients $\Sigma_{ii}/(2t)$ fitted with $\ln t$ and variances Σ_{ii} with $t \ln t - t$ are shown in Figs. 5.19 and 5.20. At a first sight these fitting results are also acceptable. Nevertheless, a more detailed analysis is required

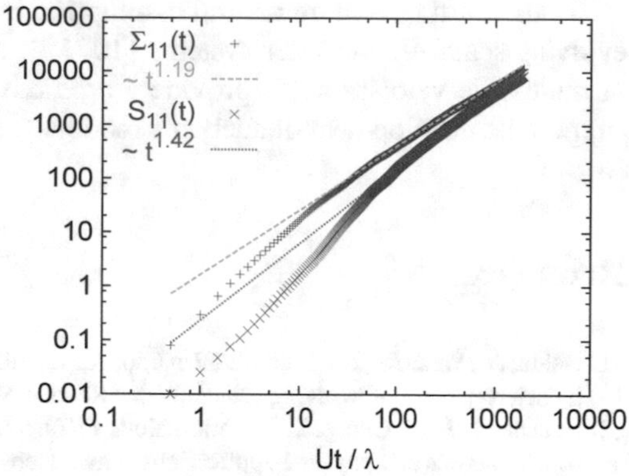

Fig. 5.18 Power-law fit between 10 and 500 days of longitudinal ensemble and effective variances; standard fit-errors for power-law exponents of 0.14% and 0.08%, respectively ($\lambda_7 = 28$ m)

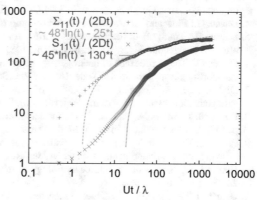

Fig. 5.19 Fit of longitudinal ensemble and effective coefficients with $a \ln t - b$ between 10 and 500 days; standard fit-errors for the coefficients $\{a, b\}$ of $\{0.30\%, 3.16\%\}$ and $\{0.36\%, 0.68\%\}$, respectively ($\lambda_7 = 28$ m)

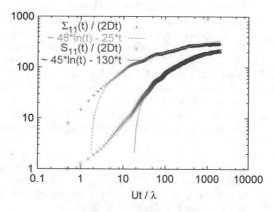

Fig. 5.20 Fit of longitudinal ensemble and effective variances with $(at \ln t - bt)$ between 10 and 500 days; standard fit-errors for the coefficients $\{a, b\}$ of $\{0.70\%, 8.46\%\}$ and $\{0.22\%, 0.40\%\}$, respectively ($\lambda_7 = 28$ m)

to decide which time behavior, the power-law one or the $\ln t$-type behavior provides the better fit with the results of the numerical experiment.

There is a long-standing attempt to introduce power law correlations and anomalous dispersion in modeling groundwater transport by rather abstract mathematical approaches [1, 3, 6, 7, 9, 17, 25]. Nevertheless, a superposition of a finite number of scales could be a more natural assumption in many contamination scenarios for evolving scale groundwater systems [10, 15]. Numerical simulations of transport in multiscale velocity fields provide a predictive tool for such scenarios and may support the development of adequate models.

References

1. Bellin, A., Pannone, M., Fiori, A., Rinaldo, A.: On transport in porous formations characterized by heterogeneity of evolving scales. Water Resour. Res. **32**, 3485–3496 (1996)
2. Bouchaud, J.-P., Georges, A.: Anomalous diffusion in disordered media: statistical mechanisms, models and physical applications. Phys. Rep. **195**, 127–293 (1990)
3. Cintoli, S., Neuman, S.P., Di Federico, V.: Generating and scaling fractional Brownian motion on finite domains. Geophys. Res. Lett. **32**, L08404 (2005)
4. Dagan, G.: Flow and Transport in Porous Formations. Springer, Berlin (1989)
5. Dagan, G.: Transport in heterogeneous porous formations: spatial moments, ergodicity, and effective dispersion. Water Resour. Res. **26**(6), 1281–1290 (1990)
6. Dagan, G.: The significance of heterogeneity of evolving scales and of anomalous diffusion to transport in porous formations. Water Resour. Res. **30**, 3327–3336 (1994)
7. Di Federico, V., Neuman, S.P.: Scaling of random fields by means of truncated power variograms and associated spectra. Water Resour. Res. **33**, 1075–1085 (1997)
8. Eberhard, J., Suciu, N., Vamos, C.: On the self-averaging of dispersion for transport in quasi-periodic random media. J. Phys. A: Math. Theor. **40**, 597–610 (2007)
9. Fiori, A.: On the influence of local dispersion in solute transport through formations with evolving scales of heterogeneity. Water Resour. Res. **37**(2), 235–242 (2001)
10. Gelhar, L.W.: Stochastic subsurface hydrology from theory to applications. Water Resour. Res. **22**(9S), 135S–145S (1986)
11. Gelhar, L.W., Axness, C.: Three-dimensional stochastic analysis of macrodispersion in aquifers. Water Resour. Res. **19**(1), 161–180 (1983)
12. Jeon, J.-H., Metzler, R.: Fractional Brownian motion and motion governed by the fractional Langevin equation in confined geometries. Phys. Rev. E **81**, 021103 (2010)
13. Kraichnan, R.H.: Diffusion by a random velocity field. Phys. Fluids **13**(1), 22–31 (1970)
14. Liu, H.H., Molz, F.J.: Block scale dispersivity for heterogeneous porous media characterized by stochastic fractals. Geophys. Res. Lett. **24**(17), 2239–2242 (1997)
15. McLaughlin, D., Ruan, F.: Macrodispersivity and large-scale hydrogeologic variability. Transp. Porous Media **42**, 133–154 (2001)
16. Sposito, G., Jury, W.A., Gupta, V.K.: Fundamental problems in the stochastic convection-dispersion model of solute transport in aquifers and field soils. Water Resour. Res. **22**(1), 77–88 (1986)
17. Suciu, N.: Spatially inhomogeneous transition probabilities as memory effects for diffusion in statistically homogeneous random velocity fields. Phys. Rev. E **81**, 056301 (2010)
18. Suciu, N., Knabner, P.: Comment on 'Spatial moments analysis of kinetically sorbing solutes in aquifer with bimodal permeability distribution' by M. Massabo, A. Bellin, and A. J. Valocchi. Water Resour. Res. **45**, W05601 (2009)
19. Suciu, N., Vamoş, C.: Comment on "Nonstationary flow and nonergodic transport in random porous media" by G. Darvini and P. Salandin. Water Resour. Res. **43**, W12601 (2007)
20. Suciu, N., Vamoş, C.: Ergodic estimations of upscaled coefficients for diffusion in random velocity fields. In: L'Ecuyér, P., Owen, A.B. (eds.) Monte Carlo and Quasi-Monte Carlo Methods 2008, pp. 617–626. Springer, Berlin (2009)

21. Suciu, N., Vamoş, C., Vanderborght, J., Hardelauf, H., Vereecken, H.: Numerical investigations on ergodicity of solute transport in heterogeneous aquifers. Water Resour. Res. **42**, W04409 (2006)
22. Suciu, N., Vamoş, C., Eberhard, J.: Evaluation of the first-order approximations for transport in heterogeneous media. Water Resour. Res. **42**, W11504 (2006)
23. Suciu, N., Vamos, C., Vereecken, H., Sabelfeld, K., Knabner, P.: Ito equation model for dispersion of solutes in heterogeneous media. Rev. Anal. Numer. Theor. Approx. **37**, 221–238 (2008)
24. Suciu, N., Vamoş, C., Vereecken, H., Sabelfeld, K., Knabner, P.: Memory effects induced by dependence on initial conditions and ergodicity of transport in heterogeneous media. Water Resour. Res. **44**, W08501 (2008)
25. Suciu, N., Vamos, C., Radu, F.A., Vereecken, H., Knabner, P.: Persistent memory of diffusing particles. Phys. Rev. E **80**, 061134 (2009)
26. Suciu, N., Attinger, S., Radu, F.A., Vamoş, C., Vanderborght, J., Vereecken, H., Knabner, P.: Solute transport in aquifers with evolving scale heterogeneity. Preprint No. 346, Mathematics Department—Friedrich-Alexander University Erlangen-Nuremberg (2011)
27. Suciu, N., Attinger, S., Radu, F.A., Vamoş, C., Vanderborght, J., Vereecken, H. Knabner, P.: Solute transport in aquifers with evolving scale heterogeneity. An. Sti. U. Ovid. Co-Mat. **23**(3), 167–186 (2015)
28. Vamoş, C., Suciu, N., Vereecken, H., Vanderborght, J., Nitzsche, O.: Path decomposition of discrete effective diffusion coefficient. Internal Report ICG-IV 00501, Research Center Jülich (2001)
29. Yaglom, A.M.: Correlation Theory of Stationary and Related Random Functions, Volume I: Basic Results. Springer, New York (1987)

Chapter 6
Probability and Filtered Density Function Approaches

Abstract Beyond mean and variance, traditionally used in stochastic approaches, the full one-point one-time concentration probability density function (PDF) is needed to estimate exceedance probabilities in assessments of groundwater contamination. By solving PDF evolution equations one avoids the cumbersome MC simulations used to obtain statistical inferences. The PDF approach is mainly useful in case of reactive transport: because reaction terms are in a closed form, there is no need to upscale them, as in case of modeling the mean behavior of species concentrations. In a filtered density function (FDF) approach, the PDF is estimated by spatial filtering. PDF/FDF equations will be formulated as Fokker–Planck equations with solutions in the Cartesian product of physical and concentration spaces. Numerical solutions will be obtained by GRW algorithms.

6.1 PDF/FDF Evolution Equations

6.1.1 Background on PDF/FDF Methods

PDF methods were originally developed in the context of modeling turbulent reacting flows as a powerful tool to close highly nonlinear terms arising from averaging chemical reaction rates [17, 18, 38, 40, 41]. The PDF approach is based on solving evolution equations for the one-dimensional (one-point one-time) Eulerian joint PDF of sets of variables describing the state of the system, for example, velocity, chemical composition, turbulent frequency, temperature, or enthalpy. PDF equations are derived by different methods [38, 40] from local balance equations governing the flow and the evolution of the thermochemical state of the system. The PDF equations are unclosed because they contain terms that cannot be determined by the one-dimensional PDF alone [40]. But, irrespective of the complexity of the set of state variables, the chemical source terms in the PDF equations are closed, namely they have the same functional form as the unaveraged reaction rates in the local balance equations [19]. The unclosed terms which require modeling are those describing the turbulent frequency, the scalar mixing by molecular diffusion,

© Springer Nature Switzerland AG 2019
N. Suciu, *Diffusion in Random Fields*, Geosystems Mathematics,
https://doi.org/10.1007/978-3-030-15081-5_6

and, when the velocity is not included among the state variables, the effects of the turbulent velocity fluctuations [17, 19].

Filtered density function (FDF) methods provide an alternative approach using spatially filtered quantities, that is, spatial averages, instead of stochastic averages on which the PDF methods are based [8, 19, 23, 29]. The meaning of the FDF is that of a PDF of state variables at scales smaller than the filter width [13, 20]. FDF evolution equations can be derived by similar procedures and have the same structure, with source terms in closed form, as the PDF equations [8, 18, 19]. With rare exceptions, the only state variables considered in FDF methods are the scalars describing the thermochemical composition [19]. Unlike the PDF, which is a deterministic function, the FDF is a random quantity. Its expectation tends to the PDF only in the limit of a small filter width [19]. However, since, by invoking an ergodic theorem, the ensemble averaging may be seen as a filtering in space with a filter width much smaller than the domain but larger than the characteristic length of large-scale motions, the FDF should approach the PDF in the limit of large filter width [20]. Even though this assertion has not been proved theoretically, in some cases the convergence may be demonstrated numerically, as we will see below in this chapter.

In modeling transport through highly heterogeneous natural groundwater systems, the randomness is introduced by the stochastic parameterization of the hydraulic conductivity, which accounts for the parameter uncertainty due to a lack of measurements. This stochasticity implies the randomness of the Darcy flow velocity and of the dependent variables of the transport equations [46]. Randomness in modeling groundwater flows can be enhanced by considering uncertain sources in the flow equations (random recharge) [37] or, for transient flows, random parameterizations of storage coefficients [1]. However, the uncertainty of the hydraulic conductivity is an omnipresent source of randomness. For this reason, only this source of randomness will be considered in the following. While in earlier stochastic approaches the focus was mainly on the mean and in some cases the variance of the concentration, during the last decade the need to model the concentration PDF received an increased attention [7, 12, 30, 44, 46, 47, 50, 58]. Concentration PDFs of conserved scalars may be inferred without solving PDF evolution equations in case of small or moderate fluctuations of the hydraulic conductivity, modeled as a lognormal random function with finite correlation lengths. Then, a Gaussian shape of the concentration may be assumed or inferred, which is completely determined by its first two moments. The statistics of these moments, specified under various assumptions in a first order perturbation approach, is finally "mapped," using the Gaussian functional shape of the concentration, onto the concentration PDF via numerical [46] or analytical techniques [7, 12]. Another favorable situation is that of stratified transport, when the Gaussian concentration can be expressed explicitly as a function of the hydraulic conductivity, with the only assumption of negligible transverse dispersion. This leads to an explicit functional dependence of the concentration PDF on the hydraulic conductivity. Moreover, this approach provides the PDF of reacting chemical species if their concentrations are fully defined by monotonous functions of conserved scalars [44].

For advective–reactive transport, PDFs of reacting species can be computed by solving evolution equations similar to those used in turbulence [58]. Such PDF equations do not contain mixing terms, because molecular diffusion is neglected. The only closure problem concerns terms due to velocity fluctuations, which are modeled as effective, or upscaled diffusion coefficients, leading to Fokker–Planck evolution equations [50, 58]. By considering the velocity among the state variables, velocity–concentration PDF equations similar to those used in turbulence theory can be derived and no closure for velocity fluctuations is necessary. Mixing is modeled similarly to turbulence approaches and the concentration PDF is obtained by integrating the joint velocity–concentration PDF over the velocity state space [30]. Evolution equations of the concentration PDF weighted by a conserved scalar, which generalizes the mass density function used in turbulence [40], can be formulated as Fokker–Planck equations [50]. Closures are provided by stochastically upscaled diffusion coefficients [47, 50] and by mixing models, which are formulated as a diffusion in concentration space. The parameters for this diffusion process can be inferred from measured/simulated concentration time series [51].

While several PDF approaches were proposed [30, 50, 58], the FDF method of modeling transport in groundwater systems is still in an infancy stage. The feasibility of the FDF approach in stochastic subsurface hydrology proposed in [52] is supported by some similarities between large eddy simulations (LES) in turbulence [8, 13, 23] and some approaches to coarse-scale, or coarse-grained simulations (CGS) in porous media [2–4, 14–16]. In both cases, spatial averages of dependent variables are used to coarsen the grid, whereas subgrid effects are modeled. The objective is to obtain results comparable to fine-grid simulations at reduced computational costs. In case of reactive transport, the upscaled equations obtained by spatial filtering contain unclosed averaged reaction terms. In LES, the closure problem is solved by coupling the filtered equations to FDF evolution equations [13]. In subsurface hydrology, effective reaction rates needed to close the problem can only be determined for specific problems under simplifying assumptions [22]. FDF approaches can be used to avoid the need to close the filtered reaction terms and CGS can be designed, based on numerical upscaling through volume averages [14, 16], multiscale finite element methods [15], or, similarly to LES, by solving filtered equations [3, 4].

Nevertheless, there are three major differences with respect to LES-FDF modeling for turbulent flows. The first one concerns the number of parameters. While only a few parameters are required to solve filtered LES equations [8, 23], upscaling flow and transport processes in groundwater, by either spatial or stochastic averaging, requires fields of parameters: the hydraulic conductivity [4, 14] or the velocity field [22, 47]. The second difference, related to the first one, is the origin of the randomness. Turbulent flows are governed by the deterministic Navier–Stokes equations but, for large Reynolds numbers, the flow velocity behaves as a random variable due to the sensible dependence of the solution on initial and boundary conditions. This mathematical aspect corresponds to an experimental lack of reproducibility of the measurements in turbulent systems [40]. In groundwater systems, the spatial variability of the hydraulic conductivity cannot be completely described. Therefore, stochastic parameterizations by random space functions are used to account for this uncertainty. The flow equations are thus solved in a probabilistic sense [7]. In this case, randomness is caused by the uncertainty of the parameter fields propagated through the flow and transport equations, which have to be modeled as stochastic equations. The third difference is given by the available experimental data. In turbulence, detailed velocity, temperature, or concentration profiles are available from measurements. Consequently, in LES-FDF simulations the objective is to obtain a good agreement with the available measurements, at scales lying between that of the fine-scale simulations, which fully resolve the variability of the turbulent flows, and that of the ensemble averaged solutions of the Navier–Stokes and transport equations, for which the variability of the unresolved scales is modeled by the solution of the PDF equation [13, 20]. In natural porous media the measurements are sparse and subject to uncertainty. Therefore, as long as the simulation relays on fields of stochastic parameters (hydraulic conductivity or velocity) and no detailed experimental data are available, the aim of the FDF approach should be a probabilistic quantification of the uncertainty for the whole hierarchy of scales. In other words, FDF approaches for modeling transport processes in groundwater systems, besides closing reaction terms, could alleviate the computational costs through estimations of the global PDF from filtered parameters and coarse-grained FDF simulations.

6.1.2   PDF/FDF Equations for Reactive Transport

Let us consider in the following the extension to reactive transport of the model of diffusion in random fields introduced in Sect. 4.3. The groundwater flow is modeled as a statistically homogeneous random velocity field $\mathbf{V}(\mathbf{x})$ with divergence-free samples. This field is a solution of the continuity and Darcy equations (4.78) with a random parameterization of the hydraulic conductivity. Furthermore, the local hydrodynamic dispersion in saturated aquifers and the molecular diffusion are modeled by an isotropic diffusion process specified by a constant diffusion coefficient D. A system of reacting chemical species described by concentrations $C_\alpha(\mathbf{x}, t) \in Y_c, \mathbf{x} \in Y_x, t \in \mathbb{R}_+, \alpha = 1, \ldots, N_\alpha$, is transported through the aquifer according to the system of balance equations

$$\frac{\partial C_\alpha}{\partial t} + V_i \frac{\partial C_\alpha}{\partial x_i} = D \frac{\partial^2 C_\alpha}{\partial x_i \partial x_i} + S_\alpha, \tag{6.1}$$

where $S_\alpha(\mathbf{C})$ denote the reaction rates. Since the velocity field $\mathbf{V}$ is a random function, the vector concentration $\mathbf{C}(\mathbf{x}, t)$ is a random field as well.

The marginal, one-point one-time probability density function $f(\mathbf{c}; \mathbf{x}, t)$ of the random concentration $\mathbf{C}$ solving (6.1) satisfies the PDF evolution equation

$$\frac{\partial f}{\partial t} + \frac{\partial}{\partial x_i}(\mathscr{V}_i f) - \frac{\partial^2}{\partial x_i \partial x_j}(\mathscr{D}_{ij} f) = -\frac{\partial^2}{\partial c_\alpha \partial c_\beta}(\mathscr{M}_{\alpha\beta} f) - \frac{\partial}{\partial c_\alpha}(S_\alpha f), \qquad (6.2)$$

where $\mathscr{V}$ and $\mathscr{D}$ are the stochastically upscaled drift vector and the diffusion tensor, respectively, and $\mathscr{M}$ is the conditional dissipation rate accounting for mixing by diffusion [50–52]. As a common convention, a semicolon is used in writing the concentration PDF $f(\mathbf{c}; \mathbf{x}, t)$ to emphasize the distinction between the value $\mathbf{c}$ taken by the random function $\mathbf{C}(\mathbf{x}, t)$ in the state space Y_c and the values $\mathbf{x}$ and t of its independent variables in $Y_x \times \mathbb{R}_+$ [31, 40].

The mostly used approaches to derive the PDF equation (6.2) are the delta function method and the test function method. The first one starts with the definition of the PDF given by the ensemble average of some delta function depending on random fields. For instance, the concentration PDF is given by $f(\mathbf{c}; \mathbf{x}, t) = \langle \delta(\mathbf{C}(\mathbf{x}, t) - \mathbf{c}) \rangle$, where the multidimensional δ function is defined by the product $\delta(\mathbf{C}(\mathbf{x}, t) - \mathbf{c}) = \prod_{\alpha=1}^{\alpha=N_\alpha} \delta(C_\alpha(\mathbf{x}, t) - c_\alpha)$. Delta functions can be used to define consistent probability distributions (see Sect. 2.1.3) and formal calculus involving them corresponds to rigorous operations with Dirac functionals [26, 50, 51]. The PDF equation is obtained by evaluating $\partial f/\partial t$ from formal derivatives of δ functions [18, 38, 44, 51] (see Appendix E.1). In the test function approach, the ensemble average of the operator $\partial/\partial t + V_i \partial/\partial x_i$ applied to a test function of state variables Q, with suitable properties, is evaluated in two different ways: first, by interchanging derivatives and stochastic averages and using the incompressibility of the velocity field, and second, by multiplying the right-hand side of (6.1) by Q and taking the ensemble average. The PDF equation follows by equating the two expressions for $\langle \partial Q/\partial t + V_i \partial Q/\partial x_i \rangle$, performing integration by parts, and considering the vanishing of Q at the boundaries of Y_c [17, 18, 40, 50].

In LES approaches, spatial filters are used to separate the dynamics of the scales larger than the filter width from sub-filter effects. The former are obtained as solutions of filtered equations and the latter, corresponding to unresolved scales, are modeled [18–20]. The filtered value of a physical quantity Q is given by the spatial average

$$\langle Q \rangle_\lambda (\mathbf{x}, t) = \int_{Y_x} Q(\mathbf{x}', t) G(\mathbf{x}' - \mathbf{x}) d\mathbf{x}', \qquad (6.3)$$

with λ being the filter width, which is implicitly defined by the filter function $G(\mathbf{x}' - \mathbf{x})$. The filter $G(\mathbf{x}' - \mathbf{x})$ is spatially invariant, non-negative, and normalized to unity, $\int_{Y_x} G(\mathbf{x}' - \mathbf{x}) d\mathbf{x}' = 1$. Furthermore, the filtering operation commutes with differentiation [59]. Under these conditions, the FDF can be defined by $f_\lambda(\mathbf{c}; \mathbf{x}, t) = \langle \delta(\mathbf{C}(\mathbf{x}, t) - \mathbf{c}) \rangle_\lambda$ and FDF equations can be derived, analogous to PDF equations, by the delta function method [18]. The FDF equation has the form of PDF equation (6.2), but its coefficients are now obtained by spatial averages (6.3) instead of ensemble averages. Note that, technically, neither the derivation of the PDF equation nor the derivation of the FDF equation requires the statistical homogeneity of the random velocity field [50, 51]. Nevertheless, there is an important difference between PDF and FDF approaches. While in the case of the PDF approach to transport in groundwater the statistical homogeneity of the random velocity is essential for the existence of the stochastically upscaled coefficients [34], spatially filtered upscaled coefficients to be used in FDF equations can be derived by methods free of homogeneity assumptions [14, 16]. This opens the perspective for FDF methods applicable to realistic situations, such as transport simulations conditional on measurements of hydraulic conductivity [35].

Because of the high dimensionality of the PDF/FDF equations (time and space dimensions and the N_α dimensions of the concentration space Y_c) solutions by standard discretization methods (finite differences or finite elements) are impracticable [18, 40]. Therefore, numerical solutions are usually obtained by MC methods. "Eulerian particle methods" simulate the finite difference solution of the PDF equation by locating an ensemble of N notional particles at each point of an Eulerian grid. These particles have assigned representative values of the state variable (e.g., concentration), initially distributed according to the initial PDF. The particles move on the grid following rules consistent with the finite difference scheme and, as N tends to infinity, the average over ensembles of particles converges to the expectation estimated by the finite difference scheme [39]. Even if they are computationally simpler than other particle methods in use, Eulerian particle methods are numerically dissipative and have low spatial accuracy [18, 33]. Another MC approach is the "stochastic Eulerian field method" which uses a representation of the PDF by an ensemble of stochastically equivalent space–time random fields, with the same one-point one-time PDF as the solution of the PDF equation [13, 24, 36]. Stochastic fields are governed by partial differential equations with a linear multiplicative noise term, interpreted in either Itô [53] or Stratonovich [43] sense. The computational effort of solving a partial differential equation for each field renders the method of stochastic Eulerian field less competitive than other MC methods when large numbers of fields are required [18]. "Lagrangian particle methods" use systems of stochastic particles moving in continuous space, according to grid-free PT procedures. Eventually, grids are used to compute averages and to interpolate the output of averaged or filtered transport equations to the particles positions. Lagrangian methods are now the dominant numerical approach for PDF equations. The performance of the above MC approaches, with several variants, is analyzed in [18].

6.1.3   The Fokker–Planck Approach

6.1.3.1   Direct Fokker–Planck Approach

Lagrangian particle methods are based on the similarity between PDF/FDF equations of type (6.2) and Fokker–Planck equations. It is common to compare [8, 23,

30], or even to assimilate [59] PDF/FDF equations to Fokker–Planck equations and to use the associated Itô equations as a model for Lagrangian particles. Further, numerical solutions to Eq. (6.2) are constructed by imposing a uniform distribution of Lagrangian particles during the simulations. To understand this constraint, let us consider the concentration PDF problem and note that the corresponding PDF fulfills the normalization condition $\int_{Y_c} f(\mathbf{c}; \mathbf{x}, t)d\mathbf{c} = 1$. On the other side, if (6.2) were a Fokker–Planck equation, its solution would be a PDF $p(\mathbf{c}, \mathbf{x}, t)$ defined on the concentration-position state space $Y = Y_c \times Y_x$ [50], which by integration over Y_c yields $\int_{Y_c} p(\mathbf{c}, \mathbf{x}, t)d\mathbf{c} = p_x(\mathbf{x}, t)$, that is, to the position PDF of the Lagrangian particles. A uniform particle distribution, $p_x(\mathbf{x}, t) \equiv const$, would suffice to make $p(\mathbf{c}, \mathbf{x}, t)$ proportional to $f(\mathbf{c}; \mathbf{x}, t)$, which allows estimating the concentration PDF from the solution of the Fokker–Planck equation. As it will be shown in the following, a general relation between the two PDFs, $f(\mathbf{c}; \mathbf{x}, t)$ and $p(\mathbf{c}, \mathbf{x}, t)$, which renders them consistent with the same normalization condition may be established if a suitable weighting function Θ exists such that $\int_{Y_c} \Theta(\mathbf{c}) f(\mathbf{c}; \mathbf{x}, t)d\mathbf{c} = p_x(\mathbf{x}, t)$.

In operator splitting schemes the transport of the PDF in physical space is treated in separate advection and diffusion steps [39]. These two steps solve Eq. (6.2) with the right-hand side set to zero [40]. This equation has the form of a Fokker–Planck equation describing the position PDF of a passive scalar. The corresponding Itô equation, used to simulate the transport step in Lagrangian particle methods [60], has the form

$$dX_i(t) = \mathcal{V}_i(\mathbf{X}, t)dt + d\tilde{W}_i(\mathbf{X}, t),\tag{6.4}$$

where $\{X_i, i = 1, 2, 3\}$ are trajectories of an Itô diffusion process and $\tilde{W}_i$ is a Wiener process with expectation $E\{\tilde{W}_i(\mathbf{X}, t)\} = 0$ and variance $E\{\tilde{W}_i^2(\mathbf{X}, t)\} = 2\int_0^t \mathscr{D}_{ii}(\mathbf{X}, t')dt'$. The other fractional steps of the Lagrangian approach [40, 60] correspond to the transport in concentration space and may be formulated in a general way as

$$dC_\alpha(t) = M_\alpha(C_\alpha(t))dt + S_\alpha(\mathbf{C}(t))dt,\tag{6.5}$$

where $C_\alpha(t) = C_\alpha(\mathbf{X}(t))$ and the coefficients M_α are provided by mixing models for the term containing the dissipation rate $\mathscr{M}_{\alpha\beta}$ in Eq. (6.2) [50].

To design a MC method based on Eqs. (6.4) and (6.5), one needs a correspondence between the two mathematical objects involved, namely: the stochastic process $\{\mathbf{C}(t), \mathbf{X}(t)\}$, indexed by a single index, t, and the multi-index random function $\mathbf{C}(\mathbf{x}, t)$. To proceed, one remarks that the Itô equation (6.4) and the corresponding Fokker–Planck equation [27] for the position PDF p_x,

$$\frac{\partial p_x}{\partial t} + \frac{\partial}{\partial x_i}(\mathcal{V}_i p_x) = \frac{\partial^2}{\partial x_i \partial x_j}(\mathscr{D}_{ij} p_x),\tag{6.6}$$

do not depend on concentrations (i.e., on the process $C_\alpha(t)$ or on the state space variable $\mathbf{c}$) and may be used to describe any conserved scalar transported in the same system (with the same parameters $\mathcal{V}_i$ and $\mathcal{D}_{ij}$), under the same initial conditions. Equation (6.6) also coincides with the equation satisfied by the ensemble averaged [30] or by the filtered scalar [59], which can be derived by multiplying the PDF/FDF equation (6.2) without the reaction term by the scalar and by taking the ensemble average, or the filter average. It follows that for any conserved scalar $\Theta(\mathbf{x}, t)$ solving Eq. (6.1) without reaction terms, the ensemble averaged $\langle \Theta \rangle$ (or the filtered scalar $\langle \Theta \rangle_\lambda$) solves the Fokker–Planck equation (6.6). Thus, we have,

$$\langle \Theta \rangle (\mathbf{x}, t)/\Theta^* = p_x(\mathbf{x}, t), \quad \text{where} \quad \Theta^* = \int_{Y_x} \langle \Theta \rangle (\mathbf{x}, t) d\mathbf{x}. \tag{6.7}$$

Equation (6.7) also holds for $\langle \Theta \rangle$ replaced by $\langle \Theta \rangle_\lambda$.

Next, let us consider a conserved scalar depending on $(\mathbf{x}, t)$ through concentrations of chemical species, $\Theta(\mathbf{x}, t) = \Theta(\mathbf{C}(\mathbf{x}, t))$. Making use of the definition of the PDF via delta functions (see Sect. 2.1.3), Eq. (6.7) can be rewritten as

$$\langle \Theta(\mathbf{C}(\mathbf{x}, t)) \rangle /\Theta^* = \frac{1}{\Theta^*} \left\langle \int_{Y_c} \Theta(\mathbf{c})\delta(\mathbf{C}(\mathbf{x}, t) - \mathbf{c}) d\mathbf{c} \right\rangle$$

$$= \frac{1}{\Theta^*} \int_{Y_c} \Theta(\mathbf{c}) f(\mathbf{c}; \mathbf{x}, t) d\mathbf{c}$$

$$= \int_{Y_c} p(\mathbf{c}, \mathbf{x}, t) d\mathbf{c}. \tag{6.8}$$

The last equality in (6.8) is ensured by choosing

$$\Theta(\mathbf{c}) f(\mathbf{c}; \mathbf{x}, t) = \Theta^* p(\mathbf{c}, \mathbf{x}, t). \tag{6.9}$$

Relation (6.9) provides a correspondence between the one-point statistics of the random concentration $\mathbf{C}(\mathbf{x}, t)$ and that of the stochastic process $\{\mathbf{C}(t), \mathbf{X}(t)\}$. Hence, the normalized scalar Θ/Θ^* is the weighting function we were looking for in the beginning. Such a correspondence has been introduced in a somewhat heuristic manner in [50] and further analyzed in [51, 52].

With $p(\mathbf{c}|\mathbf{x}, t) = p(\mathbf{c}, \mathbf{x}, t)/p_x(\mathbf{x}, t)$ being the conditional PDF of the concentration given the position of the stochastic process (6.4) and (6.5) and using (6.7), relation (6.9) becomes equivalent to

$$\frac{\Theta(\mathbf{c})}{\langle \Theta \rangle (\mathbf{x}, t)} f(\mathbf{c}; \mathbf{x}, t) = p(\mathbf{c}|\mathbf{x}, t). \tag{6.10}$$

Thus, relation (6.9) not only solves the normalization issue, but it also determines the one-point statistics of the weighted concentration PDF by that of the stochastic process (6.4) and (6.5), according to (6.10). It is worth mentioning that the PDF $f(\mathbf{c}; \mathbf{x}, t)$ of the random concentration cannot be uniquely determined by relation

(6.9). Instead, based on (6.10), the system of Itô equations (6.4) and (6.5) can be used to compute the concentration PDF weighted by a normalized conserved scalar.

The conserved scalar Θ can be chosen as the sum of all species concentrations composing the reaction system, $\Theta = \sum_{\alpha=1}^{N_\alpha} C_\alpha$. This sum is conserved in closed systems, as a consequence of mass conservation of the total amount of chemical elements contained in the reacting species molecules [51]. This statement can be proved by a slight extension of the method used by Bilger [6] to construct conserved scalars as concentrations of chemical elements. Let $r_{\alpha k}$ be the weight (e.g., the mass fraction) of the chemical element indexed by k in the composition of the molecules α and let C_k be the total concentration of the element k. Obviously, the elemental masses sum to unity, $\sum_{k=1}^{N_k} r_{\alpha k} = 1$, and $C_k = \sum_{\alpha=1}^{N_\alpha} r_{\alpha k} C_\alpha$. It follows that

$$\sum_{k=1}^{N_k} C_k = \sum_{k=1}^{N_k} \sum_{\alpha=1}^{N_\alpha} r_{\alpha k} C_\alpha = \sum_{\alpha=1}^{N_\alpha} C_\alpha \sum_{k=1}^{N_k} r_{\alpha k} = \sum_{\alpha=1}^{N_\alpha} C_\alpha,$$

that is, the sum of elemental concentrations equals the sum of species concentrations. Since elemental concentrations are conserved under chemical reactions, the sum of species concentrations is a conserved scalar. Furthermore, summing up Eq. (6.1) with species independent coefficients, one obtains the relation $\sum_{\alpha=1}^{N_\alpha} S_\alpha = 0$.

In particular, if the transport problem is formulated in terms of mass concentrations, $C_\alpha = \rho_\alpha$, and all the components of the fluid system are included among the N_α species, we have $\Theta = \sum_{\alpha=1}^{N_\alpha} \rho_\alpha = \rho$, where ρ is the fluid density. Then, according to (6.7), $\Theta^* = M$ is the total mass of fluid in Y_x, $\langle \rho \rangle(\mathbf{x}, t) = M p_x(\mathbf{x}, t)$, and the correspondence relation (6.9) takes the form

$$\rho(\mathbf{c}) f(\mathbf{c}; \mathbf{x}, t) = M p(\mathbf{c}, \mathbf{x}, t), \tag{6.11}$$

which relates the mass density function $\mathscr{F}(\mathbf{c}, \mathbf{x}; t) = \rho(\mathbf{c}) f(\mathbf{c}; \mathbf{x}, t)$ to the PDF $p(\mathbf{c}, \mathbf{x}, t)$ of the system of stochastic particles used in Lagrangian MC solution algorithms [40].

To facilitate the derivation of the FDF evolution equation, the filtered mass density function is usually written as $\mathscr{F}_\lambda(\mathbf{c}, \mathbf{x}; t) = \int_{Y_x} \rho(\mathbf{x}', t)\delta(\mathbf{C}(\mathbf{x}', t) - \mathbf{c})G(\mathbf{x}' - \mathbf{x})d\mathbf{x}'$ [18, 23]. With $\rho(\mathbf{x}, t) = \rho(\mathbf{C}(\mathbf{x}, t))$, using the FDF definition by the filtering operation (6.3) applied to $\delta(\mathbf{C}(\mathbf{x}, t) - \mathbf{c})$ and the "shifting property" $\rho(\mathbf{c})\delta(\mathbf{C}(\mathbf{x}, t) - \mathbf{c}) = \rho(\mathbf{C}(\mathbf{x}, t))\delta(\mathbf{C}(\mathbf{x}, t) - \mathbf{c})$ [51], it is easy to recognize that $\mathscr{F}_\lambda(\mathbf{c}, \mathbf{x}; t) = \rho(\mathbf{c}) f_\lambda(\mathbf{c}; \mathbf{x}, t)$. Equation (6.8) still holds when the ensemble average is replaced by the spatial filtering and the FDF version of the correspondence relation is given by (6.11), with f replaced by f_λ.

If the weighting function Θ obeys the continuity equation

$$\frac{\partial \Theta}{\partial t} + V_i \frac{\partial \Theta}{\partial x_i} = 0, \tag{6.12}$$

then, by using the correspondence (6.9), the Fokker–Planck equation governing the concentration-position PDF $p(\mathbf{c}, \mathbf{x}, t)$ of the Itô process with trajectories given by (6.4) and (6.5) may be derived directly from the transport equations (6.1). To do so, either the test function [50] approach or the delta function [51] approach, presented in Appendix E.2 can be used. This transport equation, also satisfied by weighted concentration PDFs (in particular by $\mathscr{F}$ and $\mathscr{F}_\lambda$), is

$$
\frac{\partial p(\mathbf{c}, \mathbf{x}, t)}{\partial t} + \langle V_i \rangle \frac{\partial}{\partial x_i} p(\mathbf{c}, \mathbf{x}, t) + \frac{\partial}{\partial x_i} \left[\langle U_i | \mathbf{c} \rangle \, p(\mathbf{c}, \mathbf{x}, t) \right]
$$

$$
= -\frac{\partial}{\partial c_\alpha} \left\{ \left[\left\langle D \frac{\partial^2 C_\alpha}{\partial x_i \partial x_i} \middle| \mathbf{c} \right\rangle + S_\alpha(\mathbf{c}) \right] p(\mathbf{c}, \mathbf{x}, t) \right\}. \tag{6.13}
$$

Equation (6.13) contains the conditional averages given the concentration of the velocity fluctuation $\mathbf{U} = \mathbf{V} - \langle \mathbf{V} \rangle$ and of the diffusion flux $D \frac{\partial^2 C_\alpha}{\partial x_i \partial x_i}$. The first one is traditionally closed by a gradient-diffusion hypothesis $\langle U_i \mid \mathbf{c} \rangle f = -D_{i,j}^* \partial f / \partial x_j$ [17, 40]. The second one can be related to the conditional dissipation rate, but only if the weighting factor Θ is a constant [50]. Hence, for $\Theta = const$, the two closures introduce the coefficients [50]

$$
\mathscr{V}_i = \langle V_i \rangle + \frac{\partial}{\partial x_j} D_{i,j}^*, \quad \mathscr{D}_{ij} = D + D_{i,j}^*, \quad \mathscr{M}_{\alpha\beta} = \left\langle D \frac{\partial C_\alpha}{\partial x_i} \frac{\partial C_\beta}{\partial x_i} \middle| \mathbf{c} \right\rangle. \tag{6.14}
$$

For $\Theta = const$, (6.7) implies a uniform position PDF equal to the inverse of the volume of the physical domain, $\Theta / \Theta^* = p_x(\mathbf{x}, t) \equiv 1 / \int_{Y_x} d\mathbf{x}$, and from (6.10) one obtains

$$
f(\mathbf{c}; \mathbf{x}, t) = p(\mathbf{c} | \mathbf{x}, t). \tag{6.15}
$$

Further, with (6.14) and (6.16) inserted into (6.13) one obtains the PDF equation (6.2) [50].

In turbulence applications, the $D_{i,j}^*$ component of the upscaled diffusion coefficients $\mathscr{D}_{ij}$ defined in (6.14) is an isotropic tensor, called "turbulent diffusion coefficient," which may vary in space and time [40]. If only the gradient-diffusion closure is considered in (6.13), without transforming the conditional diffusion flux, then $\mathscr{D}_{ij} = D_{i,j}^*$ and in the PDF/FDF equation (6.2) the mixing term has to be replaced by the first term in the right-hand side of (6.13) [17, 40, 41]. In models for solute plumes migrating in groundwater systems and for statistically homogeneous velocity fields, considered here, $\mathscr{D}_{ij}$ are the "ensemble dispersion coefficients" which govern the evolution of the mean concentration (see (6.6)). The corresponding term $D_{i,j}^*$ is the contribution of velocity correlations and it is an anisotropic tensor whose components are time-dependent and uniform in space [47]. The drift coefficient in (6.13) coincides then with the ensemble averaged velocity, $\mathscr{V}_i = \langle V_i \rangle$. For $\Theta = \rho = const$, Eq. (6.2) with coefficients given by (6.14) and relation (6.15) provide the framework of PDF/FDF approaches for constant density flows [8, 17, 30, 59]. It should be stressed that it is only in this simple situation of constant weighting factor Θ that the PDF equation (6.2) can be treated as a Fokker–Planck equation

Within the direct Fokker–Planck approach, the coefficients of Eq. (6.13) are determined by those of the transport equation (6.1), through unclosed conditional averages. The joint concentration-position PDF solving this equation can be numerically approximated by the associated Itô equations (6.4) and (6.5) and interpreted as a weighted concentration PDF by using the correspondence relation (6.9). Relations between joint concentration-position PDFs of Itô processes and density weighted concentration PDFs (6.11) are the core of the Lagrangian particle MC approaches. These approaches were introduced by somewhat involved arguments using the concept of "fluid particles" (rather questionable for systems undergoing diffusion [50]) [17, 31, 40].

6.1.3.2   Reverse Fokker–Planck Approach

When the balance equation for the solvent is considered together with those for $N_{\alpha-1}$ reacting species concentrations, then, summing up these equations, not only the sum of reaction rates but also that of diffusion fluxes vanishes. Then, it follows that the solution $\Theta = \sum_{\alpha=1}^{N_\alpha} C_\alpha$ of the continuity equation (6.12) is precisely the density ρ of the fluid system [21]. Considering the balance equations for all the components of the fluid is a natural choice in PDF approaches for reacting gas mixtures [40], where different components may have comparable weights. Including the carrying fluid among species components of a dilute solution transported in groundwater may cause difficulties in solving PDF/FDF problems. Even if the complicated balance equation for the solvent may be avoided by using the $N_{\alpha-1}$ Eq. (6.1) and the continuity equation (6.12) [21], the numerical solution of the system of Itô equations (6.4) and (6.5) would be complicated by the need to consider the initial condition for the carrying fluid. If one neglects the variations of the solvent concentration, which is tantamount to considering constant density flows, in a direct Fokker–Planck approach, then Eq. (6.7) imposes a uniform position PDF. In turn, a space–time constant solution to Eq. (6.6) requires some relations between the spatial derivatives of the drift and diffusion coefficients [50], which may not be fulfilled in case of statistically inhomogeneous velocity fields [34]. Such issues can be avoided in a Fokker–Planck approach which does not require the fulfillment of condition (6.12).

Such an alternative Fokker–Planck approach has been suggested in [47] and further developed in [52]. Instead of specifying the system of Itô equations by the (modeled) coefficients of the Fokker–Planck equation (6.13), modeled Itô equations may be used to derive a Fokker–Planck equation. In our case, given the system of Itô equations (6.4) and (6.5), the corresponding Fokker–Planck equation is [27]

$$\frac{\partial p}{\partial t} + \frac{\partial}{\partial x_i}(\mathcal{V}_i p) + \frac{\partial}{\partial c_\alpha}(M_\alpha p) = \frac{\partial^2}{\partial x_i \partial x_j}(\mathcal{D}_{ij} p) - \frac{\partial}{\partial c_\alpha}(S_\alpha p). \qquad (6.16)$$

According to (6.9), the solution p of the Fokker–Planck equation (6.16) coincides with the concentration PDF f weighted by a conserved scalar Θ solving (6.1) without reaction terms but not necessarily solving the continuity equation (6.12). This "reverse" Fokker–Planck approach could be a valid option in modeling groundwater systems, for which dispersion coefficients $\mathscr{D}_{ij}$ are provided by various theoretical [7, 11, 42] or numerical [16, 47] methods and mixing models may be inferred from measurements [51]. Similar reverse approaches have been used in a hydrological context to obtain the Fokker–Planck equation for the concentration PDF of a process generated by a given mixing model for fixed positions in physical space [5], and, in a general setting, to formulate the evolution equation for the velocity–concentration PDF [30]. Also similar to this approach is the derivation of the evolution equation for the mass density function from a diffusion model for the velocity of discrete particles dispersed in turbulent flows presented in [32]. The novelty of the reverse Fokker–Planck approach presented in this section is that it does not require that the conserved scalar Θ be a solution of the continuity equation.

6.1.3.3 Numerical Solutions

A straightforward solution to the system of Itô equations (6.4) and (6.5) is given by a grid-free PT in the concentration-position state space $Y_c \times Y_x$. A particle follows a trajectory which starts at an initial position $(\mathbf{c}_0, \mathbf{x}_0)$ drawn from the initial PDF $p(\mathbf{c}, \mathbf{x}, 0)$. N particles initialized in the same way are used to form a statistical ensemble. The approximation of $p(\mathbf{c}, \mathbf{x}, t)$ at $t > 0$ is given by the particle distribution in computational cells around $(\mathbf{c}, \mathbf{x})$. As an example, let us consider a two-dimensional passive transport in groundwater with initial concentration $C(\mathbf{x}, 0) = 1$ in a rectangular initial support in the (x_1, x_2) plane. The initial PDF $p(c, \mathbf{x}, 0)$ is approximated by a uniform distribution of particles in the same rectangle lifted in a plane parallel to the (x_1, x_2) plane which intersects the concentration axis at $c = 1$. The actual joint concentration-position PDF $p(c, \mathbf{x}, t)$ is then approximated at the center of the cubic cells of a regular lattice in the (c, x_1, x_2) space by the number n of particles in each cell, through the number density n/N.

In Lagrangian particle approaches, the solution algorithm differs from the straightforward solution to the system of Itô equations. A "notional particle" carries a "composition" $\mathbf{c} = \{c_1, \cdots, c_{N_\alpha}\}$ of species concentrations in physical space. Equation (6.4) is solved for N notional particles and Eq. (6.5), solved for each particle, updates its composition [17, 40]. For a fixed initial position $\mathbf{x}_0$ of the notional particle, the composition $\mathbf{c}_0$ consistent with the initial PDF $p(\mathbf{c}, \mathbf{x}, 0)$ has to be extracted from the conditional PDF $p(\mathbf{c}|\mathbf{x}, 0)$. It follows that the initial distribution of notional particles has to approximate the position PDF $p_x(\mathbf{x}, 0)$, to ensure that the joint event "extracting $\mathbf{c}_0$ and $\mathbf{x}_0$" belongs to the ensemble with the PDF $p(\mathbf{c}|\mathbf{x}, 0)p_x(\mathbf{x}, 0) = p(\mathbf{c}, \mathbf{x}, 0)$. The algorithm is mainly useful when one wants to impose a uniform position PDF p_x which allows representing the concentration PDF through conditional PDFs of the Itô process by using (6.15). In the case of two-dimensional passive transport discussed above, the notional particles are initially uniformly distributed and their concentrations are set to unity for particles inside the rectangular support of the initial concentration and to zero outside [30]. The concentration PDF $p(\mathbf{c}; \mathbf{x}, t)$ may be approximated by histograms obtained from ensembles of particles in cells, further smoothed by spline functions [40], as taking spatially constant values in cells [30], or in a weak sense, through estimations of cell averaged quantities [31].

Both PT and Lagrangian particle methods suffer from the increase of the computational costs with the number of particles and from numerical diffusion generated by interpolation of cell averages to the particles' positions [25, 50]. Such inconveniences are overcome by using a GRW algorithm equivalent to a superposition of many weak Euler schemes for systems of Itô equations projected on regular lattices (see Sect. 3.3.1). Unlike in sequential PT procedures, in GRW algorithms all the N particles used to construct a numerical estimate of the probability density solving the Fokker–Planck equation are distributed on the lattice according to the initial PDF, from the beginning of the simulation. The particles from each lattice site are then globally scattered over new positions on the lattice determined by drift and diffusion coefficients, according to binomial distributions. This numerical procedure is by construction free of numerical diffusion. Moreover, the GRW algorithm is practically insensitive to the increase of the number of particles [47, 56]. Since mean values are defined at lattice sites, the GRW solutions are not affected by the artificial diffusion caused by cell averaging and interpolation in PT and Lagrangian particle approaches. Details on implementation and convergence properties of the GRW algorithms are given in Sect. 3.3.2. The GRW algorithm used in PDF/FDF simulations is described in [50].

6.2 Spatial Coarse-Graining and FDF Simulations

6.2.1 PDF/FDF Problem for Passive Transport in Aquifers

The two-dimensional problem for passive transport in saturated aquifers formulated in Sect. 4.3.1 and considered in the MC simulations presented in Chap. 5 will now be used to illustrate the principles of the CGS and the utility of the FDF approach in subsurface hydrology. Spatial filtering will be used to smooth the velocity field and to infer upscaled dispersion coefficients, needed to parameterize FDF equations. Further, the convergence of the CGS-FDF solutions towards reference results provided by solutions of the PDF equation and MC simulations will be investigated numerically.

The passive transport in the divergence-free random velocity field $\mathbf{V}$ is governed by Eq. (6.1) with the source term set to zero, $N_\alpha = 1$, and constant local dispersion coefficient $D = 0.01\,\mathrm{m^2/day}$. The Cauchy problem is solved for an initial condition consisting of uniform concentration in a transverse slab of $1 \times 100\,\mathrm{m}$. The numerical setup of the GRW simulations is the same as in Sect. 5.1. Fine-grid and filtered velocity realizations are computed with the Kraichnan algorithm [28] described in Appendix C.2 and illustrated with a Matlab code in Appendix C.3.2.2 by considering a random hydraulic conductivity log-normally distributed, with Gaussian correlation of the $\ln K$ field. With the parameters $\langle V_1 \rangle = 1$ m/day, $\langle V_2 \rangle = 0$ m/day, $\sigma_{\ln K}^2 = 0.1$, and $\lambda_{\ln K} = 1$ m one obtains the integral scale of the $\ln K$ field $I_{\ln K} = 0.09$ m (see Appendix C.3.1.2) and asymptotic ensemble dispersion coefficients (4.58) $D_{11}^{ens} = D + \langle V_1 \rangle I_{\ln K} = 0.1$ m²/day and $D_{22}^{ens} = 0$.

The observable quantity affected by uncertainty is the cross-section concentration at the plume center of mass obtained by summing the number of particles $n(x, y, t)$ over transverse slabs $\Delta x \times L_y$, where $\Delta x = 1$ m and L_y is the transverse dimension of the two-dimensional domain. This concentration,

$$C(x, t) = \frac{1}{N \Delta x L_y} \int_0^{L_y} \int_{x-\Delta x/2}^{x+\Delta x/2} n(x', y, t) dx' dy,$$

was estimated on the trajectory $x = \langle V_1 \rangle t$ of the ensemble averaged flow velocity at 1 day intervals. Reference MC estimates for concentration PDFs and dispersion coefficients are obtained from the ensemble of 3072 transport simulations presented in [49], as well as from ensembles of 256 realizations computed during the study published in [52].

The transport of the cross-section averaged concentration C is essentially one-dimensional. Therefore, one asserts that the upscaled velocity $\mathcal{V}$ is the ensemble averaged longitudinal velocity $\langle V_1 \rangle = 1$ m/day and the upscaled diffusion coefficient $\mathcal{D}$ is the time-dependent longitudinal ensemble dispersion coefficient $D_{11}^{ens}(t)$ [47], in case of PDF simulations, and that the appropriate upscaled coefficients for FDF simulations are given by the coarse-grained coefficients estimated with filtered velocity fields by the approach described in Appendix F. The correctness of this assertion, which allows reducing by one the dimension of the PDF/FDF problem, will be demonstrated numerically by the results presented in Sect. 6.2.4. Since mixing models inspired from turbulence studies have shown a poor performance in PDF simulations for groundwater [50], the mixing term $M dt$ from equation (6.5) has been chosen as an advection–diffusion process in the concentration space, with drift and diffusion coefficients $\mathcal{V}_c$ and $\mathcal{D}_c$, respectively, derived from a stochastic analysis of simulated concentration time series [47, 50, 51]. This model, as well as a classical mixing model and a linear combination of both models are introduced in Sect. 6.2.3. Finally, the convergence with increasing filter width λ of the FDF solution based on the filtering operation (6.3) to the PDF solution is investigated numerically in Sect. 6.3.

Within the frame of the reverse Fokker–Planck approach presented in Sect. 6.1.3.2, the PDF/FDF solutions will be computed from the one-time PDF of a two-dimensional Itô process, governed by a particular form of Eqs. (6.4) and (6.5),

$$dX(t) = \mathcal{V} dt + \sqrt{2\mathcal{D}} \, dW_1(t), \tag{6.17}$$

$$dC(t) = \mathcal{V}_c dt + \sqrt{2\mathcal{D}_c} \, dW_2(t), \tag{6.18}$$

where $C(t) = C(X(t))$ and $W_1(t)$, $W_2(t)$ are two independent standard Wiener processes [50]. The PDF of the process described by (6.17) (6.18) verifies the Fokker–Planck equation

$$\frac{\partial p}{\partial t} + \mathcal{V}\frac{\partial p}{\partial x} + \mathcal{V}_c\frac{\partial p}{\partial c} = \mathcal{D}\frac{\partial^2 p}{\partial x^2} + \mathcal{D}_c\frac{\partial^2 p}{\partial c^2}. \qquad (6.19)$$

The PDF $f(c; x, t)$ and the FDF $f_\lambda(c; x, t)$ are estimated from the solution $p(c, x, t)$ of the Fokker–Planck equation (6.19) via (6.10) with $\Theta(c) = c$. The MC results presented in Figs. 6.1 and 6.2 show that there is no significant difference between the concentration PDFs $f(c; x, t)$ and the weighted PDFs $cf(c; x, t)/\langle C\rangle$, denoted by "pdf" and "w-pdf," respectively. This is a consequence of the chosen large transverse slab source, for which the transport is almost ergodic [47, 49], i.e., $C \approx \langle C\rangle$. Therefore, relation (6.10) may be approximated by (6.15). In Sect. 6.3, the concentration PDF $f(c; x, t)$ and the FDF $f_\lambda(c; x, t)$ will be estimated by the conditional PDF $p(c|x, t) = p(c, x, t)/p_x(x, t)$. The latter is obtained by two-dimensional GRW simulations consisting of superpositions of about 10^{25} solutions of the Itô equations (6.17) and (6.18) projected on a regular lattice (see Sect. 3.3.2.5). The computational particles are initially distributed on the c-axis at $x = 0$ proportionally to the concentration PDF estimated by MC simulations at $t = 1$ day. A detailed implementation of the numerical scheme is presented in [50].

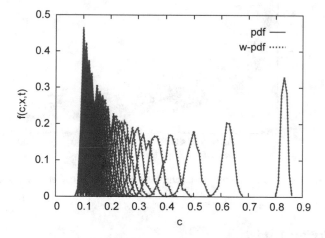

Fig. 6.1 Concentration PDFs at $x = \langle V_1\rangle t$, sampled at 2 days intervals, from $t = 0$ to $t = 100$ days, inferred from 3072 GRW simulations of the two-dimensional transport problem (from right to left) [49]

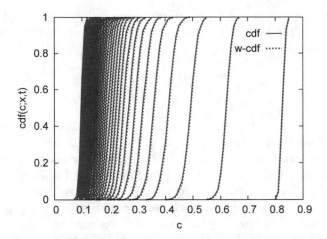

Fig. 6.2   Cumulative distribution functions (CDF), corresponding to the PDFs presented in Fig. 6.1

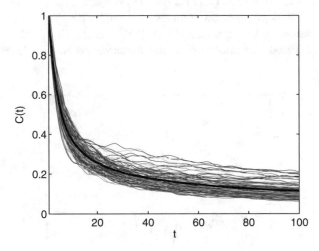

Fig. 6.3   Time series of cross-section concentrations at the plume center of mass given by MC simulations

6.2.2   Mixing Models

The coefficients $\mathscr{V}_c$ and $\mathscr{D}_c$ in Eq. (6.18) are estimated in the following by a stochastic time series analysis. An ensemble of 500 time series $C(t) = C(x = \langle V_1 \rangle t, t)$ of simulated concentrations [51] is presented in Fig. 6.3. As common in time series analysis, t is an integer and represents the count of successive terms, sampled at regular intervals of 1 day. Since the velocity field is statistically homogeneous, these time series correspond to the concentration C sampled at the

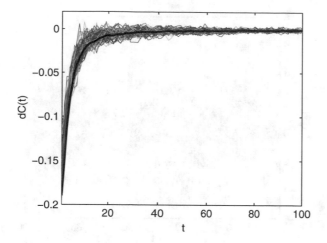

Fig. 6.4   Increments $dC(t)$ of the time series presented in Fig. 6.3

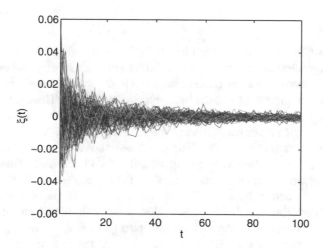

Fig. 6.5   Noisy components $\xi(t)$ of the concentration increments $dC(t)$ from Fig. 6.4

plume center of mass [49]. The corresponding increments $dC(t) = C(t+1) - C(t)$, approximating the local slope $dC(t)/dt$ of a continuous time series, are shown in Fig. 6.4. The residual noise in the time series increments, $\xi(t) = dC(t) - \langle dC(t) \rangle$, shown in Fig. 6.5 has an approximately exponentially decaying amplitude. After normalizing each ξ-sample by its maximum amplitude, $\|\xi\| = \max|\xi|$, the noise standardized in this way, $\xi/\|\xi\|$, may be approximated by a white noise (Fig. 6.6). In turn, it follows that $\xi(t)$ may be approximated by a "heteroskedastic process" [54] consisting of a white noise with rapidly decaying amplitude. The drift coefficient $\mathcal{V}_c$ in Eq. (6.18) is inferred from the local slope $\langle dC(t) \rangle$ of the ensemble mean concentration, represented by a thick line in Fig. 6.4. The maximum value of the

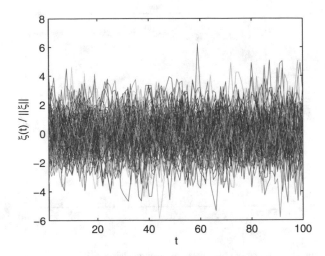

Fig. 6.6 Noisy components $\xi(t)$ of the concentration increments normalized by their maximum amplitude $\|\xi\|$

diffusion coefficient $\mathscr{D}_c$ is specified by an initial amplitude of diffusive jumps in the GRW algorithm of $5\delta c$, where $\delta c = 0.001$ is the space step in concentration space. This value is chosen as close as possible to the standard deviation 0.0053 of the maximum amplitude $\|\xi\|$ of the noise shown in Fig. 6.5. The time variation of the diffusion coefficient $\mathscr{D}_c$ which mimics the behavior of the noise in Fig. 6.5 is optimized through a crude calibration of the PDF simulations.

The time series mixing model (TS) presented above accurately describes the transport of the PDF in concentration space but it fails to reproduce the narrowing of the PDF at large times and small values of the concentration shown in Fig. 6.1 [47, 50–52]. The narrowing of the PDF is due to the local dispersion process which smooths the concentration differences, so that in the limit of uniform concentration distributions in every realization of the transport process the PDF approaches a δ function. The narrower the PDF is, the smaller are the differences between the concentration realizations and their ensemble average. An appropriate mixing model should therefore account for this "attraction" towards the mean of the concentration. For the TS model, the drift coefficient in Eq. (6.18), shown in Fig. 6.4, vanishes relatively quickly and the concentration $C(t) = C(X(t))$ will be almost constant on the trajectory $X(t) = \langle V_1 \rangle t$, with small fluctuations given by the noisy term shown in Fig. 6.5. This behavior of the TS model could be explained by the excessive smoothing in the stochastic time series analysis used to infer its parameters. An automatic algorithm to decompose time series into intrinsic components [57] could be used to check whether, besides the trend and the noise shown in Figs. 6.4 and 6.5, respectively, the concentration time series contains more components which were smoothed out by the stochastic analysis.

The attraction towards the mean of the concentration realizations is enforced by the interaction by exchange with the mean (IEM) model $\mathscr{V}_c = -\kappa(C - \langle C \rangle)/(2\tau)$, where κ is a dimensionless constant and τ is the decay time scale of the concentration fluctuations [8, 40]. For a statistically homogeneous Gaussian concentration field of a passive scalar in homogeneous and isotropic turbulence, the IEM model is precisely the expression of the conditional diffusion flux [41]. Furthermore, since in this case the density is constant and the PDF no longer depends on spatial coordinates, it follows from Eqs. (6.13) and (6.2) that the mixing terms based on conditional diffusion flux and conditional dissipation rate are equivalent. The IEM model has the drawback that it preserves the shape of the initial PDF [40, 41]. However, the simple IEM model has proved to be useful in many practical FDF problems for turbulent reacting flows [8, 23, 36]. For the purpose of the present illustration of the IEM model, the characteristic time τ is defined as a diffusion time, similarly to FDF approaches in turbulence [8, 23], in terms of effective dispersion coefficients and a characteristic length scale L. Since the transport process is almost ergodic, the effective and the ensemble coefficients are almost identical [47], so that $\tau = L^2/\mathscr{D}$. In PDF simulations, the scale L is obviously given by the correlation length $\lambda_{\ln K}$ of the log-hydraulic conductivity field. For FDF simulations the filter width λ should be accounted for, so, the length scale is defined by $L = \lambda_{\ln K} + \lambda$ [52]. Let us recall that for $\lambda > 0$, $\mathscr{D}$ is the coarse-grained diffusion coefficient, which will be defined in Sect. 6.2.4. With these parameters, the IEM model reads

$$\mathscr{V}_c^{IEM} = -\kappa \frac{\mathscr{D}}{(\lambda_{\ln K} + \lambda)^2}(C - \langle C \rangle). \qquad (6.20)$$

The model constant κ is not a universal constant and it is generally determined from comparisons with measurements or reference solutions. In turbulence studies, for instance, κ ranges between 0.6 and 3.1 [40]. In [52] it was found that $\kappa = 2$ is appropriate for PDF simulations and $\kappa = 0.6$ for FDF simulations.

Since, as we will see in the following, the IEM model is not yet satisfactory, one also considers a linear combination of TS and IEM models (hereafter referred to as the TS-IEM model),

$$\mathscr{V}_c^{TS-IEM}(t) = \frac{T-t}{T}\mathscr{V}_c^{TS}(t) + \frac{t}{T}\mathscr{V}_c^{IEM}(t), \qquad (6.21)$$

where T denotes the total simulation time. Figures 6.7 and 6.8 compare GRW-PDF simulations ($\lambda = 0$) carried out with the TS, IEM, and TS-IEM models of the drift $\mathscr{V}_c$ and the same diffusion coefficient $\mathscr{D}_c$ provided by the TS model. The reference solution and the initial condition for these simulations are those from the MC ensemble of 3072 simulations of the concentration PDF presented in Fig. 6.1. Similar FDF-GRW simulations are deferred to Sect. 6.3. As seen in Fig. 6.7, both the IEM model (6.20) and the TS-IEM model (6.21) yield narrow PDFs at small concentrations but they do not reproduce the reference solution. The PDFs are non-smooth for both models and the pure IEM model completely fails to reproduce the transport of the PDF in concentration space. The IEM solution for the cumulative distribution function shown in Fig. 6.8 is also unsatisfactory,

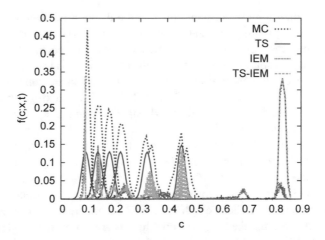

Fig. 6.7 GRW-PDF solutions $f(c; x, t)$ for different mixing models, compared to MC estimates, sampled at $x = \langle V_1 \rangle t$, for $t = 0, 5, 10, 20, 30, 50,$ and 100 days (from right to left)

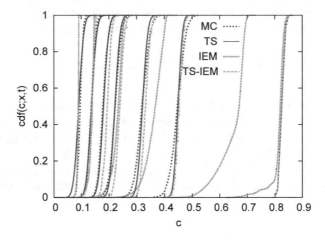

Fig. 6.8 Cumulative distribution functions corresponding to the PDF solutions presented in Fig. 6.7

the corresponding curves lagging far behind the reference solution. The TS-IEM solution for the cumulative distribution function compares satisfactorily with the TS and the reference solutions at small to intermediate times (Fig. 6.8). Since the TS model does not depend on the filter width λ, only the TS-IEM model (6.21) will be used in Sect. 6.3 to illustrate the effect of filtering in FDF simulations.

6.2.3 Upscaled Velocity Fields and Diffusion Coefficients

The computation of the filtered velocity $\langle V \rangle_\lambda$ through (6.3) is described in detail in Appendix C.2. The final expression (C.7), used the field generator code presented in Appendix C.3.2.2, reduces to the usual Kraichnan field generator when $\lambda = 0$. Therefore, the cases with $\lambda = 0$ presented in the following correspond to the fine-grained, non-filtered velocity field. The effect of filtering upon the components of the Kraichnan velocity field is illustrated in Fig. 6.9, where one can see that with increasing λ the longitudinal and transverse components of the filtered velocity, $\langle V_1 \rangle_\lambda$ and $\langle V_2 \rangle_\lambda$, approach their ensemble averages $\langle V_1 \rangle = 1$ m/day and $\langle V_2 \rangle = 0$ m/day, respectively.

Thanks to their self-averaging property, upscaled diffusion coefficients can be efficiently estimated from computations on single trajectories of the advection–diffusion process [47, 48] specified in Sect. 6.2.1. Such estimations were validated numerically for diffusion in velocity fields with finite correlation lengths [48]. One expects these estimations also to be valid in the case of power law correlated velocities as long as their covariance sampled on trajectories is ergodic (see Sect. 4.3.3). The method of self-averaging estimation is presented in Appendix F.1.

The longitudinal ensemble dispersion coefficient estimated by (F.3), as well as those given by MC-GRW simulations of two-dimensional transport [49], is defined as one-half the mean slope of the time-dependent variance of the ensemble process. The diffusion coefficient $\mathscr{D}$ in Eq. (6.17), needed in PDF/FDF simulations, is defined as one-half the local variance, which, since the process has finite first and second spatial moments at finite times, is tantamount to one-half of the derivative

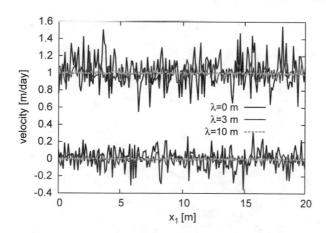

Fig. 6.9 Samples of filtered velocity fields ($\langle V_1 \rangle_\lambda$, $\langle V_2 \rangle_\lambda$) computed by applying the Gaussian spatial filter (C.6) of width λ to fine-grained velocity fields (V_1, V_2) obtained with the Kraichnan random field generator presented in Appendix C.3.2.2

of the global variance [47]. $\mathscr{D}$ is approximated in the GRW-PDF scheme from the input parameter numerically estimated by (F.3) as

$$\mathscr{D}(t) = \frac{(t + \delta t)\tilde{D}(t + \delta t) - t\tilde{D}(t)}{\delta t},$$ (6.22)

where δt is the simulation time step.

6.2.4   Coarse-Grained Simulations of Transport

Ensemble dispersion coefficients computed for diffusive transport in filtered velocity fields only account for the variability of the velocity at scales larger than the filter width λ. Therefore, as seen in Fig. 6.10, the coefficients $D_{11}^{ens}(t, \lambda)$ are smaller than the coefficients obtained for fine-grained, unfiltered velocity fields (the case $\lambda = 0$) and they approach the local dispersion coefficient D as λ increases and the velocity fluctuations are smoothed out. The sub-filter variability is modeled by fictitious diffusion coefficients. When used in a coarse-grained description, such coefficients should retrieve the statistics of the spatial moments, in theoretical stochastic approaches [10, 11, 42], or they should produce results close to those from fine-scale simulations, in numerical upscaling approaches [14, 16].

Following this paradigm, GRW-CGS of the two-dimensional transport problem formulated in Sect. 6.2.1 are conducted by using filtered velocities ($\langle V_1 \rangle_\lambda, \langle V_2 \rangle_\lambda$) and coarse-grained longitudinal diffusion coefficients obtained by adding the difference between fine-grained and filtered ensemble dispersion coefficients to the local dispersion coefficient D [52],

$$D_{11}^{cg}(t, \lambda) = D + \delta D(t, \lambda), \quad \delta D(t, \lambda) = D_{11}^{ens}(t) - D_{11}^{ens}(t, \lambda).$$ (6.23)

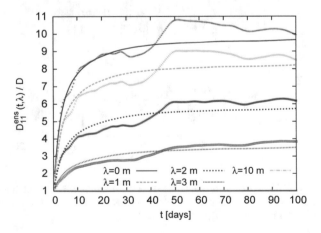

Fig. 6.10   Ensemble dispersion coefficients $D_{11}^{ens}(t, \lambda)$ derived from GRW simulations for single realizations of the filtered velocity field compared to their ensemble average (smooth curves)

Because the corrections to the transverse coefficient were found to be negligible, one chooses $D_{22}^{cg} = D$. By using expressions of dispersion coefficients in terms of trajectories of the Itô processes, it is easy to check that the corrections δD in (6.23) are entirely due to sub-filter effects only in the absence of correlations between the filtered and the sub-filter velocity fields (see Appendix F.2). Such correlations are generally non-vanishing in upscaling flow equations by spatial averages [3]. In analytical first order approximations, the condition of vanishing correlations is ensured by integrating over disjoint domains in the Fourier space [42]. In the numerical approach presented here, the lack of correlations is ensured by using independent sets of random numbers to generate filtered and fine-grained velocity fields.

The longitudinal ensemble dispersion coefficient is computed by GRW simulations from the variance of the ensemble process $X_1^{ens} = X_1 - \langle X_1 \rangle$ as $D_{11}^{ens} = \langle (X_1^{ens})^2 \rangle / (2t)$, where $\langle \cdot \rangle$ is the average over the $N = 10^{10}$ particles (realizations of the local diffusion process) and over the $R = 256$ realizations of the velocity field. Single-realization estimations of D_{11}^{ens} are computed by averaging $(X_1^{ens})^2$ for $R = 1$ over the N particles while keeping the full averaged mean $\langle X_1 \rangle$. In Fig. 6.11 it is shown that CGS, performed with filtered velocity fields and coarse-grained diffusion coefficients (6.23) determined from single-realization coefficients $D_{11}^{ens}(t, \lambda)$, recover the fine-grained ensemble coefficient computed in a single realization (corresponding to $\lambda = 0$) for large enough filter widths. Figure 6.12 shows that for the same single-realization CGS, but with coarse-grained coefficients (6.23) computed from $D_{11}^{ens}(t, \lambda)$ averaged over $R = 256$ realizations, for a filter width $\lambda = 4$ m one recovers the ensemble averaged D_{11}^{ens} given by fine-grained simulations ($\lambda = 0$). This demonstrates the usefulness of CGS, which reduces the

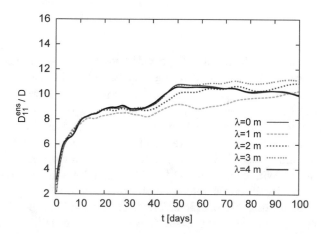

Fig. 6.11 Single-realization CGS estimations of the ensemble dispersion coefficient D_{11}^{ens} for increasing filter width λ, performed with single-realization coarse-grained diffusion coefficients $D_{11}^{cg}(t, \lambda)$

Fig. 6.12 Single-realization CGS estimations of the fine-grained ensemble dispersion coefficient D_{11}^{ens} for increasing filter width λ, performed with ensemble averaged coarse-grained diffusion coefficients $D_{11}^{cg}(t, \lambda)$

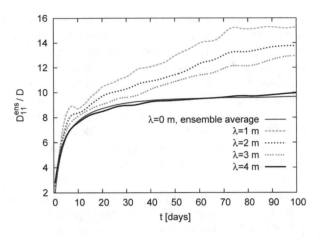

Fig. 6.13 Ensemble dispersion coefficients $D_{11}^{ens}(t, \lambda)$ for filtered velocity fields obtained by self-averaging estimations and corresponding standard deviations (represented by error bars). Solid black lines represent ensemble averaged coefficients $D_{11}^{ens}(t, \lambda)$ obtained from 256 GRW simulations

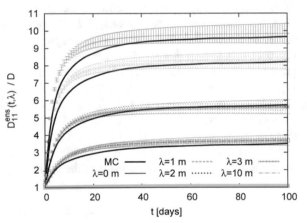

computational costs by a factor of 256, provided that the coarse-grained coefficient $D_{11}^{cg}(t, \lambda)$ can be obtained without carrying out ensembles of GRW simulations.

With the low variance $\sigma_{\ln K}^2 = 0.1$ considered in the present numerical setup, it is allowed to generate velocities at given points in space (see Appendix C.2) and to construct trajectories on which self-averaging estimations of the coefficients $D_{11}^{ens}(t, \lambda)$ for filtered velocity fields are readily available (see Appendix F.1). These self-averaging estimations, shown in Fig. 6.13, are used to infer the coarse-grained coefficients (6.23). With coefficients $D_{11}^{cg}(t, \lambda)$ obtained in this way, one finds that GRW simulations, for $R = 256$ velocity realizations, yield an acceptable recovery of the fine-grained ensemble coefficient D_{11}^{ens} (Fig. 6.14). The possibility to use self-averaging estimations of the coarse-grained coefficients $D_{11}^{cg}(t, \lambda)$ and filtered velocities to compute ensemble average estimates, demonstrated by Fig. 6.14, could significantly alleviate the costs in MC simulations by using discretization elements several times coarser than in fine-grained simulations.

Fig. 6.14 Estimations of the fine-grained ensemble averaged coefficient $D_{11}^{ens}(t)$ for increasing filter width λ, obtained from 256 GRW simulations with filtered velocities and self-averaging estimates of the coarse-grained diffusion coefficients $D_{11}^{cg}(t, \lambda)$

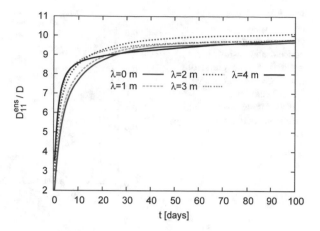

Fig. 6.15 Cross-section concentration $C(x_1, t)$ at $t = 10, 50$, and 100 days and concentration at the expected center of mass $C(x_1 = \langle V_1 \rangle t, t)$ (monotonous curves) given by single-realization GRW simulations with filtered velocities

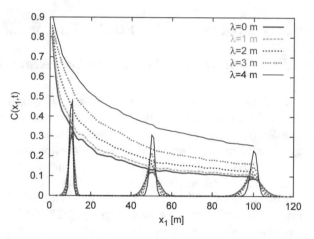

Figure 6.15 shows the cross-section concentration $C(x_1, t)$ recorded at three simulation times, together with the concentration sampled in time at the expected position of the center of mass $\langle V_1 \rangle t$, computed from a single GRW simulation with filtered velocities and with an unchanged local diffusion coefficient D. It is seen that the effect of filtering the velocity field alone is to reduce the spreading of the solute and to increase the concentration at the center of mass. Instead, GRW simulations with coarse-grained coefficients $D_{11}^{cg}(t, \lambda)$, estimated by the self-averaging method, and filtered velocities can be used to recover the fine-grained ensemble averaged cross-section concentration with a good precision (Fig. 6.16). This result indicates that the self-averaging estimations of the coarse-grained coefficients $D_{11}^{cg}(t, \lambda)$ are accurate enough for the purpose of simulating the transport of the PDF/FDF in physical space, which is governed by the same equation (6.17) as the mean concentration.

Fig. 6.16 Fine-grained
ensemble averages
corresponding to the
concentrations shown in
figure 6.15, recovered from
256 GRW simulations with
filtered velocities and
self-averaging coarse-grained
diffusion coefficients
$D_{11}^{cg}(t, \lambda)$

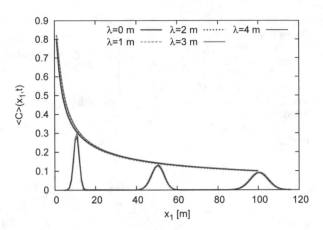

6.3   GRW Solutions for PDF/FDF Equations

6.3.1   Convergence of the Mean Concentration

GRW-FDF solutions for the problem of passive scalar transport formulated in
Sect. 6.2.1 are computed with drift coefficients for transport in physical space $\mathcal{V}$
(see Eq. (6.17)) given by filtered velocity fields $\langle V_1 \rangle_\lambda$, generated in a first order
approximation by using the randomization expression (C.7). The corresponding
diffusion coefficients, $\mathcal{D}$ in Eq. (6.17), are computed by (6.22) from coarse-grained
longitudinal ensemble dispersion coefficients $D_{11}^{cg}(t, \lambda)$. The coefficients $D_{11}^{cg}(t, \lambda)$
are obtained by (6.23) from the mean values of the self-averaging ensemble
dispersion coefficients presented in Fig. 6.13. The GRW-PDF solutions are obtained
similarly, with $\mathcal{V}$ given by the ensemble average $\langle V_1 \rangle = 1$ m/day of the fine-
grained velocity field and $\mathcal{D}$ given by the mean value of the self-averaging ensemble
dispersion coefficient $D_{11}^{ens}(t, 0)$ (case $\lambda = 0$ in Fig. 6.13). The coefficients
describing the transport in concentration space, $\mathcal{V}_c$ and $\mathcal{D}_c$ in Eq. (6.18), are those
of the TS mixing model in PDF simulations and those of the combined TS-IEM
model (6.21) in FDF simulations. The reference solution and the initial condition for
these simulations are estimated from the MC ensemble of 256 transport simulations
computed for $\ln K$ with Gaussian correlation and for $\lambda = 0$ (Sect. 6.2.4). The same
ensemble is used to estimate the drift $\mathcal{V}_c$ in the TS model while using the diffusion
coefficient $\mathcal{D}_c$ inferred in Sect. 6.2.2.

In the reverse Fokker–Planck approach to PDF/FDF solutions, the weighting
factor is the scalar concentration itself, $\Theta(c) = c$. Hence, as follows from (6.7), the
mean concentration equals the position PDF, $\langle C \rangle(x, t) = p_x(x, t)$, which is simply
obtained in the GRW-PDF/FDF simulation by summing up the joint concentration-
position PDF/FDF $p(c, x, t)$ over the c-axis. Note that in a direct approach, the
scalar concentration c cannot serve as a weighting factor Θ, because it undergoes
diffusion and does not verify the continuity equation (6.12).

Fig. 6.17 $\langle C \rangle_\lambda(x, t)$ recorded at fixed times $t = 10, 50$, and 100 days and $\langle C \rangle_\lambda(t)$ recorded at $x = \langle V_1 \rangle t$. TS-IEM results for increasing λ are compared to MC estimates and PDF solutions obtained with the TS model

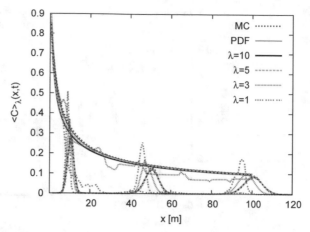

The results for the mean concentration $\langle C \rangle_\lambda$ obtained from GRW-FDF simulations using the linear combination of mixing models TS-IEM (6.21) for increasing filter width λ are compared to MC results and to GRW-PDF simulations using the TS mixing model in Fig. 6.17. One remarks that the accuracy of the GRW-FDF solutions increases with λ and for $\lambda \geq 5$ m the results obtained by the three methods practically coincide. One also remarks that the FDF solution for small λ is less smooth than similar results obtained from single-realization GRW simulations of the two-dimensional transport shown in Fig. 6.15. In both cases, one can see a misfit between $\langle C \rangle(t)$, respectively, $\langle C \rangle_\lambda(t)$, both recorded at the expected center of mass $x = \langle V_1 \rangle t$, and the peaks in the spatial distribution of the concentration located at the position of the actual center of mass, which is random and generally does not coincide with its expectation.

The mean concentration is the PDF $p_x(x, t)$ of the Itô process (6.17) which solves the Fokker–Planck equation (6.6). Thus, $\langle C \rangle_\lambda$ can be computed as well by solving Eq. (6.6), independently of the FDF problem, like in moment equations approaches [18]. As for the variance, it requires the computation of the $\langle C^2 \rangle_\lambda$, which involves (6.9) and the knowledge of the one-point FDF [51]. Since, as shown in the following, the estimated FDFs are generally inaccurate, acceptable variance estimates cannot be obtained in the present setup.

6.3.2 Convergence of FDF Solutions to PDF Solutions

Cumulative distribution functions sampled on the same trajectory $x = \langle V_1 \rangle t$ are shown in Fig. 6.18. At early times $t \leq 20$ days the cumulative distributions obtained by GRW-FDF simulations approach the PDF and the MC results with increasing filter width. Figure 6.19 shows that the FDFs are shifted towards small concentrations with increasing time, the smaller the λ, the larger the shift.

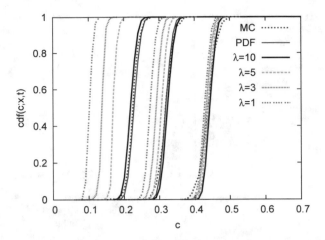

Fig. 6.18 Cumulative distribution functions (CDF) for the same cases as in Fig. 6.17, estimated at $x = \langle V_1 \rangle t$, for $t = 5$, 10, and 20 days (from right to left)

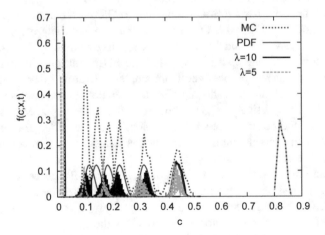

Fig. 6.19 GRW-FDF solutions $f(c; x, t)$ for the TS-IEM mixing model, compared to MC estimates and PDF solutions for the TS model, sampled at $x = \langle V_1 \rangle t$, for $t = 0, 5, 10, 20, 30, 50,$ and 100 days (from right to left)

The corresponding cumulative distributions presented in Fig. 6.20 show a similar behavior, with large deviations from the reference solution for $\lambda = 5$ m and $t \geq 20$ days. For the largest filter width, $\lambda = 10$ m, acceptable GRW-FDF estimates of the cumulative distribution can be obtained at small and intermediate times $t < 50$ days.

The behavior of the TS-IEM model at large times indicates that the left-drift, towards small concentrations, is too large. Since according to (6.21) the IEM model has a dominant contribution at large times, $\mathscr{V}_c^{TS-IEM}$ can be reduced by reducing $\mathscr{V}_c^{IEM}$. A larger time scale τ at large times is therefore needed in the

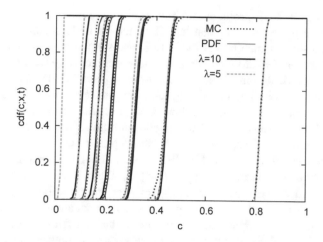

Fig. 6.20 Cumulative distribution functions corresponding to the PDFs presented in Fig. 6.19 for $t = 0, 5, 10, 20, 30$, and 50 days

IEM model. The convergent behavior of the TS-IEM model at intermediate times is ensured by the TS component $\mathscr{V}_c^{TS}$, which compensates the smallness of $\mathscr{V}_c^{IEM}$ (see Fig. 6.8). The same result can be obtained by an IEM model with a smaller time scale τ at small times. The constant time scale $\tau = (\lambda_{\ln K} + \lambda)^2/\mathscr{D}$ suggested by FDF approaches in turbulence seems to be inadequate for the PDF/FDF problem considered here.

6.4   Issues and Future Developments of PDF/FDF Approach

6.4.1   Looking for Appropriate Mixing Models

One of the most stringent problems in PDF/FDF approaches is to design the appropriate mixing model which describes the transport in concentration spaces of the system of computational particles. For the problem considered here, with the random concentration depending on a single spatial coordinate, the advection–diffusion process in concentration space, inferred from a time series analysis of simulated concentrations sampled on the mean flow trajectory, yields a correct solution for the transport of the cumulative distribution towards low concentration values. However, this time series model cannot simulate the narrowing and the raising peak of the maximum PDF/FDF values at large times and small concentrations. The classical IEM model widely used in turbulence studies, which should in principle recover this large time behavior, fails to describe the transport in concentration space of the cumulative distribution function. The IEM model is an exact mixing model, rigorously derived for homogeneous and isotropic turbulence

[41]. However, in PDF problems for transport in groundwater flows the results presented in Fig. 6.8 indicate that the IEM model underestimates the drift in concentration space which moves the PDF towards small concentration values. A linear combination TS-IEM yields an acceptable accuracy only for moderately large times. The numerical computations carried out with the TS-IEM model indicate that improvements can be achieved by choosing an appropriate variation of the characteristic time scale in the coefficient of the IEM component of the model which would result in an enhanced drift in concentration space at early times, which decreases gradually as the spreading of the solute plume progresses.

A time-dependent coefficient of the IEM model, TIEM, was heuristically inferred from the parameters of the transport problem [45]. TIEM is based on the assumption that the squared concentration gradients, involved in conditional dissipation rate $\mathcal{M}_{\alpha\beta}$ given in (6.14), evolve inversely proportional to a squared characteristic length scale λ_c, $(\nabla C)^2 \sim \lambda_c^{-2}$ as the plume spreads and its fringes and fluctuations smooth out. Since the concentration fluctuations are smoothed out by the local dispersion with the characteristic scale $\lambda_c(t) = \sqrt{2Dt}$, it is assumed that the conditional dissipation rate $\mathcal{M}_{\alpha\beta}$ decays as $\mathcal{M} \sim \lambda_c(t)^{-2} = (2Dt)^{-1}$. In addition, the coefficient $\mathscr{D}$ is chosen to be an upscaled effective diffusion coefficient, instead of upscaled ensemble coefficient (their difference is the coefficient (4.96) associated with the fluctuations of the center of mass, which do not influence the dissipation). The TIEM model is finally obtained from (6.20), for the problem considered in Sect. 6.2.1 and for coarse-graining parameter $\lambda = 0$, by choosing $\kappa = 2$ (the middle of the range reported in the literature), and by replacing the spatial scale $\lambda_{\ln K}$ by λ_c:

$$\mathscr{V}_c^{TIEM} = -\frac{\mathscr{D}}{Dt}(C - \langle C \rangle). \tag{6.24}$$

It was found that TIEM performs better than the classical IEM model, but only for small to intermediate times [45, Fig. 8].

The analysis of an ensemble of 1000 concentration time series of 2000 terms each (corresponding to a length of 2000 days), obtained by GRW simulations of the cross-section concentration at the center of mass (see Fig. 6.3), leads to a mixing model of the form

$$dC(t) \approx d\langle C(t) \rangle + \kappa(t)[C(t-1) - \langle C(t-1) \rangle] + \upsilon(t)z(t). \tag{6.25}$$

The first term in Eq. (6.25) is the slope of the mean concentration, the second term is a linear regression with variable coefficient $\kappa(t)$, and the third term consists of a noise $z(t)$ modulated by a variable amplitude $\upsilon(t)$ [9].

The first and last terms of (6.25) correspond to the TS model empirically formulated in Sect. 6.2.2, with the difference that now $z(t)$ is an AR(1) correlated noise. The linear regression (middle term) has the structure of the IEM model and is similar to the TIEM model (6.24). The difference is that the variable coefficient $\kappa(t)$ is a power law function derived from the analysis of the residual noise in the ensemble of time series. It seems that the drawback of the IEM and TIEM models

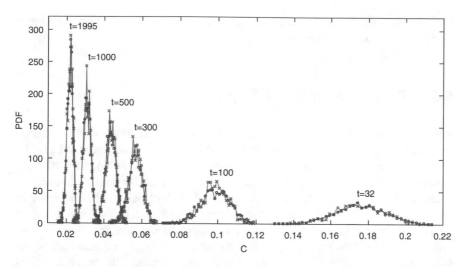

Fig. 6.21 Probability densities of the concentration $C(t)$ for $t = 32, 100, 300, 500, 1000$, and 1995 estimated from the initial GRW-MC ensemble (plotted with x markers) and by averaging on the ensemble generated with the mixing model (6.25)

is the absence of the trend increment $d\langle C(t)\rangle$ and that of the noise term. In case of the simple TS stochastic model the drawback is the absence of the regression term.

The model (6.25) reveals the structure of the mixing model which governs the transport in concentration space of the PDF of $C(t)$. The transport of the concentration PDF in concentration space presented in Fig. 6.21 shows that the mixing model (6.25) is practically exact for the PDF problem under consideration. Similarly to the classical IEM model which is appropriate for modeling a large variety of scenarios of turbulent transport, one expects that mixing models for more complex problems of transport in groundwater will have the same structure as (6.25), consisting of a deterministic trend, a noise, and a regression term generalizing the classical IEM model. Since the variable regression coefficient $\kappa(t)$ is negative, the regression term describes the time-dependent relaxation of the fluctuations $[C - \langle C\rangle]$ as the concentration approaches a spatially uniform distribution. This mechanism prevents the occurrence of negative concentrations [9, Fig. 4], which is a major drawback of diffusion-like mixing models [40].

The mixing model (6.25) requires a large number of parameters. To parameterize a model with the same structure, but for different transport conditions and physical parameters, one needs another set of real data, i.e., an ensemble of time series realizations. For practical purposes of estimating PDFs of maximum concentration by the approach proposed in [52], the mixing model (6.25) can be parameterized, for the beginning, as follows: a trend $d\langle C(t)\rangle \sim t^{3/2}$ (the slope of the Gaussian macrodispersion model of the average concentration at the plume center of mass); a white noise with variable amplitude, inferred from the analyses of individual synthetical or measured time series, for instance, by using the automatic detrending algorithm presented in [55]; the IEM model with variable coefficients introduced in

[45], specified by the local dispersion coefficient D and the longitudinal component of the effective dispersion tensor.

6.4.2 Perspectives of the FDF Approach

Although they share in common the same mathematical formalism, CGS-FDF modeling for groundwater systems and LES-FDF approach for turbulent reacting flows differ in an important aspect. Whereas in turbulence the upscaled transport parameters are not known a priori, for transport in aquifers with hydraulic conductivity modeled by statistically homogeneous random fields with low variance, the stochastically upscaled model is completely determined, at least in a first order approximation, by the statistics of the random field. In addition, the coarse-grained diffusion coefficients (6.22), estimated as ensemble averages, converge to the stochastically upscaled diffusion coefficients with increasing filter width and allow recovering the results of the stochastically upscaled model in single-realization CGS simulations (Fig. 6.12). Thus, in CGS-FDF simulations, the solution of Eq. (6.17) approaches the PDF solution for the mean concentration with increasing filter width. Provided that the same convergence holds for the solution of Eq. (6.18), which describes the transport of FDF in concentration space, the CGS-FDF solution converges to the PDF solution. The numerical results presented in this chapter partially illustrate this convergence: with high accuracy for the mean concentration, at moderately large times for the cumulative distribution function, but, in the absence of an adequate mixing model, only at small times for the filtered density. For the sake of clarity and simplicity, chemical reactions were not considered in the numerical illustration of the approach. Because the reaction terms are in closed form, they only contribute with a supplementary drift in Eq. (6.18).

As far as all the coefficients in Eqs. (6.17) and (6.18) are provided, the GRW solution to either the PDF or the FDF problem consumes almost the same amount of computational resources (e.g., computing time ranging from 0.5 to 14 s for lattices with $\sim 10^5$ and $\sim 10^6$ nodes, respectively). This was the case of small variances of the $\ln K$ field considered in Sect. 6.2.1. For larger variances of the random $\ln K$ field, the upscaled diffusion coefficients needed to parameterize Eq. (6.17) are no longer available from perturbation theories or self-averaging approximations. The advantage of the CGS-FDF would be its ability to retrieve the concentration PDF, estimated, for instance, by MC simulations, as limit for large filter widths of the simulated FDF by using filtered velocities and numerically upscaled diffusion coefficients. But the FDF approach could be even more valuable in assessing uncertainty for statistically inhomogeneous systems.

If the hydraulic conductivity is no longer statistically homogeneous, the upscaled transport parameters are not known a priori and even the existence of the upscaled transport equation is questionable. In this case, filtered velocities and single-realization estimates of coarse-grained diffusion coefficients can be used to recover the results obtained for the fully resolved velocity field (see Fig. 6.11). The

CGS-FDF modeling is now similar to the classical LES-FDF approach. Like in turbulence modeling, the ensemble average is replaced by a smoothing in space which still allows an appropriate description of the large-scale variability of the velocity field. The PDF will then be the probabilistic description obtained by the spatial smoothing of the fine-scale description of the process, and the aim of FDF modeling will be to recover this global probabilistic description on coarse grids, at affordable computational costs. Volume averaging or multi-grid numerical upscaling procedures can be used to extend the CGS-FDF strategy illustrated in this chapter to problems of reactive transport in conditions of highly variable and statistically inhomogeneous hydraulic conductivity of the groundwater system. Nonetheless, the reverse Fokker–Planck approach could also be used to design robust numerical schemes based on the GRW algorithm to solve PDF/FDF problems for other fluid systems, such as those studied in turbulence and combustion theory.

References

1. Alzraiee, A.H., Baú, D., Elhaddad, A.: Estimation of heterogeneous aquifer parameters using centralized and decentralized fusion of hydraulic tomography data from multiple pumping tests. Hydrol. Earth Syst. Sci. Discuss. **11**(4), 4163–4208 (2014)
2. Attinger, S.: Generalized coarse graining procedures for flow in porous media. Comput. Geosci. **7**(4), 253–273 (2003)
3. Beckie, R., Aldama, A.A., Wood, E.F.: Modeling the large-scale dynamics of saturated groundwater flow using spatial filtering theory: 1. Theoretical development. Water Resour. Res. **32**(5), 1269–1280 (1996)
4. Beckie, R., Aldama, A.A., Wood, E.F.: Modeling the large-scale dynamics of saturated groundwater flow using spatial filtering theory: 2. Numerical Evaluation. Water Resour. Res. **32**(5), 1281–1288 (1996)
5. Bellin, A., Tonina, D.: Probability density function of non-reactive solute concentration in heterogeneous porous formations. J. Contam. Hydrol. **94**, 109–125 (2007)
6. Bilger, R.W.: The structure of diffusion flames. Combust. Sci. Technol. **13**, 155–170 (1976)
7. Cirpka, O.A., de Barros, F.P.J., Chiogna, G., Nowak, W.: Probability density function of steady state concentration in two-dimensional heterogeneous porous media. Water Resour. Res. **47**, W11523 (2011)
8. Colucci, P.J., Jaberi, F.A., Givi, P.: Filtered density function for large eddy simulation of turbulent reacting flows. Phys. Fluids **10**(2), 499–515 (1998)
9. Crăciun, M., Vamoş, C., Suciu, N.: Analysis and generation of groundwater concentration time series. Adv. Water Resour. **111**, 20–30 (2018)
10. Dagan, G.: Upscaling of dispersion coefficients in transport through heterogeneous porous formations. In: Peters, A., et al. (eds.) Computational Methods in Water Resources X, pp. 431–439. Kluwer, Norwell (1994)
11. de Barros, F.P.J., Rubin, Y.: Modelling of block-scale macrodispersion as a random function. J. Fluid Mech. **676**, 514–545 (2011)
12. Dentz, M., Tartakovsky, D.M.: Probability density functions for passive scalars dispersed in random velocity fields. Geophys. Res. Lett. **37**, L24406 (2010)
13. Dodoulas, I.A., Navarro-Martinez, S.: Large eddy simulation of premixed turbulent flames using probability density approach. Flow Turbul. Combust. **90**, 645–678 (2013)
14. Efendiev, Y., Durlofsky, L.J.: A generalized convection-diffusion model for subgrid transport in porous media. Multiscale Model. Simul. **1**(3), 504–526 (2003)

15. Efendiev, Y., Hou, T.Y.: Multiscale Finite Element Methods. Theory and Applications. Surveys and Tutorials in the Applied Mathematical Sciences, vol. 4. Springer, New York (2009)
16. Efendiev, Y.R., Durlofsky, L.J., Lee, S.H.: Modeling of subgrid effects in coarse scale simulations of transport in heterogeneous porous media. Water Resour. Res. **36**, 2031–2041 (2000)
17. Fox, R.O.: Computational Models for Turbulent Reacting Flows. Cambridge University Press, New York (2003)
18. Haworth, D.C.: Progress in probability density function methods for turbulent reacting flows. Prog. Energy Combust. Sci. **36**, 168–259 (2010)
19. Haworth, D.C., Pope, S.B.: Transported probability density function methods for Reynolds-averaged and large-eddy simulations. In: Echekki, T., Mastorakos, E. (eds.) Turbulent Combustion Modeling. Fluid Mechanics and Its Applications, vol. 95, pp. 119–142. Springer, Dordrecht (2011)
20. Heinz, S.: Unified turbulence models for LES and RANS, FDF and PDF simulations. Theor. Comput. Fluid Dyn. **21**, 99–118 (2007)
21. Herz, M.: Mathematical modeling and analysis of electrolyte solutions. PhD thesis, Nuremberg-Erlangen University (2014). http://www.mso.math.fau.de/fileadmin/am1/projects/PhD_Herz.pdf
22. Heße, F., Radu, F.A., Thullner, M., Attinger, S.: Upscaling of the advection–diffusion–reaction equation with Monod reaction. Adv. Water Resour. **32**, 1336–1351 (2009)
23. Jaberi, F.A., Colucci, P.J., James, S., Givi, P., Pope, S.B.: Filtered mass density function for large-eddy simulation of turbulent reacting flows. J. Fluid Mech. **401**, 85–121 (1999)
24. Jones, W.P., Marquis, A.J., Prasad, V.N.: LES of a turbulent premixed swirl burner using the Eulerian stochastic field method. Combust. Flame **159**, 3079–3095 (2012)
25. Klimenko, A.Y.: On simulating scalar transport by mixing between Lagrangian particles. Phys. Fluids **19**, 031702 (2007)
26. Klimenko, A.Y., Bilger, R.W.: Conditional moment closure for turbulent combustion. Prog. Energy Combust. Sci. **25**, 595–687 (1999)
27. Kloeden, P.E., Platen, E.: Numerical Solutions of Stochastic Differential Equations. Springer, Berlin (1999)
28. Kraichnan, R.H.: Diffusion by a random velocity field. Phys. Fluids **13**(1), 22–31 (1970)
29. McDermott, R., Pope, S.B.: A particle formulation for treating differential diffusion in filtered density models. J. Comput. Phys. **226**, 947–993 (2007)
30. Meyer, D.W., Jenny, P., Tchelepi, H.A.: A joint velocity-concentration PDF method for tracer flow in heterogeneous porous media. Water Resour. Res. **46**, W12522 (2010)
31. Minier, J.-P., Peirano, E.: The PDF approach to turbulent and polydispersed two-phase flows. Phys. Rep. **352**,1–214 (2001)
32. Minier, J.-P., Chibbaro, S., Pope, S.B.: Guidelines for the formulation of Lagrangian stochastic models for particle simulations of single-phase and dispersed two-phase turbulent flows. Phys. Fluids **26**, 113303 (2014)
33. Möbus, H., Gerlinger, P., Brüggemann, D.: Comparison of Eulerian and Lagrangian Monte Carlo PDF methods for turbulent diffusion flames. Combust. Flame **124**, 519–534 (2001)
34. Morales-Casique, E., Neuman, S.P., Guadagnini, A.: Nonlocal and localized analyses of nonreactive solute transport in bounded randomly heterogeneous porous media: theoretical framework. Adv. Water Resour. **29**, 1238–1255 (2006)
35. Morales-Casique, E., Neuman, S.P., Guadagnini, A.: Nonlocal and localized analyses of nonreactive solute transport in bounded randomly heterogeneous porous media: computational analysis. Adv. Water Resour. **29**, 1399–1418 (2006)
36. Mustata, R., Valiño, L., Jiménez, C., Jones, W.P., Bondi, S.: A probability density function Eulerian Monte Carlo field method for large eddy simulations: application to a turbulent piloted methane/air diffusion flame (Sandia D). Combust. Flame **145**, 88–104 (2006)
37. Pasetto, D., Guadagnini, A., Putti, M.: POD-based Monte Carlo approach for the solution of regional scale groundwater flow driven by randomly distributed recharge. Adv. Water Resour. **34**(11), 1450–1463 (2011)

38. Pope, S.B.: The probability approach to the modelling of turbulent reacting flows. Combust. Flame **27**, 299–312 (1976)
39. Pope, S.B.: A Monte Carlo method for the PDF equations of turbulent reactive flow. Combust. Sci. Technol. **25**, 159–174 (1981)
40. Pope, S.B.: PDF methods for turbulent reactive flows. Prog. Energy Combust. Sci. **11**(2), 119–192 (1985)
41. Pope, S.B.: Turbulent Flows. Cambridge University Press, Cambridge (2000)
42. Rubin, Y., Sun, A., Maxwell, R., Bellin, A.: The concept of block effective macrodispersivity and a unified approach for grid-scale and plume-scale-dependent transport. J. Fluid Mech. **395**, 161–180 (1999)
43. Sabel'nikov, V., Soulard, O.: Rapidly decorrelating velocity-field model as a tool for solving one-point Fokker–Planck equations for probability density functions of turbulent reactive scalars. Phys. Rev. E **72**(1), 016301 (2005)
44. Sanchez-Vila. X., Guadagnini, A., Fernàndez-Garcia, D.: Conditional probability density functions of concentrations for mixing-controlled reactive transport in heterogeneous aquifers. Math. Geosci. **41**, 323–351 (2009)
45. Schüler, L., Suciu, N., Knabner, P., Attinger, S.: A time dependent mixing model to close PDF equations for transport in heterogeneous aquifers. Adv. Water Resour. **96**, 55–67 (2016)
46. Schwede, R.L, Cirpka, O.A., Nowak, W., Neuweiler, I.: Impact of sampling volume on the probability density function of steady state concentration. Water Resour. Res. **44**(12), W12433 (2008)
47. Suciu, N.: Diffusion in random velocity fields with applications to contaminant transport in groundwater. Adv. Water Resour. **69**, 114–133 (2014)
48. Suciu, N., Vamoş, C.: Ergodic estimations of upscaled coefficients for diffusion in random velocity fields. In: L'Ecuyér, P., Owen, A.B. (eds.) Monte Carlo and Quasi-Monte Carlo Methods 2008, pp. 617–626. Springer, Berlin (2009)
49. Suciu, N., Vamoş, C., Vanderborght, J., Hardelauf, H., Vereecken, H.: Numerical investigations on ergodicity of solute transport in heterogeneous aquifers. Water Resour. Res. **42**, W04409 (2006)
50. Suciu, N., Radu, F.A., Attinger, S., Schüler, L., Knabner, P.: A Fokker–Planck approach for probability distributions of species concentrations transported in heterogeneous media. J. Comput. Appl. Math. **289**, 241–252 (2015)
51. Suciu, N., Schüler, L., Attinger, S., Vamoş, C., Knabner, P.: Consistency issues in PDF methods. An. St. Univ. Ovidius Constanţa **23**(3), 187–208 (2015)
52. Suciu, N., Schüler, L., Attinger, S., Knabner, P.: Towards a filtered density function approach for reactive transport in groundwater. Adv. Water Resour **90**, :83–98 (2016)
53. Valiño, L.: A field Monte Carlo formulation for calculating the probability density function of a single scalar in a turbulent flow. Flow Turbul. Combust. **60**(2), 157–172 (1998)
54. Vamoş, C., Crăciun, M.: Separation of components from a scale mixture of Gaussian white noises. Phys. Rev. E **81**, 051125 (2010)
55. Vamoş, C., Crăciun, M.: Automatic Trend Estimation. Springer, Dordrecht (2012)
56. Vamoş, C., Suciu, N., Vereecken, H.: Generalized random walk algorithm for the numerical modeling of complex diffusion processes. J. Comput. Phys. **186**(2), 527–44 (2003)
57. Vamoş, C., Crăciun, M., Suciu, N.: Automatic algorithm to decompose discrete paths of fractional Brownian motion into self-similar intrinsic components. Eur. Phys. J. B **88**, 250 (2015)
58. Venturi, D., Tartakovsky, D.M., Tartakovsky, A.M., Karniadakis, G.E.: Exact PDF equations and closure approximations for advective–reactive transport. J. Comput. Phys. **243**, 323–43 (2013)
59. Waclawczyk, M., Pozorski, J., Minier, J.P.: New molecular transport model for FDF/LES of turbulence with passive scalar. Flow Turbul. Combust. **81**, 235–260 (2008)
60. Wang, H., Popov, P.P., Pope, S.B.: Weak second-order splitting schemes for Lagrangian Monte Carlo particle methods for the composition PDF/FDF transport equations. J. Comput. Phys. **229**, 1852–1878 (2010)

Chapter 7
Model, Scale, and Measurement

Abstract In this chapter, relations between model, scale, and measurement will be discussed. A particular attention will be paid to the perspective of using spatio-temporal upscaling to bring model output closer to the measured observable of the physical system, with emphasis on hydrological observations.

7.1 Sampling Volume and Sampling Time

7.1.1 Sampling Volume Approach

Concentrations measured in samples pumped from monitoring wells or by multi-level samplers are necessarily space averages over the support scale of the sampling procedure and measuring device [3, 12, 16, 27]. The sampling volume enters as an important parameter in analytical and numerical models because of its impact on the moments and probability distribution of the uncertain concentration transported in heterogeneous aquifers [1, 7, 18, 20, 28].

The spatial average over the sampling volume is accounted for in different ways depending on modeling approaches. In analytical and semianalytical approaches, the average concentration is an *integral of a continuous field* of concentrations [1, 18] or of related probability densities [7]. The same procedure is applied to both resident and flux-weighted concentrations [18]. When the purpose is to estimate probability distributions, the resident concentration is preferred [7, 18]. Actually resident concentration solves Eq. (6.1), which is the starting point in deriving PDF evolution equations [25] and the corresponding concentration moments equations [17]. But these equations describe the statistics of the concentration without considering a sampling volume, which corresponds to a point sampling in sampling volume approaches [20, 28]. However, by averaging the solution of Eq. (6.1) with a spatial filter, a FDF approach is obtained, which explicitly takes into account the sampling volume, determined by the width of the spatial filter [26].

In case of discrete descriptions of the transport process, such as PT [4, 20, 28] and GRW simulations [24, 26, 34], concentrations are measured by *counting particles* in the sampling volume. The relation between the two spatial averaging procedures, by

© Springer Nature Switzerland AG 2019

N. Suciu, *Diffusion in Random Fields*, Geosystems Mathematics,

https://doi.org/10.1007/978-3-030-15081-5_7

integrating continuous fields and by counting particles, is not trivial. In fact, this is equivalent to passing from a microscopic level of description to a macroscopic one. Such connections can be rigorously established if the spatial sampling is replaced by a space–time average [22, 29].

7.1.2   Spatio-Temporal Upscaling

Any measurement is also a time average and hydrological measurements are no exception [1, 8]. Even if some flow and transport conditions (e.g., small spatial scale of concentration fluctuations [1] or steady state transport [18]) justify a sampling volume approach which disregards the measurement time scale, in most cases both the spatial and temporal scales specific to the observation method [3, 8] have to be considered.

Spatial and temporal scales may be related, as, for instance, for constant water volume sampling or for constant soil sample length, when the finest observation time resolution may vary between sampling locations due to aquifer's heterogeneity [8]. But the scales may also be independent, as in case of monitoring wells, where the time scale is given by the pumping time.

A theoretical approach based on spatial and temporal averages was developed to transform microscale balance equations to useful macroscopic equations [11]. The basic assumption in this approach is the existence of differential equations which describe the transport of the volumetric density of an extensive physical quantity at the microscale. The averaging domain consists of the set product between a representative elementary volume which may vary in space and time and a variable time scale. The space–time averaging domain can thus be thought as a generalization of the classical "fluid molecule." Assuming the existence of the solution of the microscale equation [6, 11], averaging theorems which relate derivatives of averages to averages of derivatives are proved and further used to derive the macroscopic equation for multiphase transport in porous media. This equation is not closed and constitutive relations are needed to formulate well-posed problems.

Most recent papers on averaging procedures for porous media do not consider the temporal averaging, focusing mainly on volume averaging theorems for multiphase systems (see [9] and the references therein). Spatial averaging procedures are used, for instance, to formulate macroscopic equations for multiphase transport [10], to understand how the internal structure of a two-fluid porous medium system influences the macroscale state [13], or to account for dependencies between the size of the averaging volume and macroscale in derivations of Darcy's law from Navier–Stokes equations [15].

The reason that the average over the temporal domain is often disregarded could be that the micro time scales are believed to be similar to the macroscales [6]. That this assumption is not always true is shown by the following example. In Fig. 7.1 are represented noise components of concentration time series constructed by counting the number of particles at the center of mass of the plume in a two-dimensional

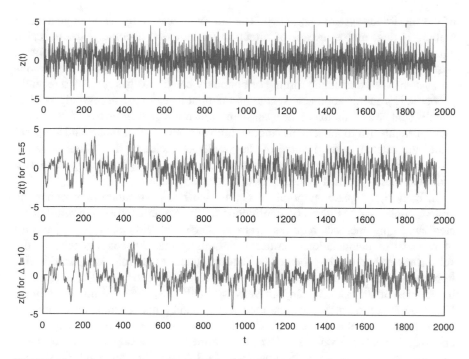

Fig. 7.1 The noise component $z(t)$ of a concentration time series obtained by an instantaneous sampling (upper panel) and noise components for time series sampled over symmetrical temporal intervals of semilengths $\Delta t = 5$ (middle panel) and $\Delta t = 10$ (lower panel)

GRW simulation of scalar transport in heterogeneous velocity fields in two different ways. First, the particles inside a slab of width equal to the transverse dimension of a rectangular domain and longitudinal dimension equal to the velocity correlation length are counted at a fixed time, and second, the particles are counted, in the same slab, over symmetrical time intervals of semilength Δt, with $\Delta t = 5$ and $\Delta t = 10$ dimensionless time units. The noise component $z(t)$ of the series is obtained after removing the additive and multiplicative deterministic components by the method introduced in [5]. One remarks that while in the first case, which corresponds to a spatial averaging, the noise is almost homogeneous, in case of sampling over finite time intervals, which corresponds to a space–time average, it is an inhomogeneous correlated noise, as indicated by the persistence of similar local trends of the noise $z(t)$ for times $t \leq 500$.

The averaging approaches mentioned above aim at upscaling processes from the microscale (pore-scale) to the macroscale of a representative elementary volume (Darcy scale) or to a kind of spatio-temporal fluid molecule. Though they can serve as models, they do not account for the scales of the measurement instruments and methods [3, 8] and for the spatial and temporal scales of the hydrological processes [19]. Upscaling continuous descriptions from Darcy scale to coarser spatial scales can be done by homogenization methods for non-periodic and random parameters of the subsurface system (e.g., [14]). The attractive feature of the homogenization method is that, unlike in case of direct spatio-temporal averaging, the upscaled equations obtained by homogenization are closed. A concept of spatio-temporal upscaling in the context of homogenization by multiple-scale asymptotic expansions for systems with fluctuating forcing or boundary conditions has been also recently proposed [2]. In this approach, spatial and temporal scales are necessarily related, which could be a limitation of the spatio-temporal homogenization from the point of view of engineering applications. Another inconvenience is that spatial and temporal scales are related to the inverse of the small parameter of the homogenization problem and the macroscopic description obtained when this parameter goes to zero cannot be directly related to the scales of the measurement.

In both direct spatio-temporal averaging and homogenization approaches the "microscopic" state is given by continuous fields. An alternative "counting particles" averaging approach requires a microscopic description given by trajectories of physical or computational particles. Assuming that the heterogeneity of the subsurface system can be modeled by random functions, a "microscopic" description corresponding to the local, Darcy scale of the transport process is given by trajectories of diffusion in random velocity fields [23] modeled numerically by PT or GRW algorithms. A general method to obtain a macroscopic description at spatio-temporal scales that can be directly related to those of the measurements and hydrological observations is the coarse-grained (CG) description of corpuscular systems through space–time (ST) averages [30]. The CGST approach also yields unclosed macroscopic equations but, since a complete microscopic description is available, it provides a frame to investigate the structure of the constitutive relations through numerical simulations [32].

7.2 Coarse-Grained Spatio-Temporal Upscaling

7.2.1 Coarse-Grained Space–Time Averages

The CGST method allows continuous modeling of corpuscular systems, without the need to use the vague notion of "fluid particle" which is small enough to be infinitesimal with respect to the size of the macroscopic system, but large enough to contain many molecules in local thermodynamic equilibrium. Continuous macroscopic fields and balance equations are derived under the only assumption of the existence of a kinematic microscopic description through piecewise smooth time functions.

If a microscopic description of a system of N particles (e.g., molecules and other physical particles, or computational particles used in GRW and PT simulations [34]) in terms of piecewise analytic time functions $\varphi_i(t) : [0, T] \longmapsto \mathbb{R}, i = 1, \ldots, N$, describing physical properties associated with each particle is available, then there exists a macroscopic description of the same system given by almost everywhere (a.e.) continuous fields through *coarse-grained space–time averages*

$$\langle\varphi\rangle(\mathbf{r}, t) = \frac{1}{2\tau\mathcal{V}} \sum_{i=1}^{N} \int_{t-\tau}^{t+\tau} \varphi_i(t') H(a^2 - (\mathbf{r}_i(t') - \mathbf{r})^2) \, dt', \tag{7.1}$$

where $\mathbf{r}_i(t)$ is the trajectory of the ith particle, $\tau < T/2$ and a are real positive parameters, $\mathcal{V} = 4\pi a^3/3$ is the volume of the sphere $S(\mathbf{r}, a)$, and H is the left continuous Heaviside function. Moreover, $\langle\varphi\rangle(\mathbf{r},t)$ has continuous partial derivatives a.e. in $\mathbb{R}^3 \times (\tau, T - \tau)$ and satisfies the identity

$$\partial_t\langle\varphi\rangle + \partial_\alpha\langle\varphi\xi_\alpha\rangle = \langle\frac{d}{dt}\varphi\rangle + \delta\varphi, \tag{7.2}$$

where $\xi_{i\alpha} = dr_{i\alpha}/dt$, $\alpha = 1, \ldots, 3$, are the velocity components of the ith particle and $\delta\varphi$ accounts for discontinuous variations of the integrand in (7.1) [30] .

For $\varphi_i \equiv 1$ the CGST average (7.1) counts the number of particles inside the sphere $S(\mathbf{r}, a)$ during a time interval of length 2τ, that is, $c = \langle 1 \rangle$ is an a.e. continuous concentration field. The velocity field, which is not only a volume density but also an average over the number of particles, is defined by $u_\alpha = \langle\xi_\alpha\rangle/\langle 1 \rangle$ [30, 33]. For $\varphi_i \equiv 1$ the first source term in (7.2), $\langle\frac{d}{dt}\varphi\rangle$, vanishes. The term $\delta\varphi$ vanishes as well, because in the absence of chemical reactions the number of particles is conserved. Thus, the identity (7.2) takes the form of the continuity equation,

$$\partial_t c + \partial_\alpha(c u_\alpha) = 0.$$

Considering the velocity v_α of the time averaged center of mass $\langle r_\alpha \rangle$ of the particles inside the sphere $S(\mathbf{r}, a)$, related to u_α by $c u_\alpha = c v_\alpha - \partial_\beta(c D_{\alpha\beta})$, according to (7.2), one obtains a diffusion equation describing the *self-diffusion* process, with a coefficient $D_{\alpha\beta}$ completely determined by the microscopic description of the system of particles [21, Eq. (4.12)]. In multi-component systems, the CG relative velocity with respect to the barycentric velocity u_α defines the unclosed particle flux of the chemical *inter-diffusion* process [29, Chp. 4, Eq. (4.37)]. Balance equations for velocity and kinetic energy can be derived from (7.2) in a similar way. These equations are unclosed, but they can be used to test various phenomenological constitutive laws in hydrodynamics [32].

For $\tau \to 0$ the CG average (7.1) with $\varphi_i \equiv 1$ gives the estimation of the concentration by the usual volume sampling formula. The essential advantage of (7.1) is that the time average yields a.e. continuous fields obeying the identity (7.2), which provides a frame to close various balance equations. Moreover, while the usual volume sampling approach requires a large number of particles for good concentration estimates, (7.1) yields concentration and velocity fields with non-vanishing support even for a single particle. This has been illustrated by the derivation of a hydrodynamic description for the record of a single stock price

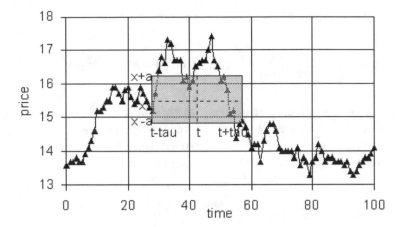

Fig. 7.2 Stock price recording and averaging window in the price-time domain

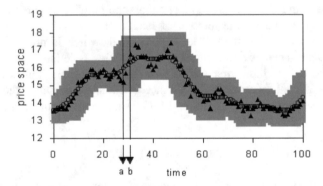

Fig. 7.3 The support of the concentration field is a price-time band that follows the trend of the mean (open circle) and of the recorded price (triangle)

in a financial market [33]. The price-time averaging procedure and the resulting concentration and velocity fields are presented in Figs. 7.2, 7.3, 7.4, 7.5, and 7.6 below.

Figure 7.2 shows the trajectory of the price x recorded at temporal intervals Δt and illustrates the CGST averaging in a price-time window of dimensions $2a$ by 2τ. The concentration defined by $c(x, t) = \langle 1 \rangle(x, t)$ verifies the continuity equation

$$\partial_t c + \partial_x (c u) = 0,$$

where $u = \langle \xi \rangle / c$ is the velocity of the price defined by the CGST procedure applied to stepwise constant velocities of the price $\xi = \Delta x / \Delta t$.

Figure 7.3 presents the support of the concentration $c(x, t)$ for an averaging window with $a = 0.5$ and $\tau = 5$. The mean price shown by circles in Fig. 7.3

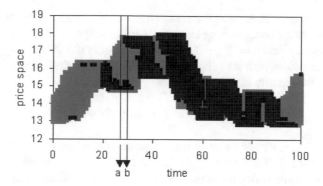

Fig. 7.4 The support of the velocity field. Grey zones correspond to positive velocity (rising price) and black zones to negative velocity (lowering price)

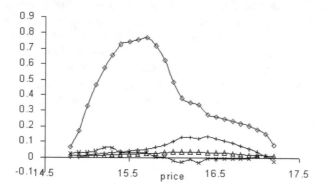

Fig. 7.5 STCG averages at t=27: concentration (open diamond), velocity (plus symbol), velocity fluctuation (open triangle), and source term (times symbol)

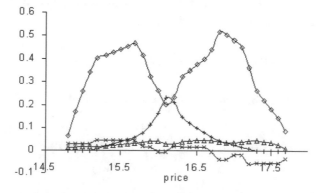

Fig. 7.6 The same STCG averages as in Fig 7.3 computed at t=30

is a spatial average weighted by the concentration, $\int_{-\infty}^{+\infty} x\, c(x,t)\, dx$. If $a \ll 1$, this mean reduces to the usual moving average over the temporal interval $[t - \tau, t + \tau]$. Since the velocity field $u = \langle \xi \rangle / c$ is non-vanishing only if $c \neq 0$, its support, shown in Fig. 7.4, coincides with the support of the concentration field shown in Fig. 7.3.

If $\varphi = \xi$, the identity Eq. (7.1) becomes

$$\partial_t\,(c\,u) + \partial_x\,(c\,u^2) + \partial_x\,(c\,\theta) = \delta_d\,\xi,$$

where $\theta = \langle (\xi - u)^2 \rangle / c$ is an average square fluctuation of the "microscopic" velocity ξ. Note that the term $\langle \frac{d}{dt}\xi \rangle$ in Eq. (7.2) vanishes because the velocity is piecewise constant. But the source term $\delta \xi$ is in general non-vanishing due to discontinuous variations of the integrand in (7.1) when particles with different velocities enter/leave the averaging domain. This term can be estimated numerically from identity (7.2) and averages computed by the CGST average (7.1). The procedure is illustrated in Figs. 7.5 and 7.6 which show the concentration c, the velocity u, the average velocity fluctuation θ, and the source term $\delta \xi$ computed in the "cuts" $t = a$ and $t = b$ through the support of the fields $c(x, t)$ and $u(x, t)$ shown in Figs. 7.3 and 7.4. One remarks that, under the influence of a nonuniform velocity (for $t = b$) the concentration develops a bimodal shape. The velocity fluctuation has a rather complex variation, and the source term is positive at low prices and negative at high prices, so that it retains the stock price in a bounded domain.

7.2.2 Continuous Fields

In the framework of stochastic modeling of transport in groundwater [23], the continuous concentration field is defined as an average M over the ensemble of realizations of the transport process, $c(\mathbf{r}, t) = M[\delta(\mathbf{r} - \mathbf{R}(t, \omega))]$, where $\mathbf{R}(t, \omega)$ denotes the realization ω of the random trajectory (also see Sect. 4.1.2). With $\mathbf{r}(t) = \mathbf{R}(t, \omega)$ and expressing the Heaviside function in (7.1) with the aid of the indicator function $1_{S(\mathbf{r},a)}(\mathbf{R}(t, \omega))$ of the sphere $S(\mathbf{r}, a)$,

$$H(a^2 - (\mathbf{r}_i(t') - \mathbf{r})^2) = 1_{S(\mathbf{r},a)}(\mathbf{R}(t, \omega)) = \int_{\mathbb{R}^3} 1_{S(\mathbf{r},a)}(\mathbf{r}')\delta(\mathbf{r}' - \mathbf{R}(t, \omega))d\mathbf{r}',$$

the averaging operator M applied to (7.1), for $\varphi_i \equiv 1$, yields

$$M[\langle 1 \rangle](\mathbf{r}, t; a, \tau) = \frac{1}{2\tau \mathscr{V}} \int_{t-\tau}^{t+\tau} dt' \int_{S(\mathbf{r},a)} c(\mathbf{r}', t')d\mathbf{r}'. \tag{7.3}$$

Equation (7.3) is a particular case of the general relation between the smooth continuous field and the CGST average of its microscopic description [21, Eq. (4.6)].

For $\tau \to 0$, Eq. (7.3) gives the relation between the "integrating continuous fields" and "counting particles" averaging procedures in sampling volume approaches.

The use of CGST averages (7.1) in numerical simulations is conditioned by the piecewise smoothness of $\varphi(t)$. Random walks on lattices are eligible because the corresponding trajectories $\mathbf{r}_i(t)$ are piecewise linear and the velocities $\xi_{i\alpha}$ are piecewise constant functions with numerable discontinuity points at the jump moments [31]. In GRW simulations, the sum over the number of particles in (7.1) is equivalent to summing up the contributions of the groups of particles jumping between the same lattice sites in the interior of the sphere $S(\mathbf{r}, a)$, in the time interval $(t - \tau, t + \tau)$.

A kinematic microscopic description of reactive transport in random velocity fields can be designed as an integrated numerical procedure, by computing realizations of the velocity field with the BGRW method described in Sect. 3.3.4.1 on the same lattice as that used to compute GRW transport solutions. Such an integrated GRW approach avoids interpolation errors, inherent when flow solutions provided by different solvers are imported into the GRW lattice. At the same time, GRW solutions to reactive transport do not need scaling factors to relate concentration and particle densities, because the number of computational particles in the GRW procedure can be as large as the number of molecules involved in chemical reactions (see Sect. 3.3.4.2).

7.2.3 CGST Average Concentration

Equations for CGST average concentration are obtained from the general identity (7.2) verified by the CGST averages $\langle \varphi \rangle$. Under local equilibrium conditions, with $\varphi_{\nu i} = 1$ if $i \leq N_\nu$ and $\varphi_{\nu i} = 0$ if $i > N_\nu$, the average $\langle \varphi_\nu \rangle$ estimates the concentration c_ν of the molecular species ν. If the relative velocity with respect to the barycentric velocity $u_\alpha = \langle \xi_\alpha \rangle / \langle 1 \rangle$, computed for $N = \sum_\nu^{\mathcal{N}} N_\nu$ particles, is introduced in (7.1), one obtains a general transport equation which, according to (7.3), governs the CGST averaged solution of the local balance equation (6.1),

$$\partial_t \langle \varphi_\nu \rangle + \partial_\alpha \langle \varphi_\nu u_\alpha \rangle + \partial_\alpha \langle \varphi_\nu (\xi_{\nu\alpha} - u_\alpha) \rangle = \delta \varphi_\nu. \tag{7.4}$$

The right-hand side of (7.4) is an upscaled reaction term which can be evaluated by computing the sum of terms in the left-hand side, as shown in Sect. 7.2.1. The sum of reaction terms vanishes as a consequence of mass conservation and, if the carrying fluid is not included among the $\mathcal{N}$ chemical species, $\sum_{\nu=1}^{\mathcal{N}} \varphi_\nu$ verifies an advection–diffusion equation (Eq. (7.4) with right-hand side set to zero) [26, Sect. 3.1]. The departure from zero of the sum of reaction terms $\sum_{\nu=1}^{\mathcal{N}} \delta \varphi_\nu$ can serve as a global error estimation of the CGST average of the integrated GRW flow and transport solution.

Equation (7.4) verified by CGST concentrations could be a starting point for various investigations, such as:

(a) check the validity of the Fick's law for the diffusion flux of the CGST average concentration and assess space–time upscaled diffusion coefficients,
(b) compare self-diffusion and inter-diffusion coefficients (see Sect. 7.2.1) and check whether the assumption of species independent diffusion coefficients holds at the measurement scale (a, τ),
(c) investigate numerically the structure of the upscaled reaction rates and their relation with the microscopic rates used in integrated GRW simulations,
(d) investigate the possibility to formulate closed macroscopic balance equations for reactive transport valid at the measurement scale (a, τ).

The space–time averaging generalizes the spatial filtering in (6.3) used in the FDF approach described in Chap. 6. However, assessing probability densities through CGST averages is more challenging than the FDF approach. The CGST average (7.1) does not commute with the time derivative and a CGST density function obtained similarly to the approach in Appendix E.1 by averaging the delta-function $\delta(c - c(\mathbf{r}, t))$ does not solve the classical PDF/FDF equations. Assessing the feasibility of such a CGST approach to model concentration probability distributions requires further theoretical investigations.

References

1. Andričević, R.: Effects of local dispersion and sampling volume on the evolution of concentration fluctuations in aquifers. Water Resour. Res. **34**(5), 1115–1129 (1998)
2. Battiato, I.: Multiscale dynamics of reactive fronts in heterogeneous media with fluctuating forcings. In: SIAM Conference on Mathematical and Computational Issues in the Geosciences September 11–14, 2017, Erlangen. http://www.siam-gs17.de/program/invited-talks-prize-presentations
3. Bayer-Raich, M., Jarsjö, J., Liedl, R., Ptak, T., Teutsch, G.: Average contaminant concentration and mass flow in aquifers from time-dependent pumping well data: analytical framework. Water Resour. Res. **40**(8), W08303 (2004)
4. Caroni, E., Fiorotto, V.: Analysis of concentration as sampled in. Transp. Porous Media **59**(1), 19–45 (2005)
5. Crăciun, M., Vamoș, C., Suciu, N.: Analysis and generation of groundwater concentration time series. Adv. Water Resour. **111**, 20–30 (2018)
6. Cushman, J.H.: Multiphase transport equations: I-general equation for macroscopic statistical, local space-time homogeneity. Transp. Theory Stat. Phys. **12**(1), 35–71 (1983)
7. de Barros, F.P.J., Fiori, A.: First-order based cumulative distribution function for solute concentration in heterogeneous aquifers: theoretical analysis and implications for human health risk assessment. Water Resour. Res. **50**(5), 4018–4037 (2014)
8. Destouni, G., Graham, W.: The influence of observation method on local concentration statistics in the subsurface. Water Resour. Res. **33**(4), 663–676 (1997)
9. Gray, W.G., Miller, C.T.: A generalization of averaging theorems for porous medium analysis. Adv. Water Res. **62**, 227–327 (2013)

10. Gray, W.G., Dye, A.L., McClure, J.E., Pyrak-Nolte, L.J., Miller, C.T.: On the dynamics and kinematics of two-fluid-phase flow in porous media. Water Resour. Res. **51**(7), 5365–5381 (2015)

11. He, Y., Sykes, J.F.: On the spatial-temporal averaging method for modeling transport in porous media. Transp. Porous Media **22**(1), 1–51 (1996)

12. Mailloux, B.J., Fuller, M.E., Rose, G.F., Onstott, T.C., DeFlaun, M.F., Alvarez, E., Hemingway, C., Hallet, R.B., Phelps, T.J., Griffin, T.: A modular injection system, multilevel sampler, and manifold for tracer tests. Ground Water **41**(6), 816–827 (2003)

13. McClure, J.E., Dye, A.L., Miller, C.T., Gray, W.G.: On the consistency of scale among experiments, theory, and simulation. Hydrol. Earth Syst. Sci. **21**(2), 1063 (2017)

14. Nolen, J., Papanicolaou, G., Pironneau, O.: A framework for adaptive multiscale methods for elliptic problems. Multiscale Model. Simul. **7**, 171–196 (2008)

15. Nordbotten, J.M., Celia, M.A., Dahle, H.K., Hassanizadeh, S.M.: Interpretation of macroscale variables in Darcy's law. Water Resour. Res. **43**(8), W08430 (2007)

16. Ptak, T., Schwarz, R., Holder, T., Teutsch, G.: Ein neues Verfahren zur Quantifizierung der Grundwasserimmission: I. Theoretische Grundlagen. Grundwasser **4**(5), 176–183 (2000)

17. Schüler, L., Suciu, N., Knabner, P., Attinger, S.: A time dependent mixing model to close PDF equations for transport in heterogeneous aquifers. Adv. Water Resour. **96**, 55–67 (2016)

18. Schwede, R.L., Cirpka, O.A., Nowak, W., Neuweiler, I.: Impact of sampling volume on the probability density function of steady state concentration. Water Resour. Res. **44**(12), W12433 (2008)

19. Skøien, J.O., Blöschl, G., Western, A.W.: Characteristic space scales and timescales in hydrology. Water Resour. Res. **39**(10), 1304 (2003)

20. Srzic, V., Cvetkovic, V., Andricevic, R., Gotovac, H.: Impact of aquifer heterogeneity structure and local-scale dispersion on solute concentration uncertainty. Water Resour. Res. **49**(6), 3712–3728

21. Suciu, N.: Mathematical background for diffusion processes in natural porous media. Forschungszentrum Jülich/ICG-4, Internal Report No. 501800 (2000)

22. Suciu, N.: On the connection between the microscopic and macroscopic modeling of the thermodynamic processes (in Romanian). Ed. Univ. Piteşti, Piteşti (2001)

23. Suciu, N.: Diffusion in random velocity fields with applications to contaminant transport in groundwater. Adv. Water. Resour. **69**, 114–133 (2014)

24. Suciu, N., Vamoş, C., Vanderborght, J., Hardelauf, H., Vereecken, H.: Numerical investigations on ergodicity of solute transport in heterogeneous aquifers. Water Resour. Res. **42**, W04409 (2006)

25. Suciu, N., Radu, F.A., Attinger, S., Schüler, L., Knabner, P.: A Fokker–Planck approach for probability distributions of species concentrations transported in heterogeneous media. J. Comput. Appl. Math. **289**, 241–252 (2015)

26. Suciu, N., Schüler, L., Attinger, S., Knabner, P.: Towards a filtered density function approach for reactive transport in groundwater. Adv. Water Resour. **90**, 83–98 (2016)

27. Teutsch, G., Ptak, T., Schwarz, R., Holder, T.: Ein neues Verfahren zur Quantifizierung der Grundwasserimmission: I. Theoretische Grundlagen. Grundwasser **4**(5), 170–175 (2000)

28. Tonina, D., Bellin, A.: Effects of pore-scale dispersion, degree of heterogeneity, sampling size, and source volume on the concentration moments of conservative solutes in heterogeneous formations. Adv. Water Resour. **31**(2), 339–354 (2008)

29. Vamoş, C.: Continuous fields associated to corpuscular systems (in Romanian). Risoprint, Cluj-Napoca (2007). English version http://ictp.acad.ro/vamos/cv-thesis/

30. Vamoş, C., Georgescu, A., Suciu, N., Turcu, I.: Balance equations for physical systems with corpuscular structure. Phys. A: Stat. Mech. Appl. **227**(1), 81–92 (1996)

31. Vamoş, C., Suciu, N., Peculea, M.: Numerical modelling of the one-dimensional diffusion by random walkers. Rev. Anal. Numér. Théor. Approx. **26**(1), 237–247 (1997)

32. Vamoş, C., Suciu, N., Georgescu, A.: Hydrodynamic equations for one-dimensional systems of inelastic particles. Phys. Rev. E **55**(5), 6277–6280 (1997)

33. Vamoş, C., Suciu, N., Blaj, W.: Derivation of one-dimensional hydrodynamic model for stock price evolution. Phys. A: Stat. Mech. Appl. **287**(3), 461–467 (2000)
34. Vamoş, C., Suciu, N., Vereecken, H.: Generalized random walk algorithm for the numerical modeling of complex diffusion processes. J. Comput. Phys. **186**(2), 527–244 (2003)

Appendix A
Numerical Simulation of Diffusion Processes

Numerical solutions of diffusion processes will be constructed with programming tools presented in Sect. 3.1.1. To ensure the reproducibility of the results, the seed of the random number generators is fixed by the script "initstate.m",

```
1  % Save as script "initstate.m" in the folder containing
2  % Matlab codes to be executed.
3  state = 137; rand('state', state); randn('state', state);
```

which is called as `initstate; state;` in Matlab simulation codes presented in the following.

A.1 Diffusion Processes Constructed with i.i.d. Variables

```
1   % Discrete diffusion processes generated with Eqs. ...
        (3.4)-(3.6) for different i.i.d. random variables
2   clear all;
3   tic
4   ex=1; %Poisson
5   %ex=2 %rand U(0.1)
6   %ex=3 %randn N(0,1)
7   N=1000; T=100; % N samples of length T
8   initstate; state;
9   D=zeros(T,3); M=zeros(T,3); V=zeros(T,3);
10  for n=1:N
11      if (ex==1) m=2.0; s=1.4142; end %[m,v]=poisstat(2);
12      if (ex==2) m=0.5; s=0.2886; end %[m,v]=unifstat(0,1);
13      if (ex==3) m=0; s=1; end %[m,v]=normstat(0,1); s=sqrt(v);
14      Z=randn(1,T);
15      S=0; R=0; X=zeros(T,3);
16      for t=1:T
17          if (ex==1) S=S+poissrnd(2); X(t,1)=S; end % Eq. (3.4)
```

© Springer Nature Switzerland AG 2019
N. Suciu, *Diffusion in Random Fields*, Geosystems Mathematics,
https://doi.org/10.1007/978-3-030-15081-5

```
18        if (ex==2) S=S+rand;         X(t,1)=S; end
19        if (ex==3) S=S+Z(t);         X(t,1)=S; end
20        X(t,2)=t*m+s*sqrt(t)*Z(t); % Eq. (3.5)
21        R=R+m+s*Z(t); X(t,3)=R;     % Eq. (3.6)
22        for i=1:3
23            M(t,i)=M(t,i)+X(t,i)/N;           %mean
24            V(t,i)=V(t,i)+X(t,i)*X(t,i)/N; %variance
25        end
26     end
27     if n==1
28        Xt=X;
29     end
30  end
31  for t=1:T
32     for i=1:3
33        V(t,i)=V(t,i)-M(t,i)*M(t,i);
34        D(t,i)=V(t,i)/(2*t); %diffusion coefficient
35     end
36  end
37  t = 1:1:T;
38  figure;
39  subplot(2,2,1)
40  plot(t,Xt(t,1),'ok',t,Xt(t,2),'b',t,Xt(t,3),'r', ...
       'MarkerSize',2)
41  legend('sum of i.i.d.','CLT','Euler ...
       scheme','Location','northwest');legend('boxoff')
42  xlabel('n'); ylabel('X_n'); set(gca, 'XTick', [0:20:100])
43  subplot(2,2,2)
44  plot(t,M(t,1),'ok',t,M(t,2),'b',t,M(t,3),'r','MarkerSize',2);
45  xlabel('n'); ylabel('\mu_{X_n}'); set(gca, 'XTick', ...
       [0:20:100])
46  subplot(2,2,3)
47  plot(t,V(t,1),'ok',t,V(t,2),'b',t,V(t,3),'r','MarkerSize',2);
48  xlabel('n'); ylabel('\sigma^{2}_{X_{n}}'); set(gca, ...
       'XTick', [0:20:100])
49  subplot(2,2,4)
50  plot(t,D(t,1),'ok',t,D(t,2),'b',t,D(t,3),'r','MarkerSize',2);
51  xlabel('n'); ylabel('D'); set(gca, 'XTick', [0:20:100])
52  if (ex==1) ylim([0 2]); end;
53  if (ex==2) ylim([0 0.1]); end;
54  if (ex==3) ylim([0.25 0.75]); end
55  toc
```

A.2   Itô–Euler Schemes

```
1  % Strong and weak Euler schemes for Ito equation
2  clear all; close all;
3  tic
4  %ex=1 %Ito linear w. cst. coefs.: dX=aXdt+bXdW
```

```
 5    %ex=2 %Ito exponential          : dX=0.5Xdt+XdW
 6    %ex=3 %Ito drift-free           : dX=XdW
 7    ex=4; %Ornstein-Uhlenbeck       : dX=-Xdt+dW
 8    if (ex==1) a=1.5;  b=1.0;   end
 9    if (ex==2) a=0.5;  b=1.0;   end
10    if (ex==3) a=0.0;  b=1.0;   end
11    if (ex==4) a=-1.0; b=1.0;   end
12    X0=1.0;
13    N=10^6; d1=3; dd=6; de=zeros(1,dd);
14    eX=zeros(dd,2); eM=zeros(dd,2);
15    eV=zeros(dd,2); eD=zeros(dd,2);
16    for d=1:dd
17        initstate; state;
18        dt=2^-(d+d1); s=sqrt(dt); T=1/dt;
19        D=zeros(T,3); M=zeros(T,3); V=zeros(T,3);
20        M(1,1)=X0; M(1,2)=X0; M(1,3)=X0;
21        V(1,1)=X0^2; V(1,2)=X0^2; V(1,3)=X0^2;
22        for n=1:N
23            W=0; X=zeros(T,3);
24            X(1,1)=X0; X(1,2)=X0; X(1,3)=X0;
25            for t=1:(T-1)
26                Z=s*randn; W=W+Z;
27                %exact analytical solution:
28                X(t+1,1)=X(1,1)*exp((a-b*b/2)*t*dt+b*W);
29                %strong Euler scheme:
30                X(t+1,2)=X(t,2)+a*X(t,2)*dt+b*X(t,2)*Z;
31                if (rand≤0.5) r=-s;
32                else r=s;
33                end
34                %weak Euler scheme:
35                X(t+1,3)=X(t,3)+a*X(t,3)*dt+b*X(t,3)*r;
36                for i=1:3
37                    M(t+1,i)=M(t+1,i)+X(t+1,i)/N; %mean
38                    V(t+1,i)=V(t+1,i)+X(t+1,i)*X(t+1,i)/N; %var
39                end
40                %error averaged over the time interval [0,1]:
41                eX(d,1)=eX(d,1)+abs(X(t+1,1)-X(t+1,2))/N*dt;
42                eX(d,2)=eX(d,2)+abs(X(t+1,1)-X(t+1,3))/N*dt;
43            end
44            if (n==1 && d==dd-1)
45                t=1:T;
46                figure(1); subplot(2,2,1)
47                plot(t*dt,X(t,1),t*dt,X(t,2),t*dt,X(t,3));
48                legend('analytical solution','strong Euler ...
                        scheme','weak Euler scheme');
49                legend('boxoff')
50                xlabel('t'); ylabel('X(t)');
51                set(gca, 'XTick', [0:0.25:1])
52            end
53        end
54        for t=1:T
55            for i=1:3
56                V(t,i)=V(t,i)-M(t,i)*M(t,i); %variance
57                D(t,i)=V(t,i)/(2*t*dt); %diffusion coefficent
```

```
58          end
59          %error averaged over the time interval [0,1]:
60          eM(d,1)=eM(d,1)+abs(M(t,1)-M(t,2))*dt;
61          eM(d,2)=eM(d,2)+abs(M(t,1)-M(t,3))*dt;
62          eV(d,1)=eV(d,1)+abs(V(t,1)-V(t,2))*dt;
63          eV(d,2)=eV(d,2)+abs(V(t,1)-V(t,3))*dt;
64          eD(d,1)=eD(d,1)+abs(D(t,1)-D(t,2))*dt;
65          eD(d,2)=eD(d,2)+abs(D(t,1)-D(t,3))*dt;
66      end
67      if d==dd
68          t = 1:1:T;
69          figure(1); subplot(2,2,2)
70          plot(t*dt,M(t,1),t*dt,M(t,2),t*dt,M(t,3));
71          xlabel('t'); ylabel('\mu(t)');
72          set(gca, 'XTick', [0:0.25:1])
73          figure(1); subplot(2,2,3)
74          plot(t*dt,V(t,1),t*dt,V(t,2),t*dt,V(t,3));
75          xlabel('t'); ylabel('\sigma^2(t)');
76          set(gca, 'XTick', [0:0.25:1])
77          figure(1); subplot(2,2,4)
78          plot(t*dt,D(t,1),t*dt,D(t,2),t*dt,D(t,3));
79          xlabel('t'); ylabel('D(t)');
80          set(gca, 'XTick', [0:0.25:1])
81      end
82      de(d)=dt;
83  end
84  d=1:dd; te=de(d)';
85  fX=fit(te,eX(d,1),'power1')
86  figure(2); subplot(2,2,1)
87  loglog(de(d),eX(d,1),de(d),eX(d,2),de(d),fX(de(d)));
88  legend('strong Euler scheme','weak Euler ...
            scheme','Location','best');
89  legend('boxoff');
90  xlabel('\Delta'); ylabel('\epsilon_{X} (\Delta)');
91  fM=fit(te,eM(d,1),'power1')
92  figure(2); subplot(2,2,2)
93  loglog(de(d),eM(d,1),de(d),eM(d,2),de(d),fM(de(d)));
94  xlabel('\Delta'); ylabel('\epsilon_{\mu} (\Delta)');
95  fV=fit(te,eV(d,1),'power1')
96  figure(2); subplot(2,2,3)
97  loglog(de(d),eV(d,1),de(d),eV(d,2),de(d),fV(de(d)));
98  xlabel('\Delta'); ylabel('\epsilon_{\sigma^2} (\Delta)');
99  fD=fit(te,eD(d,1),'power1')
100 figure(2); subplot(2,2,4)
101 loglog(de(d),eD(d,1),de(d),eD(d,2),de(d),fD(de(d)));
102 xlabel('\Delta'); ylabel('\epsilon_{D} (\Delta)');
103 toc
```

A.3   Global Random Walk

A.3.1   *One-Dimensional Unbiased GRW Algorithms*

A.3.1.1   GRW Algorithm for Wiener Processes

```
1   % GRW - a superposition of weak Euler schemes on a lattice
2   clear all; close all;
3   tic
4   initstate; state;
5   %N=10^306
6   %N=10^24
7   N=300;
8   L=200; T=200; n=zeros(1,L); nn=zeros(1,L);
9   x0=L/2; n(x0)=N; dt=0.5;
10  D=zeros(1,T); eD=zeros(1,T); M=zeros(1,T); V=zeros(1,T);
11  for t=1:T
12      ntot=0;
13      if (t==1)
14          figure(1);subplot(2,2,1);
15          bar(n); xlim([95 105]);
16          title('t=0','FontSize',8,'FontWeight','normal')
17          xlabel('x'); ylabel('n');
18      end
19      for x=2:(L-1)
20          if n(x) > 0
21              if rem(n(x),2) ≠ 0
22                  nr=floor(n(x)/2);
23                  nn(x-1)=nn(x-1)+nr; nn(x+1)=nn(x+1)+nr;
24                  if (rand≤0.5) nn(x-1)=nn(x-1)+1;
25                  else nn(x+1)=nn(x+1)+1;
26                  end
27              else
28                  nn(x-1)=nn(x-1)+n(x)/2; nn(x+1)=nn(x+1)+n(x)/2;
29              end
30          end
31      end
32      for x=1:L
33          xr=x-x0;
34          n(x)=nn(x); nn(x)=0; ntot=ntot+n(x);
35      end
36      if (t==1 || t==2 || t== 3)
37          figure(1); subplot(2,2,t+1); bar(n); xlim([95 105]);
38          title(['t=',num2str(t)],...
39          'FontSize',8,'FontWeight','normal')
40          xlabel('x'); ylabel('n');
41      end
42  end, ntot;
43  figure; bar(n); xlim([60 140]);
44  title('t=200','FontSize',8,'FontWeight','normal');
45  xlabel('x'); ylabel('n');
46  toc
```

A.3.1.2   One-Dimensional Unbiased GRW Algorithm
for Advection-Diffusion Processes

```
1
2   % Unbiased GRW algorithm for advection-diffusion
3   close all; clear all;
4   tic
5   initstate; state;
6   N=10^24;
7   L=400; T=200;
8   D0=0.25; d=1; dx=1;
9   V0=1; s=1; dt=s*dx/V0; r=2*D0*dt/(d^2*dx^2);
10  n=zeros(1,L); nn=zeros(1,L);
11  x0=L/4; n(x0)=N;
12  D=zeros(1,T); M=zeros(1,T); V=zeros(1,T);
13  restr=0; restjump=0; restsarI=0; rest2=0;
14  for t=1:T
15      ntot=0;
16      if (t==1)
17          figure(1);subplot(2,2,1); bar(n);
18          xlabel('x'); ylabel('n'); xlim([x0-5 x0+10]);
19          title('t=0','FontSize',8,'FontWeight','normal');
20      end
21      for x=2:(L-1)
22          if n(x) > 0
23              xa=x+s;
24              restr=n(x)*(1-r)+restr; nsta=floor(restr);
25              restr=restr-nsta; njump=n(x)-nsta;
26              nn(xa)=nn(xa)+nsta;
27              if(njump)>0
28                  restjump=njump/2+restjump;
29                  nj(1)=floor(restjump); ...
                        restjump=restjump-nj(1);
30                  nj(2)=njump-nj(1);
31                  for i=1:2
32                      if nj(i)>0
33                          xj=xa+(2*i-3)*d;
34                          nn(xj)=nn(xj)+nj(i);
35                      end
36                  end
37              end
38          end
39      end
40      for x=1:L
41          xr=(x-x0)*dx;
42          n(x)=nn(x); nn(x)=0; ntot=ntot+n(x);
43          M(t)=M(t)+xr*n(x); V(t)=V(t)+xr^2*n(x);
44      end
45      if (t==1 || t==2 || t== 3)
46          figure(1);subplot(2,2,t+1); bar(n);
47          xlabel('x'); ylabel('n'); xlim([x0-5 x0+10]);
```

```
48             title(['t=',num2str(t)],'FontSize',8,...
49             'FontWeight','normal');
50        end
51  end, ntot
52  figure; bar(n);
53  title('t=200','FontSize',8,'FontWeight','normal');
54  xlabel('x'); ylabel('n');
55  xlim([x0+V0*t*dt-50 x0+V0*t*dt+50]);
56  for t=1:T
57      M(t)=M(t)/ntot;
58      V(t)=V(t)/ntot-M(t)*M(t);
59      D(t)=V(t)/(2.0*t*dt); M(t)=M(t)/(t*dt);
60  end
61  t=1:T;
62  figure; plot(t,D,t,M); ...
         legend('D(t)','V(t)','Location','best');
63  toc
```

A.3.1.3   One-Dimensional FD Schemes for Advection-Diffusion Processes

```
1   %% Splitting FD-GRW schemes
2   close all; clear ;
3   tic
4   %% domain and parameters
5   L=3000; Tt=200;
6   D0=0.01; d=1; dx=0.1
7   V0=1; Cr=1
8   dt=Cr*dx/V0
9   r=2*D0*dt/(d^2*dx^2)
10  Nt=round(Tt/dt); tsel=1:round(Nt/100):round(Nt/20);
11  nn=zeros(1,L); M=zeros(1,Nt); V=zeros(1,Nt);
12  %% Gaussian initial condition
13  x0=1; ti=1; tt=ti; xx=(1:L)*dx;
14  uE = @(tt,xx) ...
         (1./sqrt(4*pi*D0*tt)).*exp(-((xx-x0-V0*tt).^2)./ ...
         (4*D0*tt));
15  gss = uE(tt,xx); n=gss/sum(gss);
16  figure(1); plot(xx,n,'r'); xlim([0 2*(x0+V0*tt)]);
17  %% splitting scheme
18  for t=1:Nt
19      %% adcvection step
20      for x=1+round(Cr+0.5):L
21          % nn(x)=n(x-Cr); % FD-GRW
22          % nn(x)=n(x)-Cr*(n(x)-n(x-1)); % backward
23          nn(x)=n(x-1)+(1-Cr)/(1+Cr)*(n(x)-nn(x-1)); % box
24      end
25      n=nn; nn=zeros(1,L);
26      %% diffusion step
27      for x=2:L-1
28          nn(x)=(1-r)*n(x)+(r/2)*(n(x-1)+n(x+1));
29      end
```

```
30      n=nn; nn=zeros(1,L);
31      %% compute first two moments M and V
32      X=(1:L)*dx; Xr=X-x0; Xr2=Xr.^2;
33      M(t)=M(t)+sum(Xr.*n); V(t)=V(t)+sum(Xr2.*n);
34      %% results 1: comparison with analytical solution
35      if sum(ismember(t,tsel))
36          tt=t*dt+ti;
37          X=(1:L)*dx;
38          gss = uE(tt,X)/sum(uE(tt,X));
39          figure(2);
40          plot(X,n,'.-b',X,gss,'k');
41          if t==tsel(1);hold
42          end
43          xlabel('x'); legend('n(x,t)','Gaussian(x,t)');
44          xlim([0 (x0+V0*(1.3*tsel(end)*dt))]);
45          figure(3);
46          plot(X,abs(gss-n),'k');
47          if t==tsel(1);hold;
48          end
49          xlabel('x'); legend('|Gaussian(x,t)-n(x,t)|');
50          xlim([0 (x0+V0*(1.3*tsel(end)*dt))]);
51      end
52
53  end
54  norm=sum(n)
55  %% results 2: estimation of V0 and D0 from numerical solution
56  t=(1:Nt)*dt;
57  M=M/norm; V=V/norm;
58  V=V-M.*M; D=V./(2*(t+ti)); M=M./(t+ti);
59  figure(4); hold all
60  subplot(1,2,1)
61  semilogy(t,M,t,D);
62  xlabel('t'); legend('V(t)','D(t)'); xlim([0 Tt]); ...
        ylim([0.001 V0+10])
63  subplot(1,2,2)
64  semilogy(t,abs(M-V0)/V0,'.',t,abs(D-D0)/D0,'.');
65  xlabel('t'); xlim([0 Tt]);
66  legend('|V(t) - V_{0}| / V_{0}','|D(t) - D_{0}| / ...
        D_{0}','Location','best');
67  %%
68  toc
```

A.3.2 Two-Dimensional Unbiased GRW Algorithm for Advection-Diffusion Processes

```
1  %2-dim GRW algorithm for advection-diffusion processes
2  close all; clear all;
3  tic
4  global state;
```

```
 5   initstate; state;
 6   N=10^24;
 7   kdisp=1;% Kdisp=0: constant coefficients;
 8           % else: diffusion in random velocity field
 9   Lx=3000; Ly=1000; T=400;
10   D=0.01; d=2; stepU=5;
11   NMOD=100; varK=0.1; U_MEAN=1; lambda=0.0; ZC1=1.0; ZC2=1.0;
12   dx=0.1; dy=dx; dt=stepU*dx/U_MEAN;
13   rx=2*D*dt/(d^2*dx^2); ry=2*D*dt/(d^2*dy^2); r=rx+ry;
14   n=zeros(Lx,Ly); nn=zeros(Lx,Ly);
15   u=zeros(Lx,Ly);v=zeros(Lx,Ly);
16   Dx=zeros(1,T); Mx=zeros(1,T); Vx=zeros(1,T);
17   Dy=zeros(1,T); My=zeros(1,T); Vy=zeros(1,T);
18   restr=0; restjump=0; restjumpx=0; restjumpy=0;
19   x0=10; y0=Ly/2; n(x0,y0)=N;
20   tsel=20:20:100;
21   [wavenum,phi,amplitude]= ...
         V_Kraichnan_Gauss_param(NMOD,varK,ZC1,ZC2,U_MEAN,lambda);
22   for i=1:Lx
23       for j=1:Ly
24           if kdisp==0
25               ur=0; vr=0;
26           else
27               x=i*dx; y=j*dy;
28               [ur,vr]=V_Kraichnan_func(x,y,wavenum, ...
29                       phi,amplitude);
30           end
31           u(i,j)=floor((ur+1)*stepU+0.5);
32           v(i,j)=floor(vr*stepU+0.5);
33       end
34   end
35   for t=1:T
36       ntot=0;
37       for x=1:(Lx-1)
38           for y=1:(Ly-1)
39               if n(x,y) > 0
40                   xa=x+u(x,y); ya=y+v(x,y);
41                   restr=n(x,y)*(1-r)+restr; nsta=floor(restr);
42                   restr=restr-nsta; njump=n(x,y)-nsta;
43                   nn(xa,ya)=nn(xa,ya)+nsta;
44                   restjump=njump*rx/r+restjump;
45                   njumpx=floor(restjump);
46                   restjump=restjump-njumpx;
47                   njumpy=njump-njumpx;
48                   if(njumpx)>0
49                       restjumpx=njumpx/2+restjumpx;
50                       nj(1)=floor(restjumpx);
51                       restjumpx=restjumpx-nj(1);
52                       nj(2)=njumpx-nj(1);
53                       for i=1:2
54                           if nj(i)>0
55                               xd=xa+(2*i-3)*d;
56                               nn(xd,ya)=nn(xd,ya)+nj(i);
57                           end
```

```
58                      end
59                  end
60                  if (njumpy) >0
61                      restjumpy=njumpy/2+restjumpy;
62                      nj (1) =floor (restjumpy) ;
63                      restjumpy=restjumpy-nj (1) ;
64                      nj (2) =njumpy-nj (1) ;
65                      for i=1:2
66                          if nj (i) >0
67                              yd=ya+ (2*i-3) *d;
68                              nn (xa, yd) =nn (xa, yd) +nj (i) ;
69                          end
70                      end
71                  end
72              end
73          end
74      end
75      for x=2: (Lx-1)
76          for y=2: (Ly-1)
77              n (x, y) =nn (x, y) ;  nn (x, y) =0;  ntot=ntot+n (x, y) ;
78              xr= (x-x0) *dx;  yr= (y-y0) *dy;
79              Mx (t) =Mx (t) +xr*n (x, y) ;  Vx (t) =Vx (t) +xr^2*n (x, y) ;
80              My (t) =My (t) +yr*n (x, y) ;  Vy (t) =Vy (t) +yr^2*n (x, y) ;
81          end
82      end
83      if sum (ismember (t, tsel))
84          [lin, col] =find (n>0) ;
85          [nl, nc] =size (n) ;
86          linind=sub2ind ([nl, nc], lin, col) ;
87          figure (1) ;plot3 (lin*dx, col*dx, n (linind)) ;
88          if t==tsel (1) ;hold
89          end
90          grid on; view (40,20)
91          xlabel ('x') ; ylabel ('y') ; zlabel ('n (x,y) ') ;
92          xlim ([0 80]) ; ylim ([35 65]) ;
93          set (gca, 'XTick', [0:20:80]) ;
94          set (gca, 'YTick', [35:10:65])
95      end
96  end
97  %%
98  for t=1:T
99      Mx (t) =Mx (t) /ntot;  My (t) =My (t) /ntot;
100     Vx (t) =Vx (t) /ntot-Mx (t) *Mx (t) ;Vy (t) =Vy (t) /ntot-My (t) *My (t) ;
101     Dx (t) =Vx (t) / (2.0*t*dt) ;  Dy (t) =Vy (t) / (2.0*t*dt) ;
102 end
103 t=1:T;
104 figure;
105 subplot (1,2,1)
106 plot (t (1:T-1) *dt, diff (Mx) /dt, t (1:T-1) *dt, diff (My) /dt) ;
107 ylim ([-0.2 1.4]) ;
108 legend ('Mx (t) ', 'My (t) ', 'Location', 'best', 'FontSize', 5) ;
109 legend ('boxoff') ;
110 xlabel ('t') ; set (gca, 'XTick', [0:50:200])
111 subplot (1,2,2)
```

```
112  plot(t*dt,Dx,t*dt,Dy);   ylim([0 0.1]);
113  legend('Dx(t)','Dy(t)','Location','best','FontSize',8);
114  legend('boxoff');
115  xlabel('t'); set(gca, 'XTick', [0:50:200])
116  toc
```

A.3.3 Biased GRW Algorithms

A.3.3.1 Two-Dimensional BGRW Algorithm for Advection-Diffusion Processes

```
1   %2-dim BGRW algorithm for advection-diffusion processes
2   close all; clear all;
3   tic
4   global state;
5   initstate; state;
6   N=10^24;
7   kdisp=1;% Kdisp=0: constant coefficients;
8           % else: diffusion in random velocity field
9   if kdisp==0
10      Lx=500; Ly=200; T=250;
11  else
12      Lx=10000; Ly=400; T=200;
13  end
14  d=1; Dfactor=1.2;
15  dx=0.01; dy=dx; D1=0.01; D2=D1; U=1;
16  dt=Dfactor*(2*D1/(dx*dx)+2*D2/(dy*dy)); dt=1./dt;
17  NMOD=100; varK=0.1; lambda=0.0; ZC1=1.0; ZC2=1.0;
18  n=zeros(Lx,Ly); nn=zeros(Lx,Ly);
19  Dx=zeros(1,T); Mx=zeros(1,T); Vx=zeros(1,T);
20  u=zeros(Lx,Ly); v=zeros(Lx,Ly);
21  ru=zeros(Lx,Ly);rv=zeros(Lx,Ly);
22  Dy=zeros(1,T); My=zeros(1,T); Vy=zeros(1,T);
23  restr=0; restjump=0; restjumpx=0; restjumpy=0;
24  x0=Lx/10; y0=Ly/2; n(x0,y0)=N;
25  tsel=20:20:100;
26  [wavenum,phi,amplitude]= ...
27  V_Kraichnan_Gauss_param(NMOD,varK,ZC1,ZC2,U_MEAN,lambda);
28  for i=1:Lx
29      for j=1:Ly
30          if kdisp==0
31              ur=0; vr=0;
32          else
33              x=i*dx; y=j*dy;
34              [ur,vr]=V_Kraichnan_func(x,y,wavenum, ...
35              phi,amplitude);
36          end
37          u(i,j)=(ur+U)*dt/dx; v(i,j)=vr*dt/dy;
38          ru(i,j)=2*D1*dt/(dx*dx); rv(i,j)=2*D2*dt/(dy*dy);
```

```
39        end
40   end
41   for t=1:T
42        ntot=0;
43        for x=1:(Lx-1)
44            for y=1:(Ly-1)
45                if n(x,y) > 0
46                    rx=ru(x,y); ry=rv(x,y); r=rx+ry;
47                    restr=n(x,y)*(1-r)+restr; nsta=floor(restr);
48                    restr=restr-nsta; njump=n(x,y)-nsta;
49                    nn(x,y)=nn(x,y)+nsta;
50                    restjump=njump*rx/r+restjump;
51                    njumpx=floor(restjump);
52                    restjump=restjump-njumpx;
53                    njumpy=njump-njumpx;
54                    if(njumpx)>0
55                        restjumpx=njumpx*0.5*(1-u(x,y)/rx) ...
56                                +restjumpx;
57                        nj(1)=floor(restjumpx);
58                        restjumpx=restjumpx-nj(1);
59                        nj(2)=njumpx-nj(1);
60                        for i=1:2
61                            if nj(i)>0
62                                xd=x+(2*i-3)*d;
63                                nn(xd,y)=nn(xd,y)+nj(i);
64                            end
65                        end
66                    end
67                    if(njumpy)>0
68                        restjumpy=njumpy*0.5*(1-v(x,y)/ry) ...
69                                +restjumpy;
70                        nj(1)=floor(restjumpy);
71                        restjumpy=restjumpy-nj(1);
72                        nj(2)=njumpy-nj(1);
73                        kdiscr=0;
74                        for i=1:2
75                            if nj(i)>0
76                                yd=y+(2*i-3)*d;
77                                nn(x,yd)=nn(x,yd)+nj(i);
78                            end
79                        end
80                    end
81                end
82            end
83        end
84        for x=2:(Lx-1)
85            for y=2:(Ly-1)
86                n(x,y)=nn(x,y); nn(x,y)=0; ntot=ntot+n(x,y);
87                xr=(x-x0)*dx; yr=(y-y0)*dy;
88                Mx(t)=Mx(t)+xr*n(x,y); Vx(t)=Vx(t)+xr^2*n(x,y);
89                My(t)=My(t)+yr*n(x,y); Vy(t)=Vy(t)+yr^2*n(x,y);
90            end
91        end
92        if sum(ismember(t,tsel))
```

```
93      [lin,col]=find(n>0);
94      [nl,nc]=size(n);
95      linind=sub2ind([nl,nc],lin,col);
96      figure(1);plot3(lin*dx,col*dy,n(linind));
97      if t==tsel(1);hold
98      end
99      grid on; view(40,20);
100     xlabel('x'); ylabel('y'); zlabel('n(x,y)');
101   end
102 end
103 for t=1:T
104     Mx(t)=Mx(t)/ntot; My(t)=My(t)/ntot;
105     Vx(t)=Vx(t)/ntot-Mx(t)*Mx(t); ...
            Vy(t)=Vy(t)/ntot-My(t)*My(t);
106     Dx(t)=Vx(t)/(2.0*t*dt); Dy(t)=Vy(t)/(2.0*t*dt);
107 end
108 t=1:T;
109 figure;
110 subplot(1,2,1)
111 plot(t(1:T-1)*dt,diff(Mx)/dt,t(1:T-1)*dt,diff(My)/dt);
112 legend('Mx(t)','My(t)','Location','best','FontSize',5);
113 legend('boxoff');
114 xlabel('t');
115 subplot(1,2,2)
116 plot(t*dt,Dx,t*dt,Dy);
117 legend('Dx(t)','Dy(t)','Location','best','FontSize',8);
118 legend('boxoff');
119 xlabel('t');
120 toc
```

A.3.3.2 One-Dimensional BGRW Algorithm for Flow in Porous Media

```
1  % BGRW solution of the flow equation (3.37)
2  % Uses analytical solution given by 'fsolution(I,omega,b)'
3  clear all; close all
4  tic
5  I=501;
6  dx=2/(I-1);
7  x=[-1:dx:1]; x2=(x(1:I-1)+x(2:I))/2;
8  omega=5; b=0.5;
9  D=1+b*sin(2*pi*omega*x);
10 csol=fsolution(I,omega,b);
11 T=400000; past=20000; maxr=0.8;
12 dt=dx^2*maxr/max(D)/2
13 r=dt*(1+b*sin(2*pi*omega*x2))/dx^2;
14 rloc=[1-2*r(1),1-(r(1:I-2)+r(2:I-1)),1-2*r(I-1)];
15 rapr=[0.5,r(1:I-2)./(r(1:I-2)+r(2:I-1))];
16 dN=10^6; Nmax=dN*(I-1);
17 n0=floor(Nmax-dN*[0:I-1]);
18 c0=n0/Nmax; cinit=c0+0*(csol-c0);
```

```
19   sumcinit=sum(cinit);
20   n=n0; nrand=rand(1,I); sumn=zeros(1,T);
21   tgraf=0; igraf=0;
22   sumc=zeros(1,floor(T/past));
23   tvect=zeros(1,floor(T/past));
24   flux1=zeros(1,floor(T/past));
25   fluxI=zeros(1,floor(T/past));
26   restr=0; restsar1=0; restsarI=0; rest2=0;
27   for t=1:T
28       restr=floor(rloc.*n)+restr; nn=floor(restr);
29       restr=restr-nn; nsta=nn;
30       nsar=n-nsta;
31       restsar1=nsar(1)/2+restsar1; nsar1=floor(restsar1);
32       restsar1=restsar1-nsar1;
33       nn(2)=nn(2)+nsar1;
34       for i=2:I-1
35           rest2=nsar(i)*rapr(i)+rest2; nsarleft=floor(rest2);
36           rest2=rest2-nsarleft;
37           nn(i-1)=nn(i-1)+nsarleft;
38           nn(i+1)=nn(i+1)+nsar(i)-nsarleft;
39       end
40       restsarI=nsar(I)/2+restsarI; nsarI=floor(restsarI);
41       restsarI=restsarI-nsarI;
42       nn(I-1)=nn(I-1)+nsarI;
43       nn(1)=n(1); nn(I)=n(I);
44       n=nn; cc=nn/Nmax;
45       c=n/Nmax;
46       if mod(t,past)==0
47           display(t);
48           tgraf=tgraf+1;
49           tvect(tgraf)=t;
50           sumc(tgraf)=sum(c);
51           flux1(tgraf)=nsar(1)/2-nsar(2)*rapr(2);
52           fluxI(tgraf)=nsar(I-1)*(1-rapr(I-1))-nsar(I)/2;
53       end
54       if t==1000 || t==10000 || t==100000
55           igraf=igraf+1;
56           tevol(igraf)=t; cgraf(igraf,:)=c;
57       end
58   end
59   figure;
60   plot(tvect,flux1,'o',tvect,fluxI,'*');
61   legend('in-flux(x=-1)','out-flux (x=1)')
62   figure;
63   plot(tvect,flux1-fluxI,'o') title('total flux')
64   figure;
65   plot(x,csol-c0,'k','LineWidth',1.5); hold all;
66   for iigraf=1:igraf
67       plot(x,cgraf(iigraf,:)-c0);
68   end
69   plot(x,c-c0,'b');
70   x2=(x(1:I-1)+x(2:I))/2; dx=2/(I-1);
71   for i=2:I-2
72       fluxnum(i)=(c(i)-c(i+1))*r(i);
```

```
73  end
74  figure;
75  plot(x2,0.5*sqrt(1-b^2)*ones(1,I-1),'k'); hold;
76  plot(x2(2:I-2),fluxnum(2:I-2)*dx/dt,'g')
77  toc
```

```
1   %Analytical solution of the one-dimensional flow equation
2   function csol=fsolution(I,omega,b)
3   dx=2/(I-1); %%% space step
4   x=[-1:dx:1]; D=1+b*sin(2*omega*pi*x);
5   domega=1/(2*omega);
6   xmin=min(x);
7   if xmin<0
8       kmin=-floor((domega-xmin)*omega);
9   end
10  xmax=max(x);
11  if xmax>0
12      kmax=ceil((xmax-domega)*omega);
13  end
14  xsol=[]; csol=[];
15  for k=kmin:kmax
16      xinf=(k-0.5)/omega;
17      if k==kmin xinf=xmin;
18      else xinf=xinf+dx;
19      end
20      xsup=(k+0.5)/omega-dx;
21      if k==kmax
22          xsup=xmax; xsol=[xsol,[xinf:dx:xsup]];
23          csol=[csol,atan((tan(pi*omega*[xinf:dx:xsup])+b) ...
24          /sqrt(1-b^2))+k*pi];
25      else
26          xsol=[xsol,[xinf:dx:xsup+dx]];
27          csol=[csol,atan((tan(pi*omega*[xinf:dx:xsup])+b) ...
28          /sqrt(1-b^2))+k*pi, (k+0.5)*pi];
29      end
30  end
31  csol=(omega*pi+asin(b)-csol)/(2*pi*omega);
32  end
```

Appendix B
GRW Solutions of Fokker–Planck Equations

B.1 GRW Approximations for Continuous Diffusion Processes

The definition of a diffusion process can be reformulated in terms of transition probabilities and conditional expectations, without requiring that the transition probability has a density (e.g., [12, p. 68]).

The continuity property (i) (Sect. 2.2.2.1), reformulated as $\lim_{\delta t \to 0} \frac{1}{\delta t} \text{Prob}\{|\hat{X}_{k+1} - \hat{X}_k| > \epsilon\} = 0$, is fulfilled if for every $\epsilon > 0$ there exists a small δt such that $\text{Prob}\{|\hat{X}_{k+1} - \hat{X}_k| > \epsilon\} = 0$. According to (3.23), $|\hat{X}_{k+1} - \hat{X}_k| = |v\delta x + \xi|$ takes on a maximum value of $(|v| + d)\delta x$. Using (3.24), one finds that if $\delta t \leq \delta t^* = \frac{rd^2\epsilon^2}{2D(|v|+d)^2}$, then $(|v| + d)\delta x \leq \epsilon$ with probability 1. Thus, the condition (i) is fulfilled because $\text{Prob}\{|\hat{X}_{k+1} - \hat{X}_k| > \epsilon\} = 1 - \text{Prob}\{|\hat{X}_{k+1} - \hat{X}_k| \leq \epsilon\} = 0$. Moreover, since transitions outside the interval $(-\epsilon, \epsilon)$ have probability zero if $\delta t \leq \delta t^*$, continuity conditions similar to (i) are also fulfilled by the higher order moments of $\hat{X}_{k+1} - \hat{X}_k$ [22]. This ensures the finiteness of the first two moments of $\hat{X}$ at finite times [6, 25].

Condition (ii) of Sect. 2.2.2.1 fulfilled by diffusion processes, reformulated as an expectation for fixed $\hat{X}_k$ and $\delta t \leq \delta t^*$, yields

$$\lim_{\delta t \to 0} \frac{1}{\delta t} E\{\hat{X}_{k+1} - \hat{X}_k\} = \lim_{\delta t \to 0} \frac{1}{\delta t} E\{V\delta t + \Delta v\delta x + \xi\} = V + \lim_{\delta t \to 0} \frac{1}{\delta t} \Delta v\delta x,$$

(B.1)

where $E\{\xi\} = 0$ according to (3.24) and $\Delta v = v - V\frac{\delta t}{\delta x}$ defines the truncation error of the advective displacement.

Condition (iii) of Sect. 2.2.2.1 is verified exactly:

$$\frac{1}{2} \lim_{\delta t \to 0} \frac{1}{\delta t} E\{(\hat{X}_{k+1} - \hat{X}_k)^2\} = \frac{1}{2} \lim_{\delta t \to 0} \frac{1}{\delta t} E\{V^2\delta t^2 + 2V\xi\delta t + \xi^2\} = D, \quad \text{(B.2)}$$

© Springer Nature Switzerland AG 2019
N. Suciu, *Diffusion in Random Fields*, Geosystems Mathematics,
https://doi.org/10.1007/978-3-030-15081-5

where one uses (3.24), which implies $E\{\xi\} = 0$ and $E\{\xi^2\} = 2D\delta t$.

Thus, the discrete process $\hat{X}$ approximates, up to truncation errors, a continuous diffusion process $X(t)$ with finite first two moments at finite times. Accordingly, the distribution of the computational particles on the GRW lattice approximates the solution of the Fokker–Planck equation (3.19).

B.2   Strict Equivalence Between GRW and the Weak Euler Scheme for Constant Velocity

In case of a constant velocity, there are no truncation errors at all if one chooses $V\delta t = v\delta x$, which cancels Δv in (B.1). The first three moments of the random walk ξ with jump probabilities given by (3.24) satisfy

$$|E(\xi)| + \left|E(\xi^3)\right| + \left|E(\xi^2) - \delta t\right| = 0 \le C\delta t^2,$$

for any positive constant C, condition required for a consistent first order truncation of the Itô–Taylor expansion [12, Section 5.12].

Thus, if V is a real constant, the discrete process (3.23) is a weak Euler scheme with convergence order $\mathcal{O}(\delta t)$ for the Itô equation $dX(t) = V dt + dW(t)$, $E(dW) = 0$ and $E((dW)^2) = 2D\delta t$.

B.3   Biased GRW Approximations for Continuous Diffusion Processes

Consider without loss of generality the one-dimensional BGRW with jump probabilities $P\{\delta x\} = r + v$, $P\{-\delta x\} = r - v$, $P\{0\} = 1 - r$, where $r = 2D\delta t/\delta x^2$ and $v = V\delta t/\delta x$.

Since in BRGW algorithms only jumps to neighbor lattice sites are allowed, the continuity condition (i) and the similar conditions for higher order moments are verified for all $\epsilon = \delta x > 0$.

Similarly to (B.1) and (B.2), one obtains

$$\lim_{\delta t \to 0} \frac{1}{\delta t} E\{\hat{X}_{k+1} - \hat{X}_k\} = \lim_{\delta t \to 0} \frac{1}{\delta t} \left[\frac{1}{2}(r + v)\delta x + \frac{1}{2}(r - v)(-\delta x)\right]$$

$$= \lim_{\delta t \to 0} \frac{1}{\delta t} v\delta x = V,$$

$$\frac{1}{2}\lim_{\delta t \to 0}\frac{1}{\delta t}E\{(\hat{X}_{k+1} - \hat{X}_k)^2\} = \frac{1}{2}\lim_{\delta t \to 0}\frac{1}{\delta t}\left[\frac{1}{2}(r+v)\delta x^2 + \frac{1}{2}(r-v)\delta x^2\right]$$

$$= \lim_{\delta t \to 0}r\frac{\delta x^2}{2\delta t} = D.$$

Thus, BGRW fulfills exactly the conditions (ii) and (iii) of Sect. 2.2.2.1.

Appendix C
Numerical Generation of Random Fields

C.1 Homogeneous Gaussian Random Fields

C.1.1 Randomization Method

A random field $u(\mathbf{x})$, $\mathbf{x} \in \mathbb{R}^d$, is said to be *homogeneous in a wide sense* (or second order homogeneous) if the mean is constant, $M(u) = const$, and the correlation function depends only on the difference vector $\rho = \mathbf{x}_1 - \mathbf{x}_2$, $M(u(\mathbf{x}_1)u(\mathbf{x}_2)) = C(\mathbf{x}_1, \mathbf{x}_2) = C(\rho)$. Since only the fluctuating part of the homogeneous field is random, in the following we consider $M(u) = 0$.

Homogeneous random fields have a *spectral representation* given by a stochastic Fourier integral

$$u(\mathbf{x}) = \int_{\mathbb{R}^d} e^{-i\mathbf{k}\mathbf{x}} Z(d\mathbf{k}). \tag{C.1}$$

$Z(d\mathbf{k})$ is a complex random measure on $\mathbb{R}^d$ with property $M(|Z(d\mathbf{k})|^2) = F(d\mathbf{k})$, where F is the *spectral distribution function* of the homogeneous random field. F is defined as the Fourier transform of the correlation function

$$C(\rho) = \frac{1}{(2\pi)^d} \int_{\mathbb{R}^d} e^{i\mathbf{k}\rho} F(d\mathbf{k}), \tag{C.2}$$

so that the spectral representation (C.1) preserves the correlation [29, Chap. 4.21]. In case of Gaussian homogeneous random fields $Z(d\mathbf{k})$ is a complex white noise random measure [14].

The standard Fourier method to generate homogeneous random fields consists of numerical evaluations of the spectral representation (C.1). The evaluation through a MC integration approach is called *randomization method* [18]. The simplest variant of the randomization method has the form of a superposition of N random sine

© Springer Nature Switzerland AG 2019

N. Suciu, *Diffusion in Random Fields*, Geosystems Mathematics,
https://doi.org/10.1007/978-3-030-15081-5

modes,

$$u(\mathbf{x}) = \frac{\sigma}{\sqrt{N}} \sum_{j=1}^{N} \left[\zeta_j \cos(2\pi \mathbf{k}_j \mathbf{x}) + \eta_j \sin(2\pi \mathbf{k}_j \mathbf{x}) \right], \tag{C.3}$$

where $\zeta_j, \eta_j, j = 1, \ldots, N_p$ are independent normal (Gaussian) random variables $N(0, \sigma^2)$ with variance $\sigma^2 = \int_{\mathbb{R}^d} F(d\mathbf{k})$, according to (C.2), and $\mathbf{k}_j$ are independent random variables, and also independent of ζ_j and η_j, sampled according to an arbitrary probability density. The accuracy of the method is influenced by the sampling of the wave numbers $\mathbf{k}_j$. Improvements are achieved if one divides the range of wave numbers into a finite number of bins, of uniform or logarithmic distributed dimensions, and one samples randomly wave numbers in each bin [14].

C.1.2   Analytic Properties of the Samples

Let us consider a one-dimensional homogeneous random field $u(x)$, $x \in \mathbb{R}$. The homogeneity of the correlation function implies

$$M\{[u(x+h) - u(x)]^2\} M\{[u(x+h)]^2\} + M\{[u(x)]^2\} - 2M\{[u(x+h)u(x)]^2\}$$
$$= 2[C(0) - C(h)].$$

Hence, if the correlation function $C(h)$ is continuous at the point $h = 0$, then the field $u(x)$ is mean square continuous for all x,

$$\lim_{h \to 0} M\{[u(x+h) - u(x)]^2\} = 0.$$

If the sequence of random variables $u_{h_i}(x) = [u(x+h_i) - u(x)]/h_i$ converges in mean square for all sequences $\{h_i\}$ converging to zero for $i \to \infty$ to a limit $u'(x)$ independent of $\{h_i\}$,

$$\lim_{i \to \infty} M\{[u_{h_i}(x) - u'(x)]^2\} = 0,$$

the one-dimensional field $u(x)$ is mean square differentiable. If the correlation function $C(h)$ has a second derivative at $h = 0$, $C''(0)$, then the field $u(x)$ is mean square differentiable for all x [29, Chap. 1.4]. Moreover, it has been proved that the existence of a finite second derivative of the correlation function at $h = 0$, $C''(0)$ is a necessary and sufficient condition for the mean square differentiability of $u(x)$ and, generally, $u(x)$ is n-times mean square differentiable if and only if $C^{(2n)}(0)$ exists and is finite [21, Section 2.6]. An example for this instance is the correlation with a Gaussian shape $C(h) \sim e^{-h^2}$. A counterexample is the exponential correlation $C(h) \sim e^{-h}$, for which $u(x)$ is nowhere mean square differentiable.

Fig. C.1 Exponential correlation: estimation of Lipschitz constant for a smooth sinusoidal velocity sample

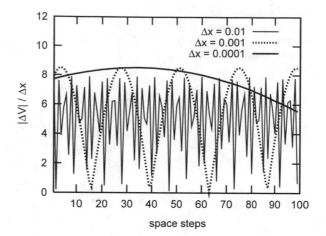

Mean square continuity and differentiability do not necessarily imply sample continuity or differentiability. Sufficient conditions for sample smoothness are given by the existence of higher order derivatives of the correlation function. For instance, if $C(h)$ has a second order derivative at $h = 0$ then the one-dimensional field $u(x)$ is equivalent to a field with samples that are continuous with probability one in every finite interval. If moreover $C^{(4)}(0)$ exists, $u(x)$ is equivalent to a field the sample functions of which have, with probability one, continuous derivatives in every finite interval. These sufficient conditions can hardly be relaxed and one expects that they are very close to the necessary conditions [4].

Lipschitz continuity, which can still hold if $u(x)$ is not differentiable, is essential for the existence and uniqueness of the solution of the Itô equation (see A2 in Sect. 3.2.1) and has to be investigated as well. A hint on the Lipschitz continuity of the drift coefficients of the Itô equation governing diffusion in random velocity fields is given by the numerical investigations presented in the following.

Figures C.1, C.2, C.3, C.4, C.5, C.6, C.7, C.8, C.9, and C.10 shows estimations of the Lipschitz constant for velocity samples generated with the randomization method for Gaussian and exponential correlations of the hydraulic conductivity (see Sects. C.2 and C.3.2 below). One can see that Lipschitz continuity is expected for Gaussian correlation but not for exponential correlation. For large numbers of modes in the randomization formula (C.3), the Lipschitz constant for samples of exponentially correlated fields behaves similarly to that of a diffusion process, which is almost everywhere non-differentiable (compare Figs. C.7, C.8 and C.9, C.10).

Fig. C.2 Exponential correlation: estimation of Lipschitz constant for a noisy velocity sample with 6400 modes

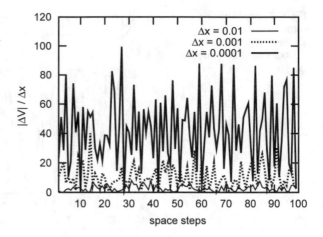

Fig. C.3 Gaussian correlation: estimation of Lipschitz constant for a smooth sinusoidal velocity sample

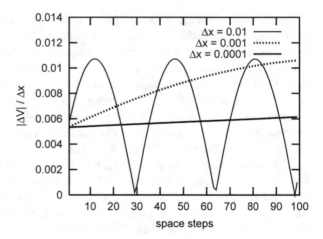

Fig. C.4 Gaussian correlation: estimation of Lipschitz constant for a noisy velocity sample with 6400 modes

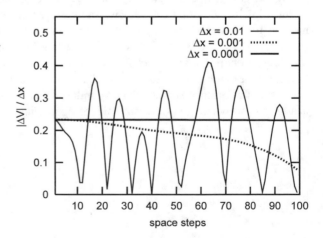

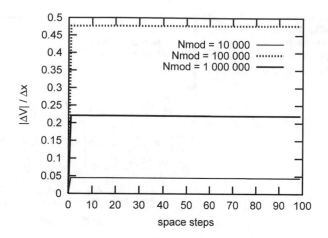

Fig. C.5 Gaussian correlation: estimation of Lipschitz constant for velocity samples with increasing numbers of modes and fixed $\Delta x = 10^{-5}$

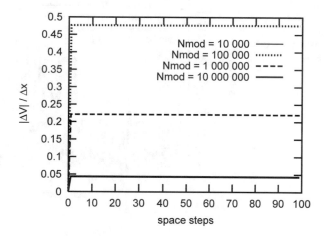

Fig. C.6 Gaussian correlation: estimation of Lipschitz constant for velocity samples with increasing numbers of modes and fixed $\Delta x = 10^{-6}$

C.2 Filtered Kraichnan Fields

In a linear approximation of the flow equations, the spectral representation of the incompressible Darcy flow velocity is proportional to that of the log-hydraulic conductivity. Hence, the velocity field and can be simulated "on a demand" at given points in space by randomization methods of form (C.3). The randomization representation also allows straightforward computations of spatially filtered velocity fields. This approach is illustrated in detail for the case of a Gaussian-correlated log-hydraulic conductivity.

Fig. C.7 Exponential correlation: estimation of Lipschitz constant for velocity samples with increasing numbers of modes and fixed $\Delta x = 10^{-5}$

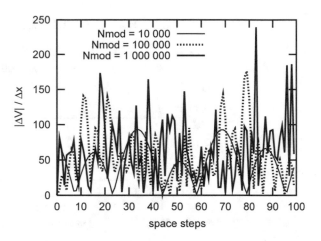

Fig. C.8 Exponential correlation: estimation of Lipschitz constant for velocity samples with increasing numbers of modes and fixed $\Delta x = 10^{-6}$

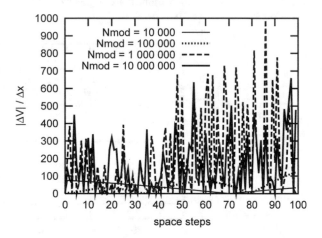

Fig. C.9 Strong diffusion process: estimation of Lipschitz constant for different Δt

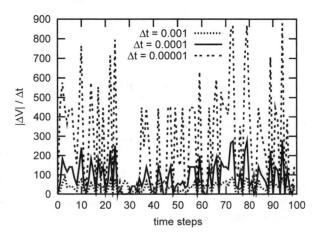

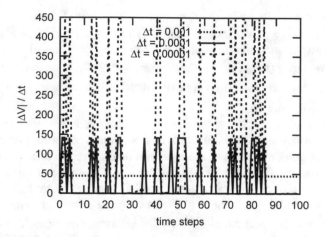

Fig. C.10 Weak diffusion process: estimation of Lipschitz constant for different Δt

A Gaussian random velocity field, corresponding to a log-normal hydraulic conductivity, can be generated in a first order approximation by the Kraichnan method [13] as a randomized spectral representation given by the well-known formula

$$V_i(\mathbf{x}) = \langle V_1 \rangle \delta_{i1} + \sigma_{\ln K} \langle V_1 \rangle \sqrt{\frac{2}{N}} \sum_{j=1}^{N} p_i(\mathbf{k}^{(j)}) \cos(\mathbf{k}^{(j)} \cdot \mathbf{x} + \phi^{(j)}), \qquad (C.4)$$

where the angular brackets stand for ensemble average, $i = 1, \ldots, d$, and d is the spatial dimension of the problem [5, 20]. The simple representation (C.4) is obtained by choosing the probability density of the random vector $\mathbf{k}^{(j)} = (k_1^{(j)}, \ldots, k_d^{(j)})$ as given by the Fourier transform of the correlation function of the statistically homogeneous Gaussian random field $\ln K$ divided by the variance $\sigma_{\ln K}^2$ (see, e.g., [15]).

Let us consider in the following a Gaussian correlation of the $\ln K$ field which, in the general anisotropic case, is given by

$$\langle \ln K(\mathbf{x}) \ln K(\mathbf{x} + \mathbf{r}) \rangle = \sigma_{\ln K}^2 \exp\left(-\sum_{i=1}^{d} \frac{r_i^2}{\lambda_{\ln K,i}^2} \right). \qquad (C.5)$$

The random vectors $\mathbf{k}^{(j)}$ are thus distributed according to the PDF

$$f(\mathbf{k}^{(j)}) = \prod_{i=1}^{d} \pi^{1/2} \lambda_{\ln K,i} \exp\left[-\pi^2 \left(k_i^{(j)} \lambda_{\ln K,i} \right)^2 \right] = \prod_{i=1}^{d} f(k_i^{(j)}),$$

that is, they have mutually independent components. With the change of variables $k_i^{(j)} = \eta/(\sqrt{2}\pi\lambda_{\ln K,i})$, where η is a normally distributed random variable of zero mean and unit variance, the corresponding probability measures of the components become Gaussian, $f(k^{(j)})dk_i^{(j)} = (2\pi)^{-1/2}\exp(\eta^2/2)d\eta$. The random components $k_i^{(j)}$ are therefore extracted from Gaussian distributions with vanishing mean and variances $1/(2\pi^2\lambda_{\ln K,i}^2)$.

The phases $\phi^{(j)}$, also being independent random numbers, are extracted from a uniform distribution in the interval $[0, 2\pi]$. Randomized spectral representations like (C.4), are particular cases of more general randomization formulas using sums of cosine and sine functions [9, 15]. In our case, the general representation is obtained when $\sqrt{2}\cos(\mathbf{k}^{(j)}\cdot\mathbf{x} + \phi^{(j)})$ in (C.4) is replaced by $\xi^{(j)}\cos(2\pi\mathbf{k}^{(j)}\cdot\mathbf{x}) + \zeta^{(j)}\sin(2\pi\mathbf{k}^{(j)}\cdot\mathbf{x})$, where $\xi^{(j)}$ and $\zeta^{(j)}$ are normal variables, mutually independent and independent of $k_i^{(j)}$(see, for instance, equations (22) and (24) in [15]). Numerical tests have shown that the two representations are equally accurate, with the only difference that the general formula is about two times slower.

The projector

$$\mathbf{p}(\mathbf{k}) = \mathbf{e}_1 - \frac{k_1\mathbf{k}}{\mathbf{k}^2}\,,$$

with $\mathbf{e}_1$ being the unit vector to which the mean flow is aligned, ensures the incompressibility of the flow. We note that when dropping the first term, as well as $\langle V \rangle$ and $p_i(\mathbf{k}^{(j)})$ from the second term, (C.4) reduces to the randomization formula for the fluctuations of the $\ln K$ field.

To illustrate the derivation of a closed form generator of filtered velocity fields, let us consider a Gaussian filter function

$$G(\mathbf{x}' - \mathbf{x}) = \left(\frac{b}{\pi}\right)^{d/2}\exp[-b(\mathbf{x}' - \mathbf{x})^2], \tag{C.6}$$

where $b = a/\lambda^2$, $a \in \mathbb{R}^+$, and λ is the filter width. Using (C.6), the filtering operation applied to Eq. (C.4) becomes

$$\langle V_i \rangle_\lambda (\mathbf{x}) = \int_{-\infty}^{\infty} d\mathbf{x}' \left(\frac{b}{\pi}\right)^{d/2}\left[\langle V_1\rangle\delta_{i1} + \sigma_{\ln K}\langle V_1\rangle\sqrt{\frac{2}{N}}\sum_{j=1}^{N} p_i(\mathbf{k}^{(j)})\cos\left(\mathbf{k}^{(j)}\cdot\mathbf{x} + \phi^{(j)}\right)\right] \cdot$$

$$\cdot \exp\left[-b(\mathbf{x}' - \mathbf{x})^2\right]$$

$$= \langle V_1\rangle\delta_{i1} + \sigma_{\ln K}\langle V_1\rangle\sqrt{2/N}\sum_{j=1}^{N} p_i(\mathbf{k}^{(j)}) \cdot$$

$$\cdot \underbrace{\int_{-\infty}^{\infty} d\mathbf{x}' \cos\left(\mathbf{k}^{(j)}\cdot\mathbf{x}' + \phi^{(j)}\right)\left(\frac{b}{\pi}\right)^{d/2}\exp\left[-b(\mathbf{x}' - \mathbf{x})^2\right]}_{=I},$$

where $\int\limits_{-\infty}^{\infty} d\mathbf{x}'$ stands for $\int\limits_{-\infty}^{\infty} \ldots \int\limits_{-\infty}^{\infty} dx_1' \ldots dx_d'$.

The integral I can be simplified by the substitution $\mathbf{q} = \mathbf{x}' - \mathbf{x}$, which results in

$$I = \left(\frac{b}{\pi}\right)^{d/2} \int\limits_{-\infty}^{\infty} d\mathbf{q} \cos\left(\mathbf{k}^{(j)} \cdot \mathbf{q} + \mathbf{k}^{(j)} \cdot \mathbf{x} + \phi^{(j)}\right) \exp\left(-b\mathbf{q}^2\right).$$

The cosine function has an argument with $2d + 1$ terms, of which d terms depend on a component of the vector $\mathbf{q}$. The goal now is to separate these d terms from the rest. This separation can be achieved by a trigonometric identity. If we take a look at this identity for fewer terms

$$\cos(u + v) = \cos(u)\cos(v) - \sin(u)\sin(v)$$

$$\cos(u_1 + u_2 + v) = \cos(u_1)\cos(u_2)\cos(v) - \sin(u_1)\sin(u_2)\cos(v)$$
$$- \sin(u_1)\cos(u_2)\sin(v) - \cos(u_1)\sin(u_2)\sin(v)$$

$$\ldots,$$

we see that there is always only one term which has no sine functions. As sine functions are antisymmetric and the exponential function with which they are multiplied under the integral is symmetric, the resulting term is antisymmetric and the evaluated integral over all these terms vanishes. Hence, only the single term with no sine functions gives a non-zero value. Thus, by using the result $\int_{-\infty}^{\infty} dq \cos(uq) \exp(-w^2 q^2) = \sqrt{\pi} \exp(-u^2/4w^2)/w$ (see [3]), the integral gives

$$I = \left(\frac{b}{\pi}\right)^{d/2} \int\limits_{-\infty}^{\infty} d\mathbf{q} \left[\prod_{\alpha=1}^{d} \cos(k_\alpha^{(j)} q_\alpha) \cos(\mathbf{k}^{(j)} \cdot \mathbf{x} + \phi^{(j)}) + \ldots\right] \exp\left(-b\mathbf{q}^2\right)$$

$$= \cos(\mathbf{k}^{(j)} \cdot \mathbf{x} + \phi^{(j)}) \exp\left[-\frac{(\mathbf{k}^{(j)})^2}{4b}\right].$$

The final expression for the Gaussian filtered velocity field is

$$\langle V_i \rangle_\lambda (\mathbf{x}) = \langle V_1 \rangle \delta_{i1} + \sigma_{\ln K} \langle V_1 \rangle \sqrt{\frac{2}{N}} \sum_{j=1}^{N} p_i(\mathbf{k}^{(j)}) \cos(\mathbf{k}^{(j)} \cdot \mathbf{x} + \phi^{(j)}) \exp\left[-\frac{(\mathbf{k}^{(j)}\lambda)^2}{4a}\right].$$

$$(\text{C.7})$$

The parameter $a \in \mathbb{R}^+$ in (C.7) is chosen to be $a = 2$ [1]. The only difference compared to the unfiltered velocity equation (C.4) is one factor for each mode, which can be precalculated and saved before the actual simulations start [19].

If instead of (C.5) one uses an exponential correlation function, filtered Kraichnan fields can be constructed similarly by using the algorithm described in [15, Sect. 5.3]. Since the resulting randomization formula is also a sum of cosine functions, the same expression (C.7) can be used to compute exponentially correlated Kraichnan fields (see codes presented in Sects. C.3.1.1–C.3.1.2 and C.3.2.1–C.3.2.2 below).

C.3   Kraichnan Field Generators

If $u(\mathbf{x})$ represents the fluctuating part of the realization of the hydraulic conductivity, of its logarithm, or of the velocity components, according to (C.7) it is evaluated as a sum of random modes

$$u(\mathbf{x}) = \sum_{j=1}^{N} a^{(j)} \cos(\mathbf{k}^{(j)} \cdot \mathbf{x} + \varphi^{(j)}), \tag{C.8}$$

where $a^{(j)}$ are constant amplitudes, $\mathbf{k}^{(j)}$ are vector wavenumbers with constant components, and $\varphi^{(j)}$ are constant phases. These parameters generated by randomization algorithms of form (C.7) are stored and used to compute the values of the realization u at all point $\mathbf{x}$ where they are needed in applications. An ensemble of realizations u is obtained by repeating the procedure for an ensemble of parameter sets $\{a^{(j)}, \mathbf{k}^{(j)}, \varphi^{(j)}\}_{j=1,N}$.

C.3.1   Hydraulic Conductivity Fields

The sum (C.8) is computed by the following Matlab function:

```
1  function Kxy = ...
2  Kraichnan_func(x,y,wavenum,phi,amplitude,K_MEAN,varK)
3  phase=2*pi*(wavenum(:,1).*x+wavenum(:,2).*y)+phi;
4  ak=sum(amplitude.*cos(phase));
5  %Kxy=ak; %lnK;
6  Kxy=K_MEAN*exp(-varK/2)*exp(ak); %hydraulic conductivity
```

The last line of the code returns the value of the hydraulic conductivity K at the point (x, y) and the commented line 5 returns the logarithm $\ln K$. The hydraulic conductivity is computed in this example by using an approximate relation between the geometric mean $K_g = \exp(\langle \ln K \rangle)$ and the arithmetic mean $\langle K \rangle$, i.e. $K_g = M(K)\exp(-\sigma^2/2)$ (see, e.g., [1, 2]). The two input parameters of this function, "K_MEAN" and "varK," stand for the mean hydraulic conductivity $\langle K \rangle$ and the variance $\sigma_{\ln K}^2$, respectively.

C.3.1.1   Two-Dimensional Log-Normal Hydraulic Conductivity Fields
with Exponential Correlation

The amplitudes, the wavenumbers, and the phases from (C.8) are computed from
the input parameters "NMOD" (number N of periodic modes in (C.7)), "varK,"
"ZC1" and "ZC2" (correlation lengths in longitudinal and transverse direction), and
"lambda" (filter width λ in (C.6)), with the following function.

```
1  function [wavenum, phi, amplitude] = ...
2  Kraichnan_Exp_param(NMOD,varK,ZC1,ZC2,lambda)
3  global state;
4  dummfac=sqrt(2*varK/NMOD);
5  gamma_1=unifrnd(0,1,NMOD,1);
6  gamma_2=unifrnd(0,1,NMOD,1);
7  wavenum(:,1)=sqrt(1./gamma_2.^2-1).*cos(2*pi*gamma_1)...
8  /(2*pi*ZC1);
9  wavenum(:,2)=sqrt(1./gamma_2.^2-1).*sin(2*pi*gamma_1)...
10 /(2*pi*ZC2);
11 norm=1./sqrt(wavenum(:,1).*wavenum(:,1)+wavenum(:,2)...
12 .*wavenum(:,2)); phi = 2.*pi*rand(NMOD,1);
13 gauss_factor=exp(-lambda*lambda./norm/2.0);
14 amplitude=dummfac*gauss_factor;
```

Correlation functions and correlation lengths of an anisotropic $\ln K$ field with
exponential correlation, estimated by using an ensemble of 1000 realizations
computed with the functions "Kraichnan_func" and "Kraichnan_Exp_param," are
shown in Fig. C.11.

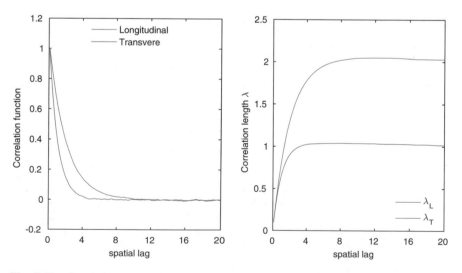

Fig. C.11   Correlations (left) and correlation lengths (right) of $\ln K$ field with anisotropic
exponential correlation

C.3.1.2   Two-Dimensional Log-Normal Hydraulic Conductivity Fields with Gaussian Correlation

Similarly to the case of exponential correlation, the parameters of the random modes in (C.8) are computed with the following function.

```
1  function [wavenum, phi, amplitude] = ...
2  Kraichnan_Gauss_param(NMOD,varK,ZC1,ZC2,lambda)
3  global state
4  sq2Pi=1/sqrt(2)/pi;
5  dummfac=sqrt(2*varK/NMOD);
6  wavenum(:,1)=sq2Pi*randn(NMOD,1)/ZC1;
7  wavenum(:,2)=sq2Pi*randn(NMOD,1)/ZC2;
8  norm=1./sqrt(wavenum(:,1).*wavenum(:,1)+wavenum(:,2)...
9  .*wavenum(:,2));
10 phi = 2.*pi*rand(NMOD,1);
11 gauss_factor=exp(-lambda*lambda./norm/2.0);
12 amplitude=dummfac*gauss_factor;
```

Correlation functions and correlation lengths of an anisotropic $\ln K$ field with Gaussian correlation, estimated by using an ensemble of 1000 realizations computed with the functions "Kraichnan_func" and "Kraichnan_Gauss_param," are shown in Fig. C.12.

Note that, unlike for exponential correlation considered in Sect. C.3.1 above, the parameter $\lambda_{\ln K}$, commonly called "correlation length" (e.g., [15]), differs from the limit of the integral of the correlation function (i.e., the *integral scale* $I_{\ln K}$) divided by the variance, which defines the correlation length in the statistical sense [29]

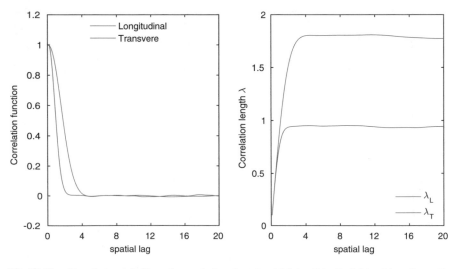

Fig. C.12 Correlations (left) and correlation lengths (right) of $\ln K$ field with anisotropic Gaussian correlation

(also see (4.57) and (2.73)). If one considers, for instance, the isotropic version of the correlation function (C.5), specified by the correlation length $\lambda_{\ln K} = 1$ m and by the variance $\sigma^2_{\ln K} = 0.1$, integrating (C.5) from zero to infinity along any straight line through the origin of the coordinate system one obtains

$$\frac{I_{\ln K}}{\sigma^2_{\ln K}} = \frac{\sqrt{\pi}}{2}\lambda_{\ln K} \approx 0.9\,\text{m}.$$

C.3.1.3   Comparison of Lognormal Hydraulic Conductivity Fields with Exponential and Gaussian Correlations

```
1   clear all; close all
2   tic
3   global state;
4   initstate; state;
5   NMOD=100;
6   varK=0.1;
7   K_MEAN=14.268;
8   lambda=0.0;
9   ZC1=2.0;
10  ZC2=1.0;
11  dx=0.1; dy=0.1; Lx=100; Ly=100;
12  Kfield=zeros(Lx,Ly);
13  for k=1:2
14      if k==1
15          [wavenum, phi, amplitude] = ...
16              Kraichnan_Exp_param(NMOD,varK,ZC1,ZC2,lambda);
17      else
18          [wavenum, phi, amplitude] = ...
19              Kraichnan_Gauss_param(NMOD,varK,ZC1,ZC2,lambda);
20      end
21      for i=1:Lx
22          for j=1:Ly
23              x=i*dx; y=j*dy;
24              Kxy=Kraichnan_func(x,y,wavenum,...
25                  phi,amplitude,K_MEAN,varK);
26              Kfield(i,j)=Kxy;
27          end
28      end
29      str='Exp'; if k==2; str='Gauss'; end
30      lgd(1:length(str))=str;
31      figure(1);
32      subplot(1,2,k)
33      surf(Kfield); colormap(jet); shading interp;
34      xlabel('x'); ylabel('y');
35      set(gca, 'XTick',0:25:100); set(gca,'YTick',0:25:100);
36      leg=legend(lgd,'Location','northwest');
37      leg.FontSize = 8;
38      figure(2);
```

```
39        subplot(1,2,k)
40        contour(1:Lx,1:Ly,Kfield/max(max(Kfield)))
41        xlim([0 100]); ylim([0 100]);
42        set(gca, 'XTick',0:20:100); set(gca,'YTick',0:20:100);
43        leg=legend(lgd,'Location','northwest');
44        leg.FontSize = 8;
45  end
46  toc
```

Realizations of K fields generated with exponential and Gaussian correlation of the $\ln K$ field are compared in Figs. C.13, C.14, and C.15.

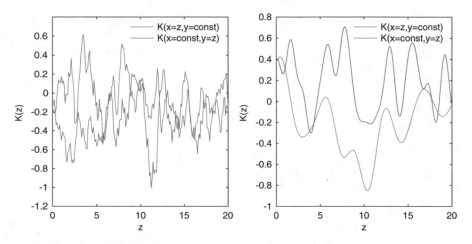

Fig. C.13   Samples of K fields with exponential (left) and Gaussian (right) correlation of $\ln K$

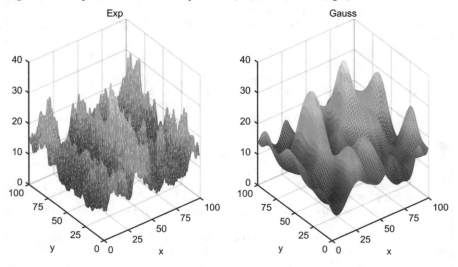

Fig. C.14 K fields with exponential (left) and Gaussian (right) correlation of $\ln K$

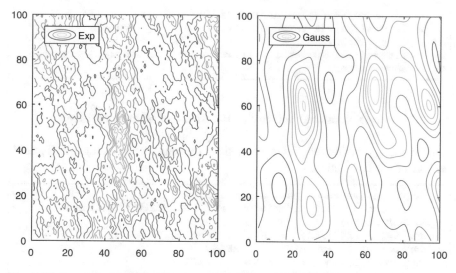

Fig. C.15 Contours of K fields with exponential (left) and Gaussian (right) correlation of ln K

C.3.2   Kraichnan Approximations of Velocity Fields

Two-dimensional velocity fields (u, v) are generated in the linear approximation (C.7) with the following function, which computes sums (C.8) for each component.

```
1  function [u, v] = ...
2  V_Kraichnan_Exp_func(x,y,wavenum,phi,amplitude)
3  phase=2*pi*(wavenum(:,1).*x+wavenum(:,2).*y)+phi;
4  u=sum(amplitude(:,1).*cos(phase));
5  v=sum(amplitude(:,2).*cos(phase));
```

C.3.2.1   Two-Dimensional Velocity Fields with Exponential Correlation

The parameters of the random modes (C.8) are computed with the following function.

```
1   function [wavenum, phi, amplitude] = ...
2   V_Kraichnan_Exp_param(NMOD,varK,ZC1,ZC2,U_MEAN,lambda)
3   global state;
4   dummfac=sqrt(2*varK/NMOD)*U_MEAN;
5   gamma_1=unifrnd(0,1,NMOD,1);
6   gamma_2=unifrnd(0,1,NMOD,1);
7   wavenum(:,1)=sqrt(1./gamma_2.^2-1).*cos(2*pi*gamma_1)...
8   /(2*pi*ZC1);
9   wavenum(:,2)=sqrt(1./gamma_2.^2-1).*sin(2*pi*gamma_1)...
10  /(2*pi*ZC2);
11  norm=1./sqrt(wavenum(:,1).*wavenum(:,1)+wavenum(:,2)...
12  .*wavenum(:,2));
13  phi = 2.*pi*rand(NMOD,1);
```

```
14  gauss_factor=exp(-lambda*lambda./norm/2);
15  amplitude(:,1)=(1-wavenum(:,1).^2.*norm.^2)*dummfac...
16  .*gauss_factor;
17  amplitude(:,2)=-wavenum(:,1).*wavenum(:,2).*norm.^2*dummfac...
18  .*gauss_factor;
```

C.3.2.2   Two-Dimensional Velocity Fields with Gaussian Correlation

The following function supplies realizations of the random modes for velocity fields
with Gaussian correlation.

```
1  function [wavenum, phi, amplitude] = ...
2  V_Kraichnan_Gauss_param(NMOD,varK,ZC1,ZC2,U_MEAN,lambda)
3  global state;
4  sq2Pi=1/sqrt(2)/pi;
5  dummfac=sqrt(2*varK/NMOD)*U_MEAN;
6  wavenum(:,1)=sq2Pi*randn(NMOD,1)/ZC1;
7  wavenum(:,2)=sq2Pi*randn(NMOD,1)/ZC2;
8  norm=1./sqrt(wavenum(:,1).*wavenum(:,1)+wavenum(:,2)...
9  .*wavenum(:,2));
10  phi = 2.*pi*rand(NMOD,1);
11  gauss_factor=exp(-lambda*lambda./norm/2);
12  amplitude(:,1)=(1-wavenum(:,1).^2.*norm.^2)*dummfac...
13  .*gauss_factor;
14  amplitude(:,2)=-wavenum(:,1).*wavenum(:,2).*norm.^2*dummfac...
15  .*gauss_factor;
```

C.3.2.3   Comparison of Velocity Fields with Exponential and Gaussian
Correlations

```
1  clear all; close all
2  tic
3  global state;
4  initstate;
5  NMOD=100;
6  varK=0.1;
7  lambda=0.0;
8  ZC1=1.0;
9  ZC2=1.0;
10  U_MEAN=1;
11  Lx=100; Ly=100; Mx=100; My=100; dx=0.1;dy=0.1;
12  Ufield=zeros(Lx,Ly); Vfield=zeros(Lx,Ly);
13  for k=1:2
14      if k==1
15          [wavenum, phi, amplitude] = ...
16          V_Kraichnan_Exp_param(NMOD,varK,ZC1,ZC2,U_MEAN,lambda);
17      else
18          [wavenum, phi, amplitude] = ...
```

```
19          V_Kraichnan_Gauss_param(NMOD,varK,ZC1,ZC2,U_MEAN, ...
               lambda);
20       end
21       for i=1:Lx
22           for j=1:Ly
23               x=i*dx; y=j*dy;
24               [u,v] = ...
25                   V_Kraichnan_func(x,y,wavenum, phi,...
26                   amplitude);
27               Ufield(i,j)=u+U_MEAN; Vfield(i,j)=v;
28           end
29       end
30       str='Exp'; if k==2; str='Gauss'; end
31       lgd(1:length(str))=str;
32       figure(1); hold
33       subplot(2,2,k)
34       surf(Ufield); colormap(jet); shading interp;
35       zlim([0 2])
36       xlabel('x','FontSize',8); ylabel('y','FontSize',8);
37       zlabel('V_x(x,y)','FontSize',8);
38       leg=legend(lgd,'Location','northwest'); leg.FontSize ...
               = 8;
39       subplot(2,2,k+2)
40       surf(Vfield); colormap(jet); shading interp;
41       zlim([-0.5 0.5])
42       xlabel('x','FontSize',8); ylabel('y','FontSize',8);
43       zlabel('V_y(x,y)','FontSize',8);
44       leg=legend(lgd,'Location','northwest');
45       leg.FontSize = 8;
46       figure(2); hold
47       subplot(1,2,k)
48       xg=1:Lx; yg=1:Ly;
49       contour(xg,yg,sqrt(Ufield.^2+Vfield.^2))
50       xlabel('x'); ylabel('y');
51       set(gca, 'XTick',0:20:100); set(gca,'YTick',0:20:100);
52       leg=legend(lgd,'Location','northwest');
53       leg.FontSize = 8;
54       xlim([0 100]); ylim([0 100]);
55       figure(3); hold all
56       subplot(1,2,k)
57       xg=1:Mx; yg=1:My;
58       quiver(xg,yg,Ufield(1:Mx,1:My),Vfield(1:Mx,1:My));
59       starty = (0:0.1:My);
60       startx = ones(size(starty));
61       streamline(xg,yg,Ufield(1:Mx,1:My),Vfield(1:Mx,1:My),...
62           startx,starty);
63       xlabel('x'); ylabel('y');
64       leg=legend(lgd,'Location','northwest');
65       leg.FontSize = 8;
66       xlim([30 70])
67       ylim([45 55])
68   end
69   toc
```

Correlation functions and correlation lengths of the velocity components generated with exponential and Gaussian correlation of the $\ln K$ field are given in Figs. C.16 and C.17, respectively. Structural features of the realizations of the velocity field for the two types of correlations are compared in Figs. C.18, C.19, C.20, and C.21.

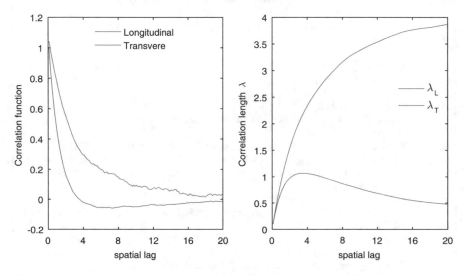

Fig. C.16   Correlations (left) and correlation lengths (right) of velocity field with anisotropic exponential correlation

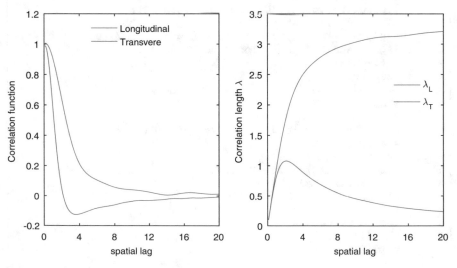

Fig. C.17   Correlations (left) and correlation lengths (right) of velocity field with anisotropic Gaussian correlation

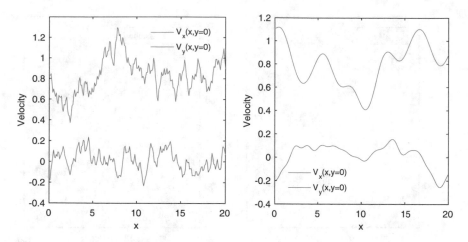

Fig. C.18   Samples of velocity fields with exponential (left) and Gaussian (right) correlation

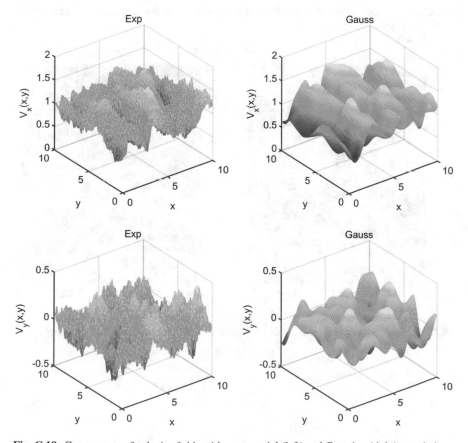

Fig. C.19   Components of velocity fields with exponential (left) and Gaussian (right) correlation

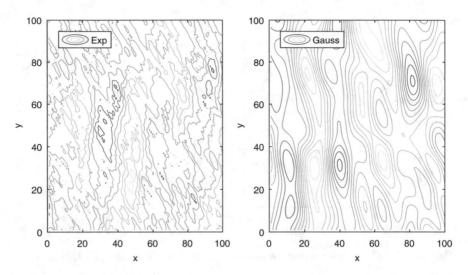

Fig. C.20   Contours of absolute values of velocity fields with exponential (left) and Gaussian (right) correlation

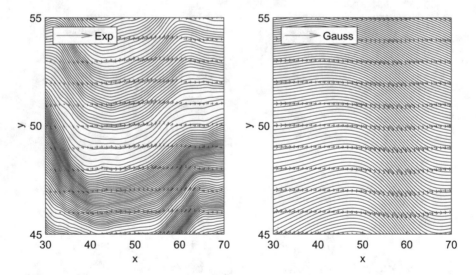

Fig. C.21   Streamlines for velocity fields with Gaussian (left) and exponential (right) correlation

Appendix D
Correlation Structure of the Itô Process

D.1 Mean Value and Covariance Components

The coefficients of the Fokker–Planck equation defined in Eqs. (4.63) and (4.64) can be related, under the conditions formulated in Eqs. (4.62), (4.65) and (4.66), to the time derivatives of the mean and covariance components. A way to derive general expressions of the mean and covariance is to start by computing their derivatives.

The derivative of the first moment can be computed by using Eq. (4.60) and the normalization property of g as follows:

$$
\frac{d}{dt}\mu_i(t) = \lim_{\Delta t \to 0} \frac{1}{\Delta t} [\mu_i(t + \Delta t) - \mu_i(t)]
$$

$$
= \lim_{\Delta t \to 0} \frac{1}{\Delta t} \left[\int x_i' c(\mathbf{x}', t + \Delta t) d\mathbf{x}' - \int x_i c(\mathbf{x}, t) d\mathbf{x} \right]
$$

$$
= \int \left[\lim_{\Delta t \to 0} \frac{1}{\Delta t} \int (x_i' - x_i) g(\mathbf{x}', t + \Delta t \mid \mathbf{x}, t) d\mathbf{x}' \right]
$$

$$
\times c(\mathbf{x}, t) d\mathbf{x}.
$$

For $\Delta t \to 0$, using Eqs. (4.62), (4.63), and (4.65) one obtains

$$
\frac{d}{dt}\mu_i(t) = \int V_i(\mathbf{x}, t) c(\mathbf{x}, t) d\mathbf{x} = \overline{V}_i(t). \tag{D.1}
$$

By integrating Eq. (D.1) one obtains Eq. (4.68) from Chap. 4.

© Springer Nature Switzerland AG 2019
N. Suciu, *Diffusion in Random Fields*, Geosystems Mathematics,
https://doi.org/10.1007/978-3-030-15081-5

To compute the derivative of the variance one proceeds like for the mean,

$$
s_{ij}(t + \Delta t) - s_{ij}(t) = \int x_i' x_j' c(\mathbf{x}', t + \Delta t) d\mathbf{x}' - \int x_i x_j c(\mathbf{x}, t) d\mathbf{x}
$$

$$
- [\mu_i(t + \Delta t)\mu_j(t + \Delta t) - \mu_i(t)\mu_j(t)]
$$

$$
= \int c(\mathbf{x}, t) d\mathbf{x} \int (x_i' - x_i)(x_j' - x_j) g(\mathbf{x}', t + \Delta t \mid \mathbf{x}, t) d\mathbf{x}'
$$

$$
+ \int x_i c(\mathbf{x}, t) d\mathbf{x} \int (x_j' - x_j) g(\mathbf{x}', t + \Delta t \mid \mathbf{x}, t) d\mathbf{x}'
$$

$$
+ \int x_j c(\mathbf{x}, t) d\mathbf{x} \int (x_i' - x_i) g(\mathbf{x}', t + \Delta t \mid \mathbf{x}, t) d\mathbf{x}'
$$

$$
- [\mu_i(t + \Delta t)\mu_j(t + \Delta t) - \mu_i(t)\mu_j(t)],
$$

and using Eqs. (4.62)–(4.66) for $\Delta t \to 0$, one obtains the derivative

$$
\frac{d}{dt} s_{ij}(t) = 2 \int D_{ij}(\mathbf{x}, t) c(\mathbf{x}, t) d\mathbf{x}
$$

$$
+ \int [x_i V_j(\mathbf{x}, t) + x_j V_i(\mathbf{x}, t)] c(\mathbf{x}, t) d\mathbf{x} - \frac{d}{dt} [\mu_i(t)\mu_j(t)],
$$

which after expressing the last term by (D.1) takes the form

$$
\frac{d}{dt} s_{ij}(t) = 2 \int D_{ij}(\mathbf{x}, t) c(\mathbf{x}, t) d\mathbf{x}
$$

$$
+ \int [x_i \left(V_j(\mathbf{x}, t) - \overline{V}_j(t) \right) + x_j \left(V_i(\mathbf{x}, t) - \overline{V}_i(t) \right)] c(\mathbf{x}, t) d\mathbf{x}. \qquad \text{(D.2)}
$$

Next, by highlighting the dependence on the initial positions of the second term in Eq. (D.2) one obtains an equivalent expression of the time derivative of $s_{ij}(t)$,

$$
\frac{d}{dt} s_{ij}(t) = 2 \int D_{ij}(\mathbf{x}, t) c(\mathbf{x}, t) d\mathbf{x}
$$

$$
+ \int c(\mathbf{x}_0, t_0) d\mathbf{x}_0 \int [(x_i - x_{0i}) \left(V_j(\mathbf{x}, t) - \overline{V}_j(t) \right)
$$

$$
+ (x_j - x_{0j}) \left(V_i(\mathbf{x}, t) - \overline{V}_i(t) \right)] g(\mathbf{x}, t \mid \mathbf{x}_0, t_0) d\mathbf{x}
$$

$$
+ \int c(\mathbf{x}_0, t_0) d\mathbf{x}_0 \int [(x_{0i} - \mu_i(t_0)) \left(V_j(\mathbf{x}, t) - \overline{V}_j(t) \right)
$$

$$
+ (x_{0j} - \mu_j(t_0)) \left(V_i(\mathbf{x}, t) - \overline{V}_i(t) \right)] g(\mathbf{x}, t \mid \mathbf{x}_0, t_0) d\mathbf{x}. \qquad \text{(D.3)}
$$

To analyze the second term of Eq. (D.3), one considers a time sequence $t_0 \leqslant t_1 \cdots < t_k < t_{k+1} \cdots \leqslant t_n = t$, $t_{k+1} - t_k = \Delta t$, and the joint probabilities p of the sequence of events $(\mathbf{x}_0, t_0), \cdots, (\mathbf{x}_{n-1}, t_{n-1}), (\mathbf{x}, t)$. The contribution of $(x_i - x_{0i})V_j(\mathbf{x}, t)$ can be expressed, using the Chapman–Kolmogorov equation and the consistency property of p that integrating over intermediate states one obtains reduced order joint probabilities, as follows

$$\int c(\mathbf{x}_0, t_0)d\mathbf{x}_0 \int (x_i - x_{0i})V_j(\mathbf{x}, t)g(\mathbf{x}, t \mid \mathbf{x}_0, t_0)d\mathbf{x}$$

$$= \int \cdots \int \left[\sum_{k=0}^{n-1}(x_{k+1,i} - x_{k,i}) \right] V_j(\mathbf{x}, t)p(\mathbf{x}, t ; \mathbf{x}_{n-1}, t_{n-1} ; \cdots ; \mathbf{x}_0, t_0)$$

$$\times d\mathbf{x}d\mathbf{x}_{n-1}\cdots d\mathbf{x}_0$$

$$= \sum_{k=0}^{n-1} \int \int \int (x_{k+1,i} - x_{k,i})V_j(\mathbf{x}, t) \, g(\mathbf{x}, t \mid \mathbf{x}_{k+1}, t_{k+1})g(\mathbf{x}_{k+1}, t_{k+1} \mid \mathbf{x}_k, t_k)c(\mathbf{x}_k, t_k)$$

$$\times d\mathbf{x}d\mathbf{x}_{k+1}d\mathbf{x}_k$$

$$= \sum_{k=0}^{n-1} \int c(\mathbf{x}_k, t_k)d\mathbf{x}_k \int (x_{k+1,i} - x_{k,i})\mathcal{V}(t, \mathbf{x}_{k+1})g(\mathbf{x}_{k+1}, t_{k+1} \mid \mathbf{x}_k, t_k)d\mathbf{x}_{k+1}, \quad \text{(D.4)}$$

where $\mathcal{V}(t; \mathbf{x}_{k+1}, t_{k+1}) = \int V_j(\mathbf{x}, t)g(\mathbf{x}, t \mid \mathbf{x}_{k+1}, t_{k+1})d\mathbf{x}$.

Because the velocity defined by Eq. (4.63) is always finite, the function $\mathcal{V}(t; \mathbf{x}_{k+1}, t_{k+1})$ is bounded, i.e. there exists a constant $M > 0$ so that $\mathcal{V}(t; \mathbf{x}_{k+1}, t_{k+1}) \leqslant M$, for all $\mathbf{x}_{k+1} \in \mathbb{R}^3$ and $t_{k+1} \in \mathbb{R}$. By using Eq. (4.64) for $i = j$ and the Cauchy–Schwarz inequality in the form

$$\left(\int_{|\mathbf{x}'-\mathbf{x}|\geqslant\epsilon} |x_i|g(\mathbf{x}', t + \Delta t \mid \mathbf{x}, t)d\mathbf{x}' \right)^2$$

$$\leqslant \int_{|\mathbf{x}'-\mathbf{x}|\geqslant\epsilon} x_i^2 g(\mathbf{x}', t + \Delta t \mid \mathbf{x}, t)d\mathbf{x}' \int_{|\mathbf{x}'-\mathbf{x}|\geqslant\epsilon} g(\mathbf{x}', t + \Delta t \mid \mathbf{x}, t)d\mathbf{x}'$$

one obtains the condition

$$\lim_{\Delta t \to 0} \frac{1}{\Delta t} \int_{|\mathbf{x}'-\mathbf{x}|\geqslant\epsilon} |x_i'| \, g(\mathbf{x}', t + \Delta t \mid \mathbf{x}, t)d\mathbf{x}' = 0. \quad \text{(D.5)}$$

Computing the last integral in Eq. (D.4) as sum between the integral over the sphere of radius ϵ and the integral outside the sphere and using the condition (D.5), one obtains

$$\left| \int\limits_{|\mathbf{x}_{k+1}-\mathbf{x}_k|\geqslant\epsilon} (x_{k+1,i}-x_{k,i})\mathcal{V}(t;\mathbf{x}_{k+1},t_{k+1})g(\mathbf{x}_{k+1},t_{k+1}\mid\mathbf{x}_k,t_k)d\mathbf{x}_{k+1}\right|$$

$$\leqslant M \int\limits_{|\mathbf{x}_{k+1}-\mathbf{x}_k|\geqslant\epsilon} \left| (x_{k+1,i}-x_{k,i})\right| g(\mathbf{x}_{k+1},t_k+\Delta t\mid\mathbf{x}_k,t_k)d\mathbf{x}_{k+1}\underset{\Delta t\to 0}{\to} 0.$$

$$(D.6)$$

Considering the negative and positive parts and applying the theorem of mean, the integral over the sphere of radius ϵ can be computed as

$$\int\limits_{|\mathbf{x}_{k+1}-\mathbf{x}_k|<\epsilon} (x_{k+1,i}-x_{k,i})\mathcal{V}(t;\mathbf{x}_{k+1},t_{k+1})g(\mathbf{x}_{k+1},t_{k+1}\mid\mathbf{x}_k,t_k)d\mathbf{x}_{k+1}$$

$$= \mathcal{V}(t;\mathbf{x}_k,t_{k+1}) \int\limits_{|\mathbf{x}_{k+1}-\mathbf{x}_k|<\epsilon} (x_{k+1,i}-x_{k,i})g(\mathbf{x}_{k+1},t_k+\Delta t\mid\mathbf{x}_k,t_k)d\mathbf{x}_{k+1}$$

$$+\mathcal{O}\left(\epsilon^2\right). \qquad\qquad (D.7)$$

For $\Delta t\to 0$, from Eq. (4.63) and Eqs. (D.4)–(D.7) one obtains

$$\int c(\mathbf{x}_0,t_0)d\mathbf{x}_0 \int (x_i-x_{0i})V_j(\mathbf{x},t)g(\mathbf{x},t\mid\mathbf{x}_0,t_0)d\mathbf{x}$$

$$= \int c(\mathbf{x}_0,t_0)d\mathbf{x}_0 \int_0^t dt' \int\int V_j(\mathbf{x},t)V_i(\mathbf{x}',t')$$

$$\times g(\mathbf{x},t\mid\mathbf{x}',t')g(\mathbf{x}',t'\mid\mathbf{x}_0,t_0)d\mathbf{x}d\mathbf{x}' + \mathcal{O}\left(\epsilon^2\right).$$

Finally, passing to the limit $\epsilon\to 0$, one obtains

$$\int c(\mathbf{x}_0,t_0)d\mathbf{x}_0 \int (x_i-x_{0i})V_j(\mathbf{x},t)g(\mathbf{x},t\mid\mathbf{x}_0,t_0)d\mathbf{x}$$

$$= \int c(\mathbf{x}_0,t_0)d\mathbf{x}_0 \int_0^t dt' \int\int V_j(\mathbf{x},t)V_i(\mathbf{x}',t')$$

$$\times g(\mathbf{x},t\mid\mathbf{x}',t')g(\mathbf{x}',t'\mid\mathbf{x}_0,t_0)d\mathbf{x}d\mathbf{x}'. \qquad (D.8)$$

Replacing in the second term of Eq. (D.3) the contribution (D.8) and the similar one obtained by permutation of the indices i and j one obtains Eq. (4.70) from Chap. 4.

D.2   Insights from Itô–Taylor Expansions

Itô–Taylor expansions provide an alternative approach to investigate the structure of the variance. Substituting the trajectory (2.46) in the expression of the second moment (2.54) one obtains

$$
m_2(t) = X_0^2 + \int_0^t b^2(t', X_{t'})dt'
$$

$$
+ 2X_0 \int_0^t E\left(a(t', X_{t'})\right) dt' + 2\int_0^t \int_0^t E\left(a(t', X_{t'})a(t'', X_{t''})\right) dt'dt''
$$

$$
+ E\left(\int_0^t a(t', X_{t'})dt' \int_0^t b(t'', X_{t''})dW_{t''}\right). \tag{D.9}
$$

The relevant terms of (D.9) are the variance of the pure diffusion process with coefficient b^2, the contributions of the mean and correlation of the drift coefficient a (second row), but also relevant could be the cross-correlation from the last row. This term can be evaluated by considering the two-dimensional process $\mathbf{X}_t = (X_t^1, X_t^2)$,

$$
X_t^1 = \int_0^t a(t', X_{t'})dt', \quad X_t^2 = \int_0^t b(t', X_{t'})dW_{t'},
$$

with vector coefficients $\mathbf{a} = (a, 0)$ and $\mathbf{b} = (0, b)^T$ (where T stands for transpose operation). For $f(\mathbf{X}_t) = X_t^1 X_t^2$, the two dimensional version of the Itô formula (2.52),

$$
f(\mathbf{X}_t) = f(\mathbf{0}) + \int_0^t dt' \sum_{k=1}^2 \left(a_k \frac{\partial f}{\partial x_k} + \frac{1}{2}b_k^2 \frac{\partial^2 f}{\partial x_k^2}\right) + \int_0^t dW_{t'} \sum_{k=1}^2 b_k \frac{\partial f}{\partial x_k},
$$

gives

$$
X_t^1 X_t^2 = \int_0^t a(t', X_{t'})dt' \int_0^{t'} b(t'', X_{t''})dW_{t''} + \int_0^t b(t', X_{t'})dW_{t'} \int_0^{t'} a(t'', X_{t''})dt''. \tag{D.10}
$$

The second term of (D.10) is an Itô integral of form $\int_0^t F(t', X_{t'})dW_{t'}$ and according to (2.49) its expectation vanishes. The expectation of the first term vanishes only if a is independent of the Wiener process, when $E(\int_0^t adt' \int_0^{t'} bdW_{t''}) = \int_0^t$

$adt'E(\int_0^{t'} bd W_{t''}) = 0$. To get an estimation of this term for the general case, one uses the first order terms in the Itô–Taylor expansions (2.56) of a and b and neglecting higher order terms one approximates the product ab as

$$b(0, X_0)[L_1 a(0, X_0) I_{(0),t'} I_{(1),t'} + L_2 a(0, X_0) I_{(1),t'} I_{(1),t'}],$$

where L_1 and L_2 are differential operators defined by

$$L_1(t, X_t) = \frac{\partial}{\partial t} + a(t, X_t) \frac{\partial}{\partial x} + \frac{1}{2} b(t, X_t) \frac{\partial^2}{\partial x^2} \text{ and } L_2(t, X_t) = b(t, X_t) \frac{\partial}{\partial x},$$

$I_{(0),t} = \int_0^t dt'$, and $I_{(1),t} = \int_0^t dW'$ (see [11] and Eq. (2.56)).

Using (D.10) for $a = b = 1$, one obtains the following decomposition of the product of integrals $I_{(0),t} I_{(1),t} = I_{(1,0),t} + I_{(0,1),t}$, which has the expectation zero, $E(I_{(1,0),t}) = \int_0^t dt' E(\int_0^{t'} d W_{t''}) = 0$, and $E(I_{(0,1),t}) = E(\int_0^t t' dW_{t'}) = 0$, according to (2.49). Proceeding to higher orders, one can see that the expectation of the multiple Itô integral vanishes if it contains at least one integration with respect to the Wiener process [12, Lemma 5.7.1]. The product of integrals $I_{(1),t} I_{(1),t} = (\int_0^t dW_{t'})^2$ has the expectation $E(I_{(1),t} I_{(1),t}) = t$, according to (2.50). It follows that

$$E(X_t^1 X_t^2) \approx \frac{1}{2} b(0, X_0) L_2 a(0, X_0) t^2 = \frac{1}{2} b^2(0, X_0) \frac{\partial}{\partial x} a(0, X_0) t^2.$$

It can be shown that from a third order Itô–Taylor expansion one obtains the following non-vanishing contributions

$$E(X_t^1 X_t^2) \approx \frac{1}{2} b L_2 a t^2 + \frac{1}{3} b L_1 L_2 a t^3 + \frac{1}{2} b L_1^2 L_2 a t^4,$$

with a and b evaluated at $(0, X_0)$. Thus, the cross-correlation contribution to the second moment (the last term of (D.9)) is in general non-vanishing.

D.3   Weak Solutions by Successive Approximations

The Itô–Taylor expansions form the previous section were constructed for a specified Wiener processes and lead to strong approximations of the Itô equation. Weak Itô–Taylor expansions can be constructed by replacing the multiple Itô integral with random variables whose moments satisfy appropriate conditions [12, Sect. 5.12]. Useful weak solutions can be also obtained by an iterative scheme consisting of successive approximations to Itô equation (2.46),

$$X_t^{(k)} = X_0 + \int_0^t a(t', X_{t'}^{(k-1)}) dt' + \int_0^t b(t', X_{t'}^{(k-1)}) d W_{t'}^{(k)}, \tag{D.11}$$

starting with some particular solution, for instance $X_t^{(0)} = X_0$, and with different Wiener processes $W_t^{(k)}$ at each order of approximation [24].

As shown in Sect. 3.2.1, if the coefficients of the Itô equation are Lipschitz continuous, then the sequence of successive approximations $X_t^{(k)}$ converges as $k \longrightarrow \infty$. Weak solutions of the equations for successive iterations k of (D.11) can be constructed by Euler scheme (3.10). If the seed of the random number generator used in the weak Euler scheme is not reset to the initial value after each iteration, then the surrogate Wiener processes (3.13) used in different orders of approximations k are independent of each other. This procedure yields a process that is not sample-path equivalent to X_t but has the same probability law.

Because the drift coefficient $a(t', X_{t'}^{(k-1)})$ is independent of $W_{t'}^{(k)}$ at each order of approximation, the expectation of the cross-correlation term of the second moment (D.9) factorizes

$$E\left(\int_0^t a(t', X_{t'}^{(k-1)})dt' \int_0^t b(t'', X_{t''}^{(k-1)})dW_{t''}^K\right) =$$

$$E\left(\int_0^t a(t', X_{t'}^{(k-1)})dt'\right) E\left(\int_0^t b(t'', X_{t''}^{(k-1)})dW_{t''}^K\right),$$

and the second factor vanishes according to (2.49). Then, from (D.9) and (2.54) one obtains the variance $s(t) = m_2(t) - m_1^2(t)$ of the Itô process,

$$s(t) = \int_0^t b^2(t', X_{t'})dt'$$

$$+ 2\int_0^t \int_0^t \left[E\left(a(t', X_{t'})a(t'', X_{t''})\right) - E\left(a(t', X_{t'})\right) E\left(a(t'', X_{t''})\right)\right] dt'dt''.$$

$$(D.12)$$

The trajectory (2.46) can be written as sum of drift and diffusion displacements,

$$X_t = X_t^a + X_t^b, \text{ where } X_t^a = X_0 + \int_0^t a(t', X_{t'})dt' \text{ and } X_t^a = \int_0^t b(t', X_{t'})dW_{t'},$$

and the variance (D.12) becomes $s(t) = s^a(t) + s^b(t)$. In case of a strong solution, one has the supplementary term $s^{ab}(t)$, the cross-correlation in (D.9). Figure D.1 shows the variance terms $s^a(t)$, $s^b(t)$, and $s^{ab}(t)$ obtained by successive approximations (D.11) and by the usual weak Euler scheme for the Itô process (2.46) with $a = 1 + \sin(x)$ and $b = \sqrt{2}$ (diffusion coefficient $D = b^2/2 = 1$). The approximation (D.11) consisted of $k = 3$ iterations and the variances were computed by averaging over 1000 trajectories.

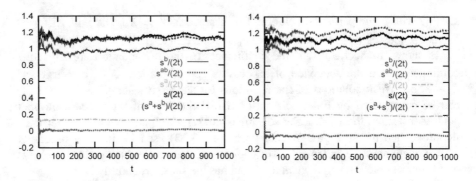

Fig. D.1 Variance terms for the solution of the Itô equation (2.46) given by 3 successive approximations (D.11) (left) and by the weak solution computed with the Euler scheme (3.10) (right)

Appendix E
Derivation of PDF Equations
by δ-Function Method

E.1 The PDF Equation

In studies on turbulent reacting flows probability densities are often defined by expectations of Dirac-δ functions [7, 8, 16, 17]. For instance, the concentration PDF is given by

$$f(\mathbf{c}; \mathbf{x}, t) = \langle \delta(\mathbf{C}(\mathbf{x}, t) - \mathbf{c}) \rangle, \tag{E.1}$$

where the angular bracket indicates stochastic average with respect to the probability measure of the random concentration vector $\mathbf{C}(\mathbf{x}, t)$ and the multidimensional δ function is defined by the product $\delta(\mathbf{C}(\mathbf{x}, t) - \mathbf{c}) = \prod_{\alpha=1}^{\alpha=N_\alpha} \delta(C_\alpha(\mathbf{x}, t) - c_\alpha)$.

The consistency of (E.1) with stochastic averaging is ensured by the definition of the δ functional. For instance, the integral weighted by the PDF (E.1) of a continuous function $Q(\mathbf{c})$ over the concentration space Y_c yields the expectation of Q as a function of the random concentration $\mathbf{C}(\mathbf{x}, t)$:

$$\int_{Y_c} Q(\mathbf{c}) f(\mathbf{c}; \mathbf{x}, t) d\mathbf{c} = \left\langle \int_{Y_c} Q(\mathbf{c}) \delta(\mathbf{C}(\mathbf{x}, t) - \mathbf{c}) d\mathbf{c} \right\rangle = \langle Q(\mathbf{C}(\mathbf{x}, t)) \rangle.$$

To derive the above relation one uses the continuity of $Q(\mathbf{c})$, which allows the permutation of the integral with the stochastic average (see Sect. 2.1.2.1), and the obvious property $\int_{Y_c} Q(\mathbf{c}) \delta(\mathbf{C}(\mathbf{x}, t) - \mathbf{c}) d\mathbf{c} = \int_{\mathbb{R}^{N_\alpha}} I_{Y_c}(\mathbf{c}) Q(\mathbf{c}) \delta(\mathbf{C}(\mathbf{x}, t) - \mathbf{c}) d\mathbf{c} = Q(\mathbf{C}(\mathbf{x}, t))$, where I_{Y_c} is the indicator function of the set Y_c. The definition (E.1) can be generalized, by using products of δ functions, to obtain a consistent hierarchy of multi-point probability distributions which completely define the random concentration $\mathbf{C}(\mathbf{x}, t)$ as a random function (see Sect. 2.1.3).

A straightforward derivation of the evolution equation for the PDF $f(\mathbf{c}; \mathbf{x}, t)$ is based on the evaluation of the time derivative of its definition (E.1) through formal manipulations of derivatives of δ functions [10]. In terms of Dirac distributions, the

© Springer Nature Switzerland AG 2019
N. Suciu, *Diffusion in Random Fields*, Geosystems Mathematics,
https://doi.org/10.1007/978-3-030-15081-5

derivative of the δ function is defined by the relation

$$\int_{-\infty}^{\infty} \delta'(y_0 - y)\varphi(y)dy = -\int_{-\infty}^{\infty} \delta(y_0 - y)\varphi'(y)dy, \tag{E.2}$$

for any smooth function φ with compact support in $\mathbb{R}$. The usual notation for the distributional derivative is $\delta'[\varphi] = -\delta[\varphi'] = -\varphi'(y_0)$. Further, (E.2) can be generalized for the case where y_0 is a given value of a composite function of some variable x, $y_0 = g(x)$. Then, the Dirac functional applied to the test function φ reads

$$\int_{-\infty}^{\infty} \delta(g(x) - y)\varphi(y)dy = \varphi(g(x)). \tag{E.3}$$

The derivative of (E.3) with respect to x follows as

$$\frac{d}{dx} \int_{-\infty}^{\infty} \delta(g(x) - y)\varphi(y)dy = \varphi'(g(x))\frac{dg(x)}{dx}$$

$$= \frac{dg(x)}{dx} \int_{-\infty}^{\infty} \delta(g(x) - y)\varphi'(y)dy$$

$$= -\frac{dg(x)}{dx} \int_{-\infty}^{\infty} \delta'(g(x) - y)\varphi(y)dy,$$

where the second equality is implied by (E.3) and the third equality by (E.2). In the common compact notation for distributions we get

$$\frac{d}{dx}\delta(g(x) - y)[\varphi] = -\frac{dg(x)}{dx}\delta'(g(x) - y)[\varphi],$$

which is often written in applications to turbulent flows as [10, 16, 17]

$$\frac{d}{dx}\delta(g(x) - y) = -\frac{dg(x)}{dx}\frac{d}{dy}\delta(g(x) - y). \tag{E.4}$$

The correctness of the results obtained with formal manipulations of the relation (E.4) can be checked by multiplying them with test functions, integrating over y and using (E.2) (see, e.g., [10, Sect. 2.2.1]).

Considering multidimensional δ functions and the vectorial function $\mathbf{g} = \mathbf{C}(\mathbf{x}, t)$ one obtains, similarly to (E.4), the formal expression of the partial derivative with respect to the time variable,

$$\frac{\partial}{\partial t}\delta(\mathbf{C}(\mathbf{x}, t) - \mathbf{c}) = -\frac{\partial C_\alpha(\mathbf{x}, t)}{\partial t}\frac{\partial}{\partial c_\alpha}\delta(\mathbf{C}(\mathbf{x}, t) - \mathbf{c}), \tag{E.5}$$

and the partial derivatives with respect to the spatial variables,

$$\frac{\partial}{\partial x_i}\delta(\mathbf{C}(\mathbf{x}, t) - \mathbf{c}) = -\frac{\partial C_\alpha(\mathbf{x}, t)}{\partial x_i}\frac{\partial}{\partial c_\alpha}\delta(\mathbf{C}(\mathbf{x}, t) - \mathbf{c}). \tag{E.6}$$

Using (E.5), the time derivative of the PDF (E.1) is computed as follows:

$$\begin{aligned}
\frac{\partial f(\mathbf{c}; \mathbf{x}, t)}{\partial t} &= \left\langle \frac{\partial}{\partial t}\delta(\mathbf{C}(\mathbf{x}, t) - \mathbf{c})\right\rangle \\
&= -\left\langle \frac{\partial C_\alpha(\mathbf{x}, t)}{\partial t}\frac{\partial}{\partial c_\alpha}\delta(\mathbf{C}(\mathbf{x}, t) - \mathbf{c})\right\rangle \\
&= -\frac{\partial}{\partial c_\alpha}\left\langle \frac{\partial C_\alpha(\mathbf{x}, t)}{\partial t}\delta(\mathbf{C}(\mathbf{x}, t) - \mathbf{c})\right\rangle.
\end{aligned} \tag{E.7}$$

Performing the stochastic average on the right-hand side of (E.7) requires more information on the statistics of the random concentration $\mathbf{C}(\mathbf{x}, t)$ than the knowledge of the one-point PDF $f(\mathbf{C}; \mathbf{x}, t)$. To see that, note that the time derivative is the limit of the increment of $\mathbf{C}$ divided by the corresponding time increment, which is a two-point quantity. Thus, a two-point PDF is needed if one wants to perform the average before taking the limit. Alternatively, the average can be performed if one knows the joint statistics of $\partial \mathbf{C}/\partial t$ and $\mathbf{C}$. To obtain meaningful expressions of these kind of stochastic averages, let us follow the approach of Fox [7] and consider a generic random function $\mathbf{Z}(\mathbf{x}, t)$ which is not described by the one-point PDF $f(\mathbf{C}; \mathbf{x}, t)$. Similarly to the average occurring in (E.7), one obtains in general, for a random function $F(\mathbf{Z}(\mathbf{x}, t))$, the average

$$\begin{aligned}
\langle F(\mathbf{Z}(\mathbf{x}, t))\delta(\mathbf{C}(\mathbf{x}, t) - \mathbf{c})\rangle &= \left\langle \delta(\mathbf{C}(\mathbf{x}, t) - \mathbf{c})\int_{Y_z} F(\mathbf{z})\delta(\mathbf{Z}(\mathbf{x}, t) - \mathbf{z})d\mathbf{z}\right\rangle \\
&= \int_{Y_z} F(\mathbf{z})\,\langle \delta(\mathbf{Z}(\mathbf{x}, t) - \mathbf{z})\delta(\mathbf{C}(\mathbf{x}, t) - \mathbf{c})\rangle\, d\mathbf{z} \\
&= \int_{Y_z} F(\mathbf{z})f(\mathbf{c}, \mathbf{z}; \mathbf{x}, t)d\mathbf{z} \\
&= f(\mathbf{c}; \mathbf{x}, t)\int_{Y_z} F(\mathbf{z})f(\mathbf{z}|\mathbf{c}; \mathbf{x}, t)d\mathbf{z},
\end{aligned}$$

where $f(\mathbf{c}, \mathbf{z}; \mathbf{x}, t) = \langle \delta(\mathbf{Z}(\mathbf{x}, t) - \mathbf{z})\delta(\mathbf{C}(\mathbf{x}, t) - \mathbf{c})\rangle$ defines, similarly to (E.1), the joint PDF in the $(\mathbf{c}, \mathbf{z})$ space and $f(\mathbf{z}|\mathbf{c}; \mathbf{x}, t) = f(\mathbf{c}, \mathbf{z}; \mathbf{x}, t)/f(\mathbf{c}; \mathbf{x}, t)$ is a conditional PDF. Finally one obtains

$$\langle F(\mathbf{Z}(\mathbf{x}, t))\delta(\mathbf{C}(\mathbf{x}, t) - \mathbf{c})\rangle = \langle F(\mathbf{Z}(\mathbf{x}, t))|\mathbf{C}(\mathbf{x}, t) = \mathbf{c}\rangle\, f(\mathbf{c}; \mathbf{x}, t), \tag{E.8}$$

which is just the expectation of the random function $F(\mathbf{Z}(\mathbf{x}, t))$ conditional on a fixed value of the concentration vector $\mathbf{c}$ multiplied by the one-point PDF $f(\mathbf{c}; \mathbf{x}, t)$. In the following, the shorthand notation $\langle \cdot | \mathbf{c} \rangle$ will be used to denote conditional averages $\langle \cdot | \mathbf{C}(\mathbf{x}, t) = \mathbf{c} \rangle$.

Now, substituting the time derivative $\partial C_\alpha(\mathbf{x}, t)/\partial t$ from the transport equation (6.1) into (E.7) and using (E.8), one obtains the evolution equation for the PDF $f(\mathbf{c}; \mathbf{x}, t)$,

$$\frac{\partial f(\mathbf{c}; \mathbf{x}, t)}{\partial t} = \frac{\partial}{\partial c_\alpha} \left\{ \left[\left\langle V_i \frac{\partial C_\alpha}{\partial x_i} \middle| \mathbf{c} \right\rangle - \left\langle D \frac{\partial^2 C_\alpha}{\partial x_i \partial x_i} \middle| \mathbf{c} \right\rangle - S_\alpha(\mathbf{c}) \right] f(\mathbf{c}; \mathbf{x}, t) \right\}. \quad \text{(E.9)}$$

The last term in (E.9) is in a closed form because the reaction term S_α in (6.1) is completely determined by the random concentration $\mathbf{C}$, which implies that its conditional expectation is the value of S_α evaluated for the sample space value $\mathbf{c}$,

$$\langle S_\alpha(\mathbf{C}(\mathbf{x}, t)) | \mathbf{c} \rangle = \int_{Y_z} S_\alpha(\mathbf{c}) f(\mathbf{z} | \mathbf{c}; \mathbf{x}, t) d\mathbf{z} = S_\alpha(\mathbf{c}) \int_{Y_z} f(\mathbf{z} | \mathbf{c}; \mathbf{x}, t) d\mathbf{z} = S_\alpha(\mathbf{c}).$$

The first two terms on the right-hand side of (E.9) are unclosed and require modeling.

The first unclosed term can be transformed as follows:

$$\frac{\partial}{\partial c_\alpha} \left[\left\langle V_i \frac{\partial C_\alpha}{\partial x_i} \middle| \mathbf{c} \right\rangle f(\mathbf{c}; \mathbf{x}, t) \right] = \frac{\partial}{\partial c_\alpha} \left\langle V_i \frac{\partial C_\alpha}{\partial x_i} \delta(\mathbf{C}(\mathbf{x}, t) - \mathbf{c}) \right\rangle$$

$$= -\frac{\partial}{\partial x_i} [\langle V_i | \mathbf{c} \rangle f(\mathbf{c}; \mathbf{x}, t)], \quad \text{(E.10)}$$

where the first equality follows from (E.8) and the final result is obtained from (E.6) and the incompressibility condition $\partial V_i/\partial x_i = 0$. Defining the velocity fluctuation $\mathbf{U} = \mathbf{V} - \langle \mathbf{V} \rangle$ and using the gradient diffusion closure $\langle \mathbf{U} | \mathbf{c} \rangle f(\mathbf{c}; \mathbf{x}, t) = -\mathbf{D}^* \nabla f(\mathbf{c}; \mathbf{x}, t)$ [7, 8, 16], one writes Eq. (E.10) in the form

$$\frac{\partial}{\partial c_\alpha} \left[\left\langle V_i \frac{\partial C_\alpha}{\partial x_i} \middle| \mathbf{c} \right\rangle f(\mathbf{c}; \mathbf{x}, t) \right] = -\frac{\partial}{\partial x_i} [\langle V_i \rangle f(\mathbf{c}; \mathbf{x}, t)] + \frac{\partial}{\partial x_i} D_{i,j}^* \frac{\partial}{\partial x_j} f(\mathbf{c}; \mathbf{x}, t).$$

$$\text{(E.11)}$$

The upscaled diffusion tensor $D_{i,j}^*$ is provided by turbulence models [17] or by stochastic upscaling of diffusion in random velocity fields [22].

The second unclosed term in (E.9) is the conditional expectation of the molecular diffusion term in the transport equation (6.1). In the following it will be shown that this term is related to the divergence, in the physical space, of the diffusive flux of the PDF $f(\mathbf{c}; \mathbf{x}, t)$, with the same diffusion coefficient D as in Eq. (6.1). Since by (E.1) $f(\mathbf{c}; \mathbf{x}, t) = \langle \delta(\mathbf{C}(\mathbf{x}, t) - \mathbf{c}) \rangle$, the diffusive flux of f is determined by the expectation of (E.6) and its divergence can be expressed as follows:

$$D\frac{\partial^2}{\partial x_i \partial x_i} f(\mathbf{c}; \mathbf{x}, t) = D\frac{\partial^2}{\partial x_i \partial x_i} \langle \delta(\mathbf{C}(\mathbf{x}, t) - \mathbf{c}) \rangle$$

$$= -D\left\langle \frac{\partial^2 C_\alpha}{\partial x_i \partial x_i}(\mathbf{x}, t)\frac{\partial}{\partial c_\alpha}\delta(\mathbf{C}(\mathbf{x}, t) - \mathbf{c}) - \frac{\partial C_\alpha}{\partial x_i}(\mathbf{x}, t)\frac{\partial}{\partial c_\alpha}\frac{\partial}{\partial x_i}\delta(\mathbf{C}(\mathbf{x}, t) - \mathbf{c}) \right\rangle$$

$$= -\frac{\partial}{\partial c_\alpha}\left\langle D\frac{\partial^2 C_\alpha}{\partial x_i \partial x_i}(\mathbf{x}, t)\delta(\mathbf{C}(\mathbf{x}, t) - \mathbf{c}) \right\rangle + \frac{\partial^2}{\partial c_\alpha \partial c_\beta}\left\langle D\frac{\partial C_\alpha}{\partial x_i}\frac{\partial C_\beta}{\partial x_i}\delta(\mathbf{C}(\mathbf{x}, t) - \mathbf{c}) \right\rangle$$

$$= -\frac{\partial}{\partial c_\alpha}\left[\left\langle D\frac{\partial^2 C_\alpha}{\partial x_i \partial x_i}(\mathbf{x}, t)\middle|\mathbf{c}\right\rangle f(\mathbf{c}; \mathbf{x}, t)\right] + \frac{\partial^2}{\partial c_\alpha \partial c_\beta}\left[\left\langle D\frac{\partial C_\alpha}{\partial x_i}\frac{\partial C_\beta}{\partial x_i}\middle|\mathbf{c}\right\rangle f(\mathbf{c}; \mathbf{x}, t)\right].$$

$$(E.12)$$

Using (E.11) and (E.12) one defines the coefficients

$$\mathscr{V}_i = \langle V_i \rangle + \frac{\partial}{\partial x_j}D^*_{i,j},\tag{E.13}$$

$$\mathscr{D}_{ij} = D + D^*_{i,j},\tag{E.14}$$

$$\mathscr{M}_{\alpha\beta} = \left\langle D\frac{\partial C_\alpha}{\partial x_i}\frac{\partial C_\beta}{\partial x_i}\middle|\mathbf{c}\right\rangle.\tag{E.15}$$

With the coefficients given by (E.13)–(E.15), Eq. (E.9) takes the form of the PDF equation (6.2).

E.2   The Fokker–Planck Equation

It has been shown in [26–28] that for a weighting function Θ which obeys the continuity equation

$$\frac{\partial \Theta}{\partial t} + V_i\frac{\partial \Theta}{\partial x_i} = 0\tag{E.16}$$

the evolution of the joint concentration-position PDF $p(\mathbf{c}, \mathbf{x}, t) = \Theta(\mathbf{c})f(\mathbf{c}; \mathbf{x}, t)$ is described by the following Fokker–Planck equation

$$\frac{\partial p(\mathbf{c}, \mathbf{x}, t)}{\partial t} + \langle V_i \rangle\frac{\partial}{\partial x_i}p(\mathbf{c}, \mathbf{x}, t) = -\frac{\partial}{\partial x_i}\left[\langle U_i|\mathbf{c}\rangle\, p(\mathbf{c}, \mathbf{x}, t)\right]$$

$$-\frac{\partial}{\partial c_\alpha}\left\{\left[\left\langle D\frac{\partial^2 C_\alpha}{\partial x_i \partial x_i}\middle|\mathbf{c}\right\rangle + S_\alpha(\mathbf{c})\right]p(\mathbf{c}, \mathbf{x}, t)\right\}.$$

$$(E.17)$$

To derive Eq. (E.17) in the δ-function approach from previous section, one uses the "shifting" property

$$\Theta(\mathbf{c})\delta(\mathbf{C}(\mathbf{x}, t) - \mathbf{c}) = \Theta(\mathbf{C}(\mathbf{x}, t))\delta(\mathbf{C}(\mathbf{x}, t) - \mathbf{c}), \qquad (E.18)$$

which can be readily checked by integrating both sides of (E.18) with respect to $\mathbf{c}$. To derive the equation for $p(\mathbf{c}, \mathbf{x}, t) = \Theta(\mathbf{c})f(\mathbf{c}; \mathbf{x}, t) = \Theta(\mathbf{c}) \langle \delta(\mathbf{C}(\mathbf{x}, t) - \mathbf{c}) \rangle$ one starts, as for (E.7), by computing its time derivative:

$$\frac{\partial p(\mathbf{c}, \mathbf{x}, t)}{\partial t} = \frac{\partial}{\partial t} \langle \Theta(\mathbf{c})\delta(\mathbf{C}(\mathbf{x}, t) - \mathbf{c}) \rangle = \frac{\partial}{\partial t} \langle \Theta(\mathbf{C}(\mathbf{x}, t))\delta(\mathbf{C}(\mathbf{x}, t) - \mathbf{c}) \rangle$$

$$= \left\langle \Theta(\mathbf{C}(\mathbf{x}, t)) \frac{\partial}{\partial t} \delta(\mathbf{C}(\mathbf{x}, t) - \mathbf{c}) \right\rangle + \left\langle \delta(\mathbf{C}(\mathbf{x}, t) - \mathbf{c}) \frac{\partial}{\partial t} \Theta(\mathbf{C}(\mathbf{x}, t)) \right\rangle$$

$$= -\frac{\partial}{\partial c_\alpha} \left\langle \Theta(\mathbf{C}(\mathbf{x}, t)) \frac{\partial C_\alpha(\mathbf{x}, t)}{\partial t} \delta(\mathbf{C}(\mathbf{x}, t) - \mathbf{c}) \right\rangle + \left\langle \delta(\mathbf{C}(\mathbf{x}, t) - \mathbf{c}) \frac{\partial}{\partial t} \Theta(\mathbf{C}(\mathbf{x}, t)) \right\rangle.$$

$$(E.19)$$

In the first equality of (E.19) $\Theta(\mathbf{c})$ can be introduced under average because it is a non-random function of the state space variable $\mathbf{c}$, the second equality follows from the shifting property (E.18), and in the first term of the last line one uses the partial spatial derivative (E.6) of the δ-function. With the time derivative of C_α from the transport equation (6.1), this latter term becomes

$$-\frac{\partial}{\partial c_\alpha} \left\langle \Theta(\mathbf{C}(\mathbf{x}, t)) \left[-V_i \frac{\partial C_\alpha}{\partial x_i} + D \frac{\partial^2 C_\alpha}{\partial x_i \partial x_i} + S_\alpha \right] \delta(\mathbf{C}(\mathbf{x}, t) - \mathbf{c}) \right\rangle$$

$$= \frac{\partial}{\partial c_\alpha} \left\langle \Theta(\mathbf{C}(\mathbf{x}, t)) V_i \frac{\partial C_\alpha}{\partial x_i} \delta(\mathbf{C}(\mathbf{x}, t) - \mathbf{c}) \right\rangle$$

$$- \frac{\partial}{\partial c_\alpha} \left\{ \left[\left\langle D \frac{\partial^2 C_\alpha}{\partial x_i \partial x_i} \Big| \mathbf{c} \right\rangle + S_\alpha(\mathbf{c}) \right] \Theta(\mathbf{c}) f(\mathbf{c}; \mathbf{x}, t) \right\}. \qquad (E.20)$$

To obtain the second term on the right-hand side of (E.20) one uses the shifting property (E.18) and the conditional average (E.8). This term is just the last term of the Fokker–Planck equation (E.17). The first term on the right-hand side of (E.20) can be rewritten by using (E.6) as

$$\frac{\partial}{\partial c_\alpha} \left\langle \Theta(\mathbf{C}(\mathbf{x}, t)) V_i \frac{\partial C_\alpha}{\partial x_i} \delta(\mathbf{C}(\mathbf{x}, t) - \mathbf{c}) \right\rangle = -\left\langle \Theta(\mathbf{C}(\mathbf{x}, t)) V_i \frac{\partial}{\partial x_i} \delta(\mathbf{C}(\mathbf{x}, t) - \mathbf{c}) \right\rangle$$

$$= \left\langle \delta(\mathbf{C}(\mathbf{x}, t) - \mathbf{c}) \frac{\partial}{\partial x_i} [\Theta(\mathbf{C}(\mathbf{x}, t)) V_i] \right\rangle - \left\langle \frac{\partial}{\partial x_i} [\Theta(\mathbf{C}(\mathbf{x}, t)) V_i \delta(\mathbf{C}(\mathbf{x}, t) - \mathbf{c})] \right\rangle.$$

$$(E.21)$$

Note that due to the incompressibility condition $\partial V_i / \partial x_i = 0$ and because Θ obeys the continuity equation (E.16), the first term in the final expression (E.21) cancels the last term of (E.19). Further, using (E.18), (E.8), the incompressibility condition, and writing the velocity as a sum of mean and fluctuations, the second term of (E.21) becomes

$$
-\left\langle \frac{\partial}{\partial x_i} [\Theta(\mathbf{C}(\mathbf{x}, t)) V_i \delta(\mathbf{C}(\mathbf{x}, t) - \mathbf{c})] \right\rangle = -\frac{\partial}{\partial x_i} \langle \Theta(\mathbf{c}) V_i \delta(\mathbf{C}(\mathbf{x}, t) - \mathbf{c}) \rangle
$$

$$
= -\langle V_i \rangle \frac{\partial}{\partial x_i} p(\mathbf{c}, \mathbf{x}, t) - \frac{\partial}{\partial x_i} [\langle U_i | \mathbf{c} \rangle \, p(\mathbf{c}, \mathbf{x}, t)] . \tag{E.22}
$$

The Fokker–Planck equation (E.17) is finally obtained by recursively substituting (E.22), (E.21), and (E.20) into (E.19). For constant weighting facto Θ, e.g. constant density, the Fokker–Planck equation (E.17) coincides with the PDF equation (6.2) with coefficients given by (E.13)–(E.15).

Appendix F
Upscaled Dispersion Coefficients

F.1 Self-averaging Estimations of Diffusion Coefficients

Let us consider a component of the ensemble process $\{X_t, t \geq 0\}$ of mean zero starting from $X_0 = 0$, obtained by subtracting the ensemble mean $\langle X_{i,t} \rangle$ from one of the components $X_{i,t}$, $i = 1, 2$, of the transport process. If the continuous time interval $[0, t]$ is partitioned into S subintervals of equal length τ, so that $t = S\tau$, a position on the trajectory X_t and its square X_t^2 can be expressed in terms of increments $\delta X_s = X_{s\tau} - X_{(s-1)\tau}$ as

$$X_t = \sum_{s=1}^{S} \delta X_s, \quad X_t^2 = \sum_{s=1}^{S} (\delta X_s)^2 + 2 \sum_{r=1}^{S-1} \sum_{s=1}^{S-r} \delta X_s \delta X_{s+r}.$$

Further, introducing the time averages

$$\rho(r) = \frac{1}{S-r} \sum_{s=1}^{S-r} \delta X_s \delta X_{s+r}, \tag{F.1}$$

X_t^2 can be rewritten as

$$X_t^2 = S\rho(0) + 2 \sum_{r=1}^{S-1} (S - r)\rho(r). \tag{F.2}$$

Since the averages ρ defined by (F.1) are similar to stochastic correlations, (F.2) is a single-realization form of Taylor formula for processes with stationary variance [22].

The increment process δX_s is determined through advective displacements by the velocity field sampled on trajectories [23], which, in a first order approximation, has

© Springer Nature Switzerland AG 2019
N. Suciu, *Diffusion in Random Fields*, Geosystems Mathematics,
https://doi.org/10.1007/978-3-030-15081-5

ergodic covariances [22]. Therefore, one expects that the ergodic estimation (F.1) converges to the correlation of δX_s, so that (F.2) provides a good approximation of the variance of X_t. Then, the diffusion coefficient can be estimated on a single sample of X_t as $\tilde{D} = X_t^2/(2S\tau)$. With this, (F.2) yields

$$\tilde{D} = \frac{1}{2\tau}\rho(0) + \frac{1}{\tau}\sum_{r=1}^{S-1}\left(1 - \frac{r}{S}\right)\rho(r). \tag{F.3}$$

Since, as r gets closer to S the time averages ρ are poor ergodic estimates of the correlations, a rigorous ergodicity statement is not possible in this case. Nevertheless, the reliability of the self-averaging estimator (F.3) has been demonstrated by numerical experiments [23]. In practice, acceptable results are obtained if the estimation time $S\tau$ is larger than the total simulation time T. For instance, for the PDF simulations over $T = 100$ days presented in Chap. 6 dispersion coefficients were estimated on samples X_t of length $S\tau = 400$ days. Also, for an increased accuracy, the ergodic estimations (F.1) of the correlations ρ were further averaged over 10^6 paths S sampled on the same trajectory. With these parameters, the computation time for the estimation (F.3) of the diffusion coefficient was of about 4 min. For comparison, a MC estimation of the of same dispersion coefficients over 100 days required about 10 min and 256 processors.

F.2   Coarse-Grained Dispersion Coefficients

A statistically homogenous velocity field of mean $\langle V \rangle$ and fluctuations $U = V - \langle V \rangle$ is related to the filtered velocity by

$$V = \langle V \rangle + \langle U \rangle_\lambda + U_\lambda, \tag{F.4}$$

where $\langle U \rangle_\lambda$ is the filtered fluctuation of the fine-grained velocity and U_λ is the residual sub-filter velocity fluctuation about the filtered velocity $\langle V \rangle_\lambda = \langle V \rangle + \langle U \rangle_\lambda$. The longitudinal component of the transport process starting from $X = 0$ is described by the Itô equation

$$X_1(t) = \int_0^t V_1(t')dt' + W(t), \tag{F.5}$$

where $V_1(t') = V_1(X(t'))$ and $W(t)$ is a Wiener process of mean zero and variance equal to $2Dt$. Since the velocity field is statistically homogeneous, the ensemble average of (F.5) is $\langle X_1 \rangle(t) = \langle V_1 \rangle t$. The longitudinal ensemble process $X_1^{ens}(t) = X_1(t) - \langle X_1(t) \rangle$ verifies the same equation (F.5), with the velocity component V_1 replaced by its fluctuation U_1, and the half derivative of its variance defines the ensemble dispersion coefficient

$$D_{11}^{ens}(t) = D + \int_0^t \langle U_1(t)U_1(t')\rangle dt'. \tag{F.6}$$

From (F.4) and (F.5) one obtains the equivalent expression

$$D_{11}^{ens}(t) = D + \int_0^t \langle \langle U_1 \rangle_\lambda(t) \langle U_1(t') \rangle_\lambda \rangle dt'$$

$$+ \int_0^t \langle U_{\lambda,1}(t)U_{\lambda,1}(t')\rangle dt'$$

$$+ \int_0^t \langle \langle U_1 \rangle_\lambda(t) U_{\lambda,1}(t')\rangle dt' + \int_0^t \langle \langle U_1 \rangle_\lambda(t') U_{\lambda,1}(t)\rangle dt'. \tag{F.7}$$

The first line in (F.7) corresponds to (F.6) with U_1 replaced by $\langle U_1 \rangle_\lambda$, that is to the coefficient $D_{11}^{ens}(t, \lambda)$ derived for a filtered velocity field. The second line is given by the integral from (F.6) written for the correlation of the sub-filter velocity fluctuations. As for the last line in (F.7), it describes a contribution to the ensemble coefficient produced by correlations between filtered and sub-filter velocity fluctuations. If this last contribution vanishes, then the contribution from the second line of (F.7) corresponds to the correction $\delta D_{11}(t, \lambda)$ of the coarse-grained diffusion coefficient defined by (6.23).

References

1. Attinger, S.: Generalized coarse graining procedures for flow in porous media. Comput. Geosci. **7**(4), 253–273 (2003)
2. Bellin, A., Salandin, P., Rinaldo, A.: Simulation of dispersion in heterogeneous porous formations: statistics, first-order theories, convergence of computations. Water Resour. Res. **28**(9), 2211–2227 (1992)
3. Bronstein, I.N., Semendjajew, K.A., Musiol, G., Mühlig, H.: Taschenbuch der Mathematik. Verlag Harri Deutsch, Frankfurt am Main (2006)
4. Cramér, H., Leadbetter, M.R.: Stationary and Related Stochastic Processes. Wiley, New York (1967)
5. Dentz, M., Kinzelbach, H., Attinger, S., Kinzelbach, W.: Temporal behavior of a solute cloud in a heterogeneous porous medium. 3. Numerical simulations. Water Resour. Res. **38**, 1118 (2002)
6. Doob, J.L.: Stochastic Processes. Wiley, New York (1990)
7. Fox, R.O.: Computational Models for Turbulent Reacting Flows. Cambridge University Press, New York (2003)
8. Haworth, D.C.: Progress in probability density function methods for turbulent reacting flows. Prog. Energy Combust. Sci. **36**, 168–259 (2010)
9. Heße, F., Prykhodko, V., Schlüter, S., Attinger, S. Generating random fields with a truncated power-law variogram: a comparison of several numerical methods. Environ. Model. Softw. **55**, 32–48 (2014)

10. Klimenko, A.Y., Bilger, R.W.: Conditional moment closure for turbulent combustion. Prog. Energy Combust. Sci. **25**, 595–687 (1999)
11. Kloeden, P.E., Platen, E.: Relations between multiple Ito and Stratonovich integrals. Stoch. Anal. Appl. **9**(3), 86–96 (1991)
12. Kloeden, P.E., Platen, E.: Numerical Solutions of Stochastic Differential Equations. Springer, Berlin (1999)
13. Kraichnan, R.H.: Diffusion by a random velocity field. Phys. Fluids **13**(1), 22–31 (1970)
14. Kramer, P.R., Kurbanmuradov, O., Sabelfeld, K.: Comparative analysis of multiscale Gaussian random field simulation algorithms. J. Comp. Phys. **226**, 897–924 (2007)
15. Kurbanmuradov, O.A., Sabelfeld, K.K.: Stochastic flow simulation and particle transport in a 2D layer of random porous medium. Transp. Porous Med. **85**, 347–373 (2010)
16. Pope, S.B.: PDF methods for turbulent reactive flows. Prog. Energy Combust. Sci. **11**(2), 119–192 (1985)
17. Pope, S.B.: Turbulent Flows. Cambridge University Press, Cambridge (2000).
18. Sabelfeld, K.: Monte Carlo Methods in Boundary Value Problems. Springer, Berlin (1991)
19. Schüler, L., Suciu, N., Knabner, P., Attinger, S.: A time dependent mixing model to close PDF equations for transport in heterogeneous aquifers. Adv. Water Resour. **96**, 55–67 (2016)
20. Schwarze, H., Jaekel, U., Vereecken, H.: Estimation of macrodispersion by different approximation methods for flow and transport in randomly heterogeneous media. Transp. Porous Med. **43**, 265–287 (2001)
21. Stein, M.L.: Interpolation of Spatial Data—Some Theory for Kriging. Springer, New York (1999)
22. Suciu, N.: Diffusion in random velocity fields with applications to contaminant transport in groundwater. Adv. Water Resour. **69**, 114–133 (2014)
23. Suciu, N., Vamoş, C.: Ergodic estimations of upscaled coefficients for diffusion in random velocity fields. In: L'Ecuyér, P., Owen, A.B. (eds.) Monte Carlo and Quasi-Monte Carlo Methods 2008, pp. 617–626. Springer, Berlin (2009)
24. Suciu, N., Vamoş, C., Vereecken, H., Sabelfeld, K., Knabner, P.: Ito equation model for disperion of solutes in heterogeneous media. Rev. Anal. Numer. Theor. Approx. **37**, 221–238 (2008)
25. Suciu, N., Vamoş, C., Radu, F.A., Vereecken, H., Knabner, P.: Persistent memory of diffusing particles. Phys. Rev. E **80**, 061134 (2009)
26. Suciu, N., Radu, F.A., Attinger, S., Schüler, L., Knabner, P.: A Fokker–Planck approach for probability distributions of species concentrations transported in heterogeneous media. J. Comput. Appl. Math. **289**, 241–252 (2015)
27. Suciu, N., Schüler, L., Attinger, S., Vamoş, C., Knabner, P.: Consistency issues in PDF methods. An. St. Univ. Ovidius Constanţa **23**(3), 187–208 (2015)
28. Suciu, N., Schüler, L., Attinger, S., Knabner, P.: Towards a filtered density function approach for reactive transport in groundwater. Adv. Water Resour. **90**, 83–98 (2016)
29. Yaglom, A.M.: Correlation Theory of Stationary and Related Random Functions, Volume I: Basic Results. Springer, New York (1987)

Index

Advection
  -diffusion, 67
  -diffusion-reaction, 4
  -dominated, 6, 73, 140
  -scheme, 73
Anomalous diffusion, 123, 125, 152
Auto-regressive of order 1, 48, 55, 186
Average
  spatio-temporal, 196
  spatial, 158, 161, 193
  spatio-temporal, 194
  stochastic, 13, 158
    variance, moments, correlation, 13

Breakthrough curves, 106, 149
Brownian motion
  continuous time, 34
  discrete time, 55, 56

Cellular automata, 7, 87
Characteristic function, 29
Coarse-grained/coarse-graining
  coefficients, 170, 175, 179, 180, 182
  simulations, 159, 160, 178
  spatial, 169
  spatio-temporal, 196
Convergence, random sequences, 50
  central limit theorem, 52, 55, 96, 100
  law of large numbers, 52
Correlation
  length, 111, 236
  time, 45, 96

Corsin's conjecture, 97
Courant number, 68, 74, 80, 84
Cross-section concentration, 149, 170, 181, 186

Darcy's law, 4, 85
Diffusion processes
  drift and diffusion coefficients, 26, 27, 54
  equivalent Itô and Fokker–Planck
    representations, 58
  inter-diffusion, 197
  in a large sense, 26
  self-diffusion, 197
  in the sense of Kolmogorov, 25
Diffusive behavior, 43, 95, 104
Dispersion
  macrodispersion, 187
  effective, 119
  ensemble, 119
  local, 102
  macrodispersion, 99
  of the center of mass, 119
  process, 107
Dispersion coefficient
  ensemble, 119
  dispersivity, 107
  effective, 119, 126
  ensemble, 126, 127, 130
  local, 107, 118
  of the center of mass, 126
  upscaled, 112
Dynamical systems, 7, 24, 92

© Springer Nature Switzerland AG 2019
N. Suciu, *Diffusion in Random Fields*, Geosystems Mathematics,
https://doi.org/10.1007/978-3-030-15081-5

Effective diffusion coefficient, 43, 61, 84, 99
Ergodicity, 52, 122, 124, 141, 143
Eulerian
    correlation, 99
    statistics, 35
Exponential correlation, 109, 123

Fick's law, 113
Filtered density function (FDF)
    approach, 158, 159
    equation, 161
Filtered mass density function, 165
Finite difference, 29, 35, 65, 70, 72
    backward scheme, 74, 211
    box scheme, 74, 211
    splitting scheme, 72
    staggered scheme, 84
Finite element, 82, 87
    GRW-MFEM approach, 84
First order approximations, 108
Flow equation, 4, 84
    GRW solution, 85
Fokker–Planck equation, 26–28, 30–36, 41–43,
            112, 118, 126
    fundamental solution, 27
Fractional Brownian motion, 124, 146
Fractional Gaussian noise, 114, 124

Gaussian
    correlation, 104, 109, 123
    diffusion, 27, 43, 96, 107, 114, 117
    distribution, 16, 44, 107
    process, 27, 31, 33, 34, 42
    random variable, 7
Global random walk, 3, 62
    biased, 77
    boundary conditions, 81
    unbiased, 67
        one-dimensional, 67
        two-dimensional, 74
Green–Kubo formula, 43, 96

Hurst exponent, 124, 146

Itô equation, 34, 55
    iterative scheme, 250
    Itô calculus, 32, 55
    Milstein scheme, 40

strong and weak
    Euler schemes, 58
    solutions, 57
Itô formula, 38, 116
Itô–Taylor expansions, 39, 116, 222, 249, 250

Kraichnan fields
    continuity and differentiability, 226
    field generators
        hydraulic conductivity, 234
        velocity, 239
    filtered fields, 229
    randomization method, 225

Lagrangian
    correlation, 95, 97
    statistics, 36
    velocity, 94
Langevin equation, 42, 45, 118, 146
Liouville equation, 28, 31, 32, 93, 95
Lipschitz continuity, 57, 227
Liuoville equation, 7

Markov processes
    integro-differential Chapman-Kolmogorov
            equation, 28
    stationary, 23
    Chapman-Kolmogorov equation, 22, 23
    conditional probability density, 22
    diffusion equation, 28
    Markov operator, 23
    Markov property, 22
    strong ergodicity, 24
    transition probability, 22
Master equation, 28, 29, 31
Measure, 11
    Lebesgue, 17, 18, 20, 21, 24, 25
    Lebesgue-Stieltjes, 13
    measurable space, 11
    measure space, 11
Memory effects, 112, 116, 144
Mixing model, 159, 163, 170, 182, 183, 185
Monte Carlo simulations, 3, 5, 7, 122, 139,
            140, 169, 171, 172, 177, 180, 188

Nonanticipating function, 37, 41
Numerical diffusion, 6, 65, 68, 74, 82, 87, 169

Ornstein–Uhlenbeck process, 42, 55, 60, 62

Particle tracking, 7, 58, 66
Peclet number, 73, 84, 140, 144
Power law, 124, 125, 133
Probability
    distribution of a random function, 14
        n-dimensional distributions, 14
        Kolmogorov Theorem, 15
        probability densities, 20, 21
    distribution of a random variable, 12
    intuitive definition, 11
    repartition function, 13
    space, 11
Probability density function (PDF)
    approach, 157
    equation, 161, 253
    Fokker–Planck equation, 166, 257

Radon–Nikodym theorem, 20, 21
Random function, 14
    equivalent representations, 17
Random variable, 12
    phase space, 12
    space of elementary events, 12
Reactive transport, 87, 160, 165, 201

Scale effect, 99, 112, 116
Self-averaging, 43, 44, 71, 125, 141, 262
Slutsky's theorem, 124
Solute plume, 149
Stratonovich equation, 5, 41, 84, 113
Subdiffusion, 124
Superdiffusion, 124

Taylor formula, 96, 148, 261
Transfer function, 106
Travel-time, 104, 149

Uncorrelated increments, 33, 117, 148
Upscaling
    stochastic, 159
    numerical, 159, 189
    spatial, 159, 179, 195
    spatio-temporal, 194, 196

White noise
    continuous, 33
    discrete, 52, 55
Wiener process, 7, 16, 34, 55, 100

Printed in the United States
By Bookmasters